THE CHEMISTRY
OF ORGANOTIN COMPOUNDS

ORGANOMETALLIC CHEMISTRY

A Series of Monographs

EDITORS

P. M. MAITLIS
MCMASTER UNIVERSITY
HAMILTON, ONTARIO
CANADA

F. G. A. STONE
UNIVERSITY OF BRISTOL
BRISTOL, ENGLAND

ROBERT WEST
UNIVERSITY OF WISCONSIN
MADISON, WISCONSIN

BRIAN G. RAMSEY: Electronic Transitions in Organometalloids, 1969.

R. C. POLLER: The Chemistry of Organotin Compounds, 1970.

Other volumes in preparation.

THE CHEMISTRY
OF ORGANOTIN COMPOUNDS

R. C. Poller

*Department of Chemistry, Queen Elizabeth College,
University of London*

1970

ACADEMIC PRESS · NEW YORK

© R. C. Poller, 1970

Published in the United States of America by
ACADEMIC PRESS, INC.
111, Fifth Avenue
New York, New York 10003

and in Great Britain by
LOGOS PRESS LIMITED

in association with

ELEK BOOKS LIMITED
2 All Saints Street, London, N.1.

CHEMISTRY

Library of Congress Catalog Number 75–11602

Printed in Great Britain by
Spottiswoode, Ballantyne & Co. Ltd., London and Colchester

CONTENTS

v

PREFACE

In writing this book the aim has been to produce a balanced and up-to-date account of organotin chemistry intended primarily for those concerned with pure and applied research in this field. It is hoped that the book will also serve as an introduction to a fascinating area of organometallic chemistry for research workers in allied subjects and for advanced undergraduates.

The various classes of organotin compounds are systematically presented in the first two-thirds of the book, the remaining space being devoted to topics which are more satisfactorily treated in a unified manner. In a work of this size it would be inappropriate to give exhaustive lists of compounds, particularly as these are available elsewhere (the most useful source, which contains references to the end of 1964, is *Organometallic Compounds, Part II*, Second Edition, edited by R. W. Weiss, Springer-Verlag, New York, 1967). The descriptions of the various classes of compounds are, therefore, illustrated by concise tables showing the physical properties of typical members of each class.

Although it is not possible, or desirable, to suppress entirely one's own interests the choice of material has been largely determined by the content of the current chemical literature. Systematic coverage of the literature extends to the end of 1968 and several references for 1969 have been included. Since the process of publication always takes a finite time, readers who wish to bring themselves comprehensively up-to-date with the literature on organotin chemistry should begin searching with the first issue of *Chemical Abstracts* for 1969 (Volume 70).

It is a pleasure to record thanks to Miss Alison Anderson of Logos Press for her valuable help and advice.

Queen Elizabeth College,   R. C. POLLER
University of London.
December 1969.

ix

1

SOME PRELIMINARY CONSIDERATIONS

1.1 NOMENCLATURE

Current practice in the naming of organotin compounds is based on the use of 'tin' as a suffix so that Me_4Sn is tetramethyltin and $(Me_3Sn)_2O$ is bis(trimethyltin) oxide[6, 28]. In the older system compounds are regarded as derivatives of the (sometimes hypothetical) stannanes Sn_nH_{2n+2} and the above compounds are called tetramethylstannane and hexamethyldistannoxane respectively. Both systems have advantages and both are used in this book. The modern system is logically based but reluctance to regard an organotin group as a substituent can result in cumbersome or inappropriate names; for example, the most felicitous name for $(Bu_3Sn)_2NEt$ is bis(tributylstannyl)ethylamine. The principal objection to the stannane system is that, for catenated compounds with more than two tin atoms, the parent compound does not exist but as this does not apply to the majority of organotin compounds it does not seem to be a serious objection.

1.2. REACTIVITY OF ORGANOTIN COMPOUNDS

Before beginning a detailed scrutiny of organotin chemistry it is appropriate to consider the extent to which the general reactivity of organotin compounds can be predicted from fundamental principles. In fact it is difficult to make valid generalisations since many factors will be involved in determining how a particular compound will react in a given set of circumstances[57]. Hence it is most profitable to discuss specific reactions of, for example, the tin-carbon bond in detail (Chapter 3) but it is, nevertheless, instructive to consider in a more general way, the more pertinent factors influencing reactivity. Discussion is mainly confined to reactions of the tin–carbon bond in comparison with corresponding bonds between carbon and other Group IVb elements.

Metal–carbon bond strengths have been reviewed by Skinner[53] who noted that mean bond dissociation energies ($\bar{D}$) fall as a subgroup is descended so that $\bar{D}(C-R) > \bar{D}(Si-R) > \bar{D}(Ge-R) >$

$\bar{D}$(Sn—R) > $\bar{D}$(Pb—R). Some figures are given in Table I which, although useful for indicating trends, are imperfect guides to reactivity since (a) they are mean values [$\bar{D}$(Sn—Me) is one quarter of the energy required to break the 4 C—Sn bonds in Me$_4$Sn, this differs from the dissociation energy D(Me$_3$Sn—Me) i.e. the energy required to break one Sn—C bond in Me$_4$Sn] and (b) they refer to homolytic bond fission, whereas the majority of the reactions of these compounds involve heterolysis. By the use of electron-impact studies a number of bond dissociation energies have been investigated for organotin compounds (Table II). These figures, when compared with those

TABLE 1

Mean bond dissociation energies (kcal. mole^{-1})

Bond	$\bar{D}^a$	Reference
Sn—Me	50–52	53, 54, 55
Sn—Et	46–47	53, 54, 55
Sn—Ph	54	54
Sn—O(Et$_3$SnOCOPh)	98	54
Sn—Sn(Ph$_3$SnSnPh$_3$)	30 ± 15	54
C—Me	83	13
Si—Me	70	13
Ge—Me	59	13
Ge—Et	57	53
Pb—Me	37	13
Pb—Et	31	53

^{a} Unless otherwise stated the error is ±1–2 kcal. mole^{-1}.

given in Table I emphasise the fact that bonds dissociate in a stepwise fashion with different energies being involved at each step. From measurements of the heats of redistribution reactions it was shown that, while 56 kcal. mole^{-1} are required to dissociate the first Sn—C bond in dimethyltin dichloride, only 32 kcal. mole^{-1} are required for the second Sn—C bond[41].

Although progress is being made in the determination of bond energies the problem of predicting reactivity in heterolytic processes remains and it is necessary to consider the relative ionic character of the tin–carbon bond.

TABLE II

Bond dissociation energies (kcal. mole^{-1}) estimated from electron impact studies

Bond	D^a	Reference	Bond	D^a	Reference
Me$_3$Sn—Me	73	61	Me$_3$Sn—C$_2$H$_3$	78	61
Me$_3$Sn—Me	60	32a	Me$_3$Sn—CH$_2$Ph	57	61
Me$_3$Sn—Me	61	33a	Me$_3$Sn—Br	94	61
Me$_3$Sn—Et	68	61	Me$_3$Sn—I	80	61
Me$_3$Sn—Et	55	32a	Me$_3$Sn—SnMe$_3$	76	61
Me$_3$Sn—Pr	68	61	Me$_3$Sn—SnMe$_3$	63	32a
Me$_3$Sn—Pr	55	32a	Me$_3$Sn—GeMe$_3$	67	33a
Me$_3$Sn—Ph	57	61	Me$_3$Sn—SiMe$_3$	66	33a
			Me$_3$Sn—CMe$_3$	60	33a

^{a} The error is $\pm$2–6 kcal. mole^{-1}.

Unfortunately the electronegativities of the Group IVb elements is a controversial subject; the results of a number of different approaches to the problem have recently been tabulated[46]. It is implicit in Pauling's definition of electronegativity 'the power of an atom in a molecule to attract electrons to itself'[44] that the electronegativity of an element may vary with its molecular environment and this, together with the fact that it is not possible to make direct electronegativity measurements, accounts for the large number of different values published[8]. The values given by Pauling[44] are unsatisfactory in that they scarcely discriminate between the elements silicon, germanium, tin and lead. Although certain measurements[15, 40] are compatible with the electronegativity sequence which would be predicted from the positions of the elements in the Periodic Table i.e. C > Si > Ge > Sn > Pb, Allred and Rochow[4] have adduced compelling chemical and physical evidence for the sequence C > Pb > Ge > Sn > Si. This order has been criticised[16, 17] but there is some theoretical justification for it and a similar alternation of electronegativity values is observed in Group IIIb[3]. The Allred-Rochow values are: C, 2·60; Si, 1·90; Ge, 2·00; Sn, 1·93; Pb, 2·45 and while these emphasise the polarity of the tin–carbon bond they offer little hope of explaining the very real differences in reactivity between organosilicon and organotin compounds. (For example, hydrogen in the alkyl groups of tetraalkylsilanes can be substituted by chlorine or bromine whereas these reagents give only tin–carbon cleavage products with tetraalkylstannanes.)

The relative inductive effects of trialkylstannyl and analogous groups are more certain. Studies of the relative rates of acid cleavage of the compounds p-$Me_3MCH_2C_6H_4SnMe_3$ (M = Si, Ge, Sn)[9] and the rates of alkaline hydrolysis of $R_3M(CH_2)_nCOOEt$ (R = Me,Et; n = 1, 2, 3; M = Si, Ge, Sn)[50] and of other compounds[18a], spectroscopic measurements on $(Me_3Si)_2N(MMe_3)$ (M = Si, Ge, Sn, Pb)[51] and calculations[27] establish that the R_3M groups have a +I effect. The order of electron release is $Me_3Pb > Me_3Sn > Me_3Ge > Me_3Si$.

One difficulty which often arises in experiments designed to measure relative inductive effects is that there may be some π-character in a bond between tin and an element possessing p electrons[39, 60]. Thus in a Sn—X bond where X is sp^2C, N, O, S or halogen it is possible that, opposing the inductive electron drift Sn $\rightarrow$ X, there may be some overlap between a filled p-orbital on X and empty $5d$ orbitals on Sn causing a transfer of electron density in the opposite direction as in I.

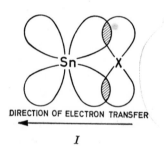

DIRECTION OF ELECTRON TRANSFER

I

Despite some controversy[19, 25, 47, 48] it is widely accepted that there are significant $(p \rightarrow d)\pi$ contributions to the Si—N bonds in compounds such as the planar trisilylamine[45]. Is it reasonable to assume that tin and the other Group IVb elements will show a similar tendency to π-bond formation? Theoretical considerations suggest that the overlap of p and d orbitals would not be particularly sensitive to the sizes of the orbitals[14] and there is no marked decrease in $p\pi$–$p\pi$ bond strengths in Group III as the atomic numbers increase[45]. Although this suggests that π-bonding should be present in suitably substituted tin compounds, examination of the literature shows disagreement regarding its extent. There is some evidence that, in phenyltin compounds, interaction occurs between the π-electrons of the phenyl groups and the $5d$ orbitals of tin. This conclusion is based on the interpretation of nmr-[38, 59], infrared-[33] and ultraviolet[33]-spectra and dipole moments[26] of phenyltin compounds and on the acid strengths of the substituted benzoic acids p-Me$_3$MC$_6$H$_4$COOH (M = C, Si, Ge, Sn)[11]. It cannot be assumed that other aromatic groups behave similarly since the pK$_a$ values for a series of pyridines having Me$_3$M (M = C, Si, Ge, Sn) substituents in the 2-position indicated the absence of $p\pi$–$d\pi$ bonding in the tin–pyridine link[5]. The existence of $p\pi$–$d\pi$ bonding in propenyltin compounds is indicated by lithium cleavage experiments[52] and differences in chemical shifts observed in the nmr spectra of tetraallyl- and tetravinyl-tin compounds suggest double bond character in the tin–vinyl bond[7] but opposite conclusions were reached from the infrared spectra of CH$_2$═CHMCl$_3$ compounds (M = Si, Ge, Sn)[34]. In the course of infrared and Raman studies of compounds R$_3$SnXSnR$_3$ (X = O, S) it was concluded that there was no π-contribution to the Sn—O and Sn—S bonds[32], similar conclusions regarding Sn—O and Sn—S bonds in other compounds were reached by other workers[1, 2, 38a]. Using diverse arguments several authors have concluded that there is no double bond character in the tin–nitrogen bond[24, 25, 30, 47, 48] but nmr measurements on methylstannylamines indicated considerable $p\pi$–$d\pi$ bonding[35]. Calculations based upon nuclear quadrupole resonance measurements indicated π-character in the Sn—I bonds of diethyltin diiodide[10] whereas it was concluded from the dipole moments of organotin chlorides that the Sn—Cl bond order is close to unity[26]. There is evidence for $d\pi$–$d\pi$ interaction in the bonds formed between tin and certain transition metals[20, 29] and the possibility of drift of d-electrons from tin into C—F antibonding orbitals in trifluoromethyltin compounds has been discussed[12]. To summarise, the extent of $p\pi$–$d\pi$ bonding in tin compounds is uncertain, the balance of evidence suggests that it may not be important except, perhaps, in some sp^2C—Sn bonds.

Another factor which can be important in influencing reactivity is the increased tendency to higher coordination numbers on passing from germanium to tin. This general subject is developed in Chapter 11 but we note here that this can lead to changes in reaction mechanisms with striking results[18].

The foregoing brief discussion indicates the difficulty of evaluating some of the factors which determine reactivity. This, together with the problem of identifying the particular factors which are relevant to a specific reaction point up the challenge and interest involved in a fundamental understanding of organotin reactions.

1.3. ISOTOPES OF TIN

Tin (atomic number 50, atomic weight 118·69) has the following isotopic composition[43]:

Isotope:	112	114	115	116	117	118	119	120	122	124
% Abundance:	1	0·68	0·34	14·4	7·54	24·1	8·62	32·5	4·7	5·9

This multiplicity of isotopes is, on the whole, advantageous to the experimentalist in the range and usefulness of the physical techniques which are consequently applicable to organotin compounds (Chapter 13). For studies utilising radioactive tracer techniques the ^{113}Sn isotope is employed; ^{113}Sn metal, some inorganic compounds, and a few organotin derivatives containing this isotope are available commercially. Details of the synthesis of $Bu_3^{113}SnCl$ [58], $(Bu_3^{113}Sn)_2O$ [56] and $Bu_3^{113}SnOCOPh$ [42] have been published.

1.4. ANALYSIS OF ORGANOTIN COMPOUNDS

The best routine method for the determination of tin in organotin compounds is based on conversion to stannic oxide which is estimated gravimetrically. The following details are similar to those given by van der Kerk and Luijten[36].

The organotin compound (~0·1 g) is weighed into a silica test tube (length 130 mm, external diameter 17 mm, internal diameter 15 mm) which has been heated, with a Meker burner, to constant weight. By means of a dropping pipette, 25 drops of a mixture of concentrated nitric acid (3 volumes) and concentrated sulphuric acid (2 volumes) are added cautiously. When the reaction subsides the tube is supported horizontally and the mixture heated with a low flame. As frothing subsides the rate of heating is increased until, finally, the tube is heated strongly with a Meker burner for 2–3 hours.

Titrimetric methods[21, 49] of tin determination are usually based on a final oxidation of tin(II) to tin(IV) and suffer from the following disadvantages: (a) all traces of the oxidising agent used in the initial combusion must be removed and (b) an inert atmosphere must be maintained until the final titration is complete because of the ready oxidation of tin(II) by oxygen. Tin estimation by complexometric titration[22], polarography[23], spectrography[31] and X-ray fluorescence[37] has also been described.

Methods for the estimation of specific classes of organotin compounds are given in the appropriate chapters.

References

1. Abel, E. W. and Armitage, D. A., *Advances in Organometallic Chemistry*, Academic Press, New York and London, 1967, **5**, 1.
2. Abel, E. W., Armitage, D. A. and Brady, D. B., *Trans. Faraday Soc.*, 1966, **62**, 3459.
3. Allred, A. L., *J. Inorg. Nuclear Chem.*, 1961, **17**, 215.
4. Allred, A. L. and Rochow, E. G., *J. Inorg. Nuclear Chem.*, 1958, **5**, 269.
5. Anderson, D. G., Chipperfield, J. R. and Webster, D. E., *J. Organometallic Chem.*, 1968, **12**, 323.
6. Anon., *Chem. Abs.*, 1962, **56**, 66N.
7. Blears, D. J., Danyluk, S. S. and Cawley, S., *J. Organometallic Chem.*, 1966, **6**, 284.
8. Bott, R. W., Eaborn, C. and Swaddle, T. W., *J. Organometallic Chem.*, 1966, **5**, 233.
9. Bott, R. W., Eaborn, C. and Walton, D. R. M., *J. Organometallic Chem.*, 1964, **2**, 154.
10. Bryuchova, E. V., Semin, G. K., Goldanskii, V. I. and Khrapov, V. V., *Chem. Comm.*, 1968, 491.
11. Chatt, J. and Williams, A. A., *J. Chem. Soc.*, 1954, 4403.
12. Clark, H. C. and Tsai, J. H., *J. Organometallic Chem.*, 1967, **7**, 515.
13. Coates, G. E., Green, M. L. H., Powell, P. and Wade, K., *Principles of Organometallic Chemistry*, Methuen, London, 1968, 4.
14. Craig, D. P., MacColl, A., Nyholm, R. S., Orgel, L. E. and Sutton, L. E., *J. Chem. Soc.*, 1954, 332.
15. Dessy, R. E., Pohl, R. L. and King, R. B., *J. Amer. Chem. Soc.*, 1966, **88**, 5121.
16. Drago, R. S., *Record of Chemical Progress*, 1965, **26**, 157.
17. Drago, R. S. and Matwiyoff, N. A., *J. Organometallic Chem.*, 1965, **3**, 62.
18. Eaborn, C. and Pande, K. C., *J. Chem. Soc.*, 1960, 1566.
18a. Eaborn, C., Thompson, A. R. and Walton, D. R. M., *J. Organometallic Chem.*, 1969, **17**, 149.
19. Ebsworth, E. A. V., *Chem. Comm.*, 1966, 530.
20. Edmondson, R. C. and Newlands, M. J., *Chem. and Ind.*, 1966, 1888.
21. Farnsworth, M. and Pekola, J., *Analyt. Chem.*, 1959, **31**, 410.
22. Geyer, R. and Seidlitz, H. J., *Z. Chem.*, 1964, **4**, 468.
23. Geyer, R. and Seidlitz, H. J., *Z. Chem.*, 1967, **7**, 114.
24. Hester, R. E. and Jones, K., *Chem. Comm.*, 1966, 317.
25. Hogben, M. G., Oliver, A. J. and Graham, W. A. G., *Chem. Comm.*, 1967, 1183.
26. Huang, H. H., Hui, K. M. and Chiu, K. K., *J. Organometallic Chem.*, 1967, **11**, 515.
27. Huheey, J. E., *J. Phys. Chem.*, 1965, **69**, 3284.
28. I.U.P.A.C. Rules of Organic Nomenclature, *Handbook for Chemical Society Authors*, The Chemical Society, London, 1961, 111.
29. Jetz, W., Simons, P. B., Thompson, J. A. J. and Graham, W. A. G., *Inorg. Chem.*, 1966, **5**, 2217.
30. Jones, K. and Lappert, M. F., *J. Organometallic Chem.*, 1965, **4**, 295.

31. Kreshkov, A. P. and Kuchkarev, E. A., *Zavodsk. Lab.*, 1966, **32**, 558 (*Chem. Abs.*, 1966, **65**, 6292e).

32. Kriegsmann, H., Hoffmann, H. and Geissler, H., *Z. Anorg. Allgem. Chem.*, 1965, **341**, 24.

32a. Lampe, F. W. and Niehaus, A., *J. Chem. Phys.*, 1968, **49**, 2949.

33. Lapkin, I. I. and Dumler, V. A., *Uch. Zap. Permsk. Gos. Univ. No.* 111, 1964, 185 (*Chem. Abs.*, 1966, **64**, 12045e).

33a. Lappert, M. F., Simpson, J. and Spalding, T. R., *J. Organometallic Chem.*, 1969, **17**, P1.

34. Leites, L. A., Pavlova, I. D. and Egorov, Yu. P., *Teor. i Eksperim. Khim.*, *Akad. Nauk Ukr. S.S.R.*, 1965, **1**, 311 (*Chem. Abs.*, 1965, **63**, 13024e).

35. Lorberth, J. and Kula, M. R., *Chem. Ber.*, 1965, **98**, 520.

36. Luijten, J. G. A. and van der Kerk, G. J. M., *Investigations in the Field of Organotin Chemistry*, Tin Research Institute, Greenford, 1966, 84.

37. Mahr, C. and Stork, G., *Z. Anal. Chem.*, 1966, **221**, 1.

38. Maire, J. C., *J. Organometallic Chem.*, 1967, **9**, 271.

38a. Marchand, A., Mendelsohn, J., Lebedeff, M. and Valade, J., *J. Organometallic Chem.*, 1969, **17**, 379.

39. Matwiyoff, N. A. and Drago, R. S., *J. Organometallic Chem.*, 1965, **3**, 393.

40. Nagy, J., Réffy, J., Kuszmann-Borbély, A. and Palossy-Becker, K., *J. Organometallic Chem.*, 1967, **7**, 393.

41. Nash, G. A., Skinner, H. A. and Stack, W. F., *Trans. Faraday Soc.*, 1965, **61**, 640.

42. Otto, P. Ph. H. L., Creemers, H. M. J. C., and Luijten, J. G. A., *J. Label. Compounds*, 1967, **2**, 339.

43. Pascal, P., *Nouveau Traité de Chimie Minerale*, Masson et Cie, Paris, 1963, **8**, 285.

44. Pauling, L., *The Nature of the Chemical Bond*, 3rd Edn., Cornell Univ. Press, Ithaca, N.Y. 1960, 88.

45. Perkins, P. G., *Chem. Comm.*, 1967, 268.

46. Prokof'ev, A. K., Nechiporenko, V. P. and Kostyanovskii, R. G., *Izv. Akad. Nauk S.S.S.R. Ser. Khim.*, 1967, 794.

47. Randall, E. W. and Zuckerman, J. J., *Chem. Comm.*, 1966, 732.

48. Randall, E. W. and Zuckerman, J. J., *J. Amer. Chem. Soc.*, 1968, **90**, 3167.

49. Reverchon, R., *Chim. Anal.* (*Paris*), 1965, **47**, 70 (*Chem. Abs.*, 1965, **62**, 15428f).

50. Rijkens, F., Janssen, M. J., Drenth, W. and van der Kerk, G. J. M., *J. Organometallic Chem.*, 1964, **2**, 347.

51. Scherer, O. J. and Schmidt, M., *J. Organometallic Chem.*, 1964, **1**, 490.

52. Seyferth, D., Singh, G. and Suzuki, R., *Pure Appl. Chem.* 1966, **13**, 159.

53. Skinner, H. A., *Advances in Organometallic Chemistry*, Academic Press, New York and London, 1964, **2**, 49.

54. Tel'noi, V. I. and Rabinovich, I. B., *Russian J. Phys. Chem.*, 1966, **40**, 842.

55. Tel'noi, V. I. and Rabinovich, I. B., *Zh. Fiz. Khim.*, 1965, **39**, 2076.

56. Tenny, K. S. and Tenny, A. M., *J. Label. Compounds*, 1968, **4**, 54.

57. Tobias, R. S. and Hutcheson, S., *J. Organometallic Chem.*, 1966, **6**, 535.

58. Tucker, W. P., *Inorg. Nucl. Chem. Lett.*, 1968, **4**, 83.
59. Verdonck, L. and van der Kelen, G. P., *Bull. Soc. Chim. Belges*, 1965, **74**, 361.
60. West, R., *J. Organometallic Chem.*, 1965, **3**, 314.
61. Yergey, A. L. and Lampe, F. W., *J. Amer. Chem. Soc.*, 1965, **87**, 4204.

COMPOUNDS WITH FOUR ORGANIC GROUPS ATTACHED TO A TIN ATOM

By the year 1959 some 330 compounds having four organic groups bound to tin were known and over two thirds of these were of the type R_3SnR'. The fully symmetrical R_4Sn compounds are less common than the R_2SnR_2' types and only 10 compounds with three different organic groups $R_2R'R''Sn$ had been described[35]. A practicable method for the preparation of unsymmetrical $RR'R''R'''Sn$ compounds has only very recently been described[6a, 23a]. It is a reflection of the great interest shown in organotin chemistry during the last decade that the total number of compounds in these classes now known, at the most conservative estimate, exceeds 600.

2.1. ALKYL- AND ARYL-TIN COMPOUNDS

Organic groups attached to a wide range of metals will readily exchange with chlorine atoms bound to tin.

$$\text{>C—M} + \text{>Sn—Cl} \longrightarrow \text{>C—Sn<} + \text{MCl}$$

M is a metal such as magnesium, lithium, aluminium, sodium, potassium or zinc.

For the laboratory and industrial[38] preparation of symmetrical organotin compounds good yields are usually obtained from the reaction between stannic chloride and a Grignard reagent.

$$4RMgX + SnCl_4 \longrightarrow R_4Sn + 2MgX_2 + 2MgCl_2$$
$$(R = \text{alkyl or aryl})$$

Increased yields of tetraalkyltin compounds are obtained if the alkyl halide and stannic chloride are added simultaneously to the magnesium using heptane containing limited amounts of ether as solvent[131]. Continuous processes have been devised for the large scale preparation of tetraalkyltin compounds[126]. Good yields of the latter have been obtained by reaction between Grignard reagents and tin alkoxides[60].

There are many recent examples[19, 36, 37, 80] of the preparation of unsymmetrical compounds of the types R_3SnR' and R_2SnR_2' by reaction between an organotin mono- or di-halide and an organo-magnesium halide. When a Grignard reagent made from an equi-molecular mixture of methyl iodide and ethyl bromide was treated with stannic chloride the principal product was diethyldimethyltin [75].

The Grignard method can be used to introduce organic groups containing functional substituents providing these do not interact with organomagnesium compounds, for example [41]:

$$PhOCH(OMe)MgBr + Me_3SnCl \longrightarrow PhOCH(OMe)SnMe_3 + MgBrCl.$$

Chloromethyldiethyltin chloride (obtained by reaction between diethyltin dichloride and diazomethane) reacts with ethylmagnesium bromide to give chloromethyltriethyltin [46].

$$Et_2Sn(CH_2Cl)Cl + EtMgBr \longrightarrow Et_3SnCH_2Cl + MgBrCl$$

Reaction between perchlorophenylmagnesium chloride and tri-phenyltin chloride gave $Ph_3SnC_6Cl_5$ in 67% yield, however stannic chloride gave only a 4% yield of $(C_6Cl_5)_4Sn$ due to the intervention of an exchange reaction resulting in the preferential formation of hexachlorobenzene [24]. A number of perfluorophenyltin compounds have been made utilising perfluorophenylmagnesium bromide [32, 74].

In certain cases it is advantageous to replace the Grignard reagent by an organolithium compound, as in the preparation of $(C_6F_5)_4Sn$ where the yield of 13% obtained from stannic chloride and perfluoro-magnesium bromide was increased to 91% when perfluorophenyl-lithium was used [109]. 2-Lithioperfluorobiphenyl was used to prepare *bis*(2-perfluorobiphenylyl) dimethyltin (*I*) [21]; perchlorophenyltri-methyltin and 1,4-*bis*(trimethylstannyl)-2,3,5,6-tetrachlorobenzene

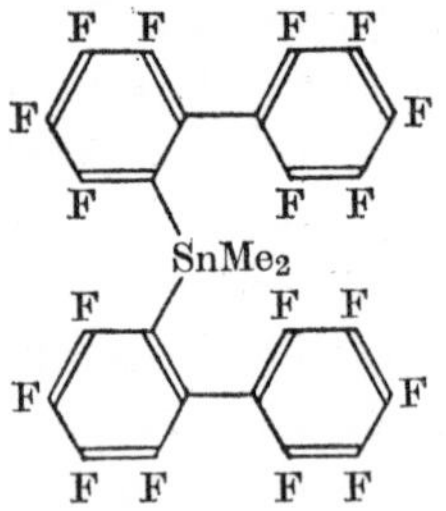

I

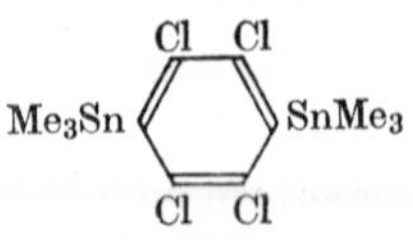

II

(*II*) were made from the appropriate lithium compounds and trimethyltin chloride [98].

For the preparation of cyclic organotin compounds dilithium reagents are preferred.

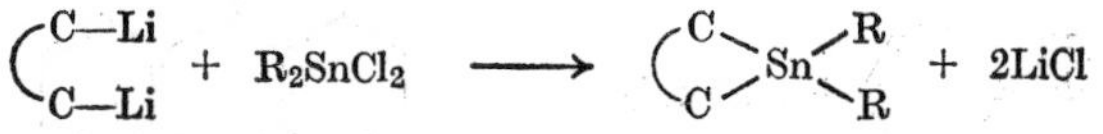

$$C\underset{C-Li}{\overset{C-Li}{\big<}} + R_2SnCl_2 \longrightarrow C\underset{C}{\overset{C}{\big<}}Sn\underset{R}{\overset{R}{\big>}} + 2LiCl$$

Some examples of this class of compound are shown in Table I; in some cases only very small amounts of the cyclic compounds are obtained due to the preferential formation of polymeric products. If a dilithium reagent is treated with stannic chloride then spirocyclic compounds result.

$$2\left(C\underset{C-Li}{\overset{C-Li}{\big<}}\right) + SnCl_4 \longrightarrow \left(C\underset{C}{\overset{C}{\big<}}Sn\underset{C}{\overset{C}{\big>}}\right) + 4LiCl$$

In this way all of the organolithium compounds listed in Table I, with the exception of 2,2'-dilithiobiphenyl, gave the corresponding spirocyclic compounds together with varying amounts of polymeric products. An alternative route to the products shown in Table I is to heat the spirocyclic compound with stannic chloride and to treat the resulting dichloride with a Grignard reagent, for example [51]:

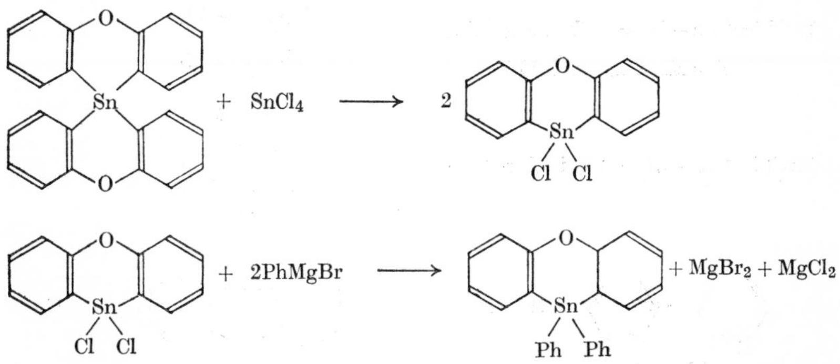

Treatment of compound *III* with stannic chloride gave the expected spirocyclic product *IV*. The latter compound was also isolated when *III* was treated with divinyltin dichloride, presumably by the intervention of a particularly ready disproportionation reaction [78].

TABLE I

The preparation of cyclic organotin compounds

Starting materials	Product	Reference
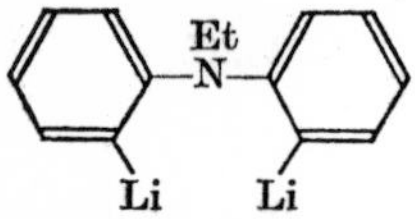 $+ \text{Ph}_2\text{SnCl}_2$	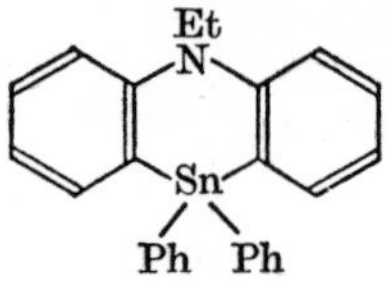	25
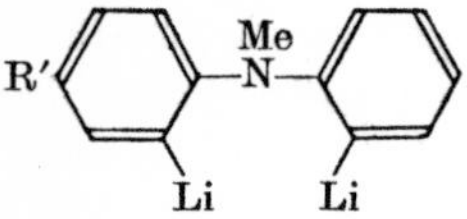 $\text{R}' + \text{R}_2\text{SnCl}_2$ (R = Me, Ph; R' = H, Me, Br)	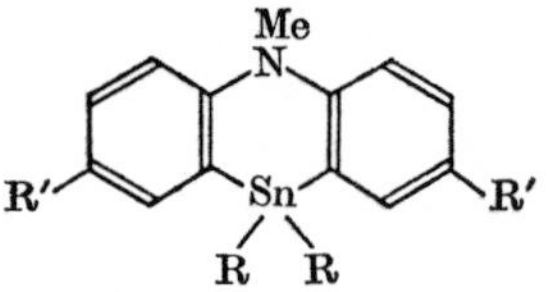	50
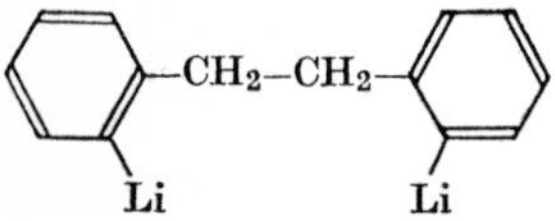 $-\text{CH}_2-\text{CH}_2-$ $+ \text{Ph}_2\text{SnCl}_2$	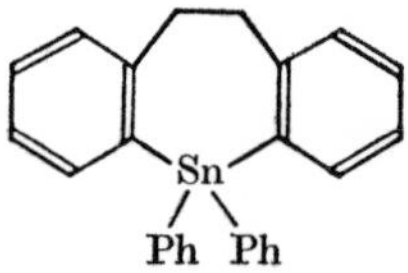	49
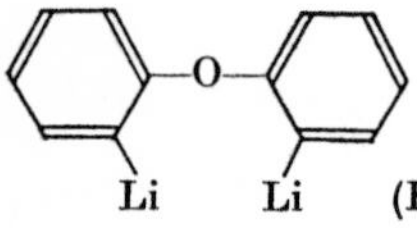 $-\text{O}-$ $+ \text{R}_2\text{SnCl}_2$ (R = Me, Et, Bu)	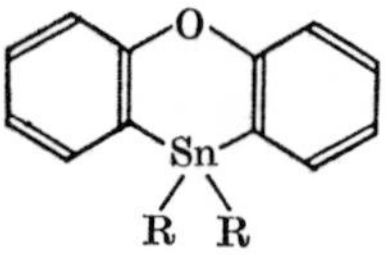	51
$\text{PhCLi}{=}\text{CPh}-\text{CPh}{=}\text{CPhLi} + \text{R}_2\text{SnCl}_2$ (R = Me, $\text{CH}_2{=}\text{CH}$, Ph)	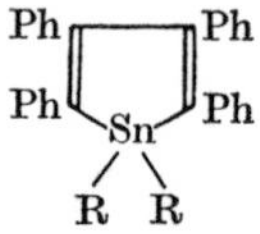	54
$\text{PhCLi}{=}\text{CH}-\text{CH}{=}\text{CPhLi} + \text{Me}_2\text{SnCl}_2$	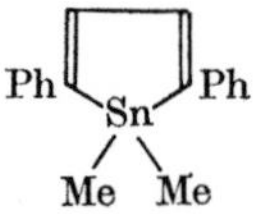	2
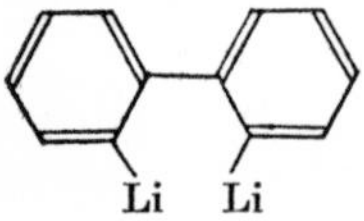 $+ \text{Me}_2\text{SnCl}_2$	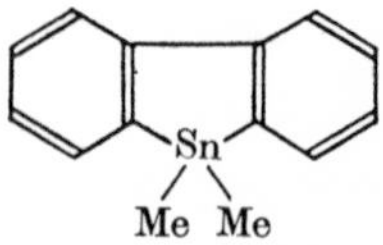	134
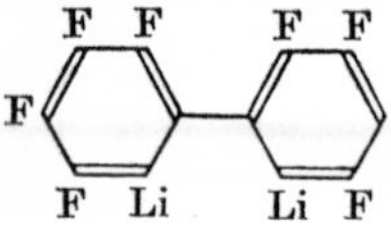 $+ \text{R}_2\text{SnCl}_2$ (R = Me, Ph)	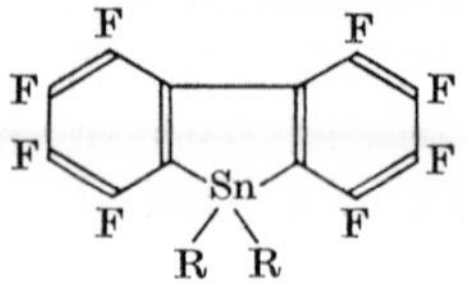	15

DiGrignard reagents have been used to prepare cyclic compounds [122] and the spirocyclic compound V was obtained, in low yield, from $BrMg(CH_2)_5MgBr$ and stannic chloride [4].

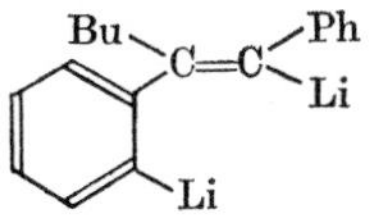

III 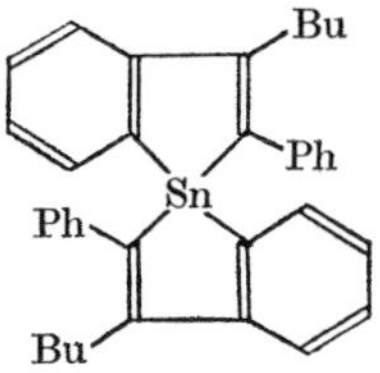 IV 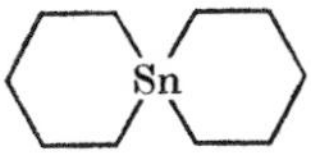   V

The dilithium derivative of 1,2-dicarborane reacts with organotin monochlorides to give 1,2-distannyl derivatives [9, 115].

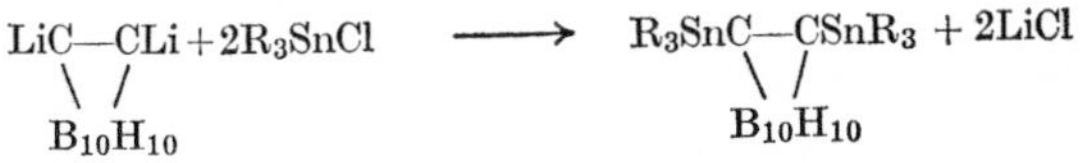

With organotin dihalides, cyclic dimeric products may be formed [85, 115]:

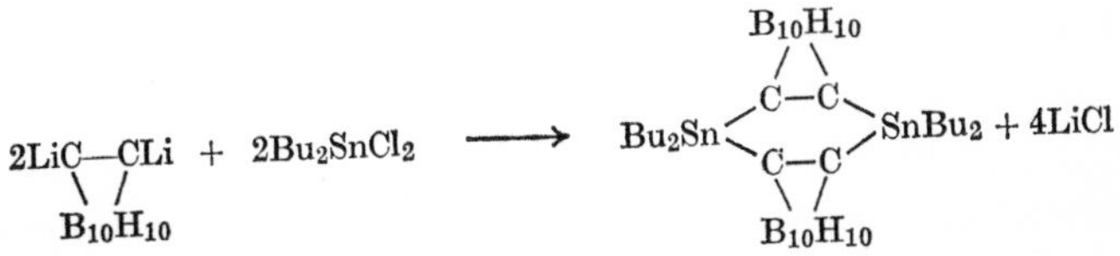

or depending upon the conditions, somewhat higher molecular weight oligomers are obtained [8, 10]. The dilithium derivatives of 1,7- and 1,12-carboranes [85] and the Grignard reagent from bromo-methylcarborane [116] have been used to prepare other stannylcarboranes.

Organolithium compounds have been used to link tin to organic groups containing phosphorus [96].

$$Ph_2P(S)CHRLi + Ph_3SnCl \longrightarrow Ph_2P(S)CHRSnPh_3 + LiCl$$

$$(R = H, Me)$$

Although an organolithium reagent is not involved it is appropriate to mention here an unusual method of forming a tin–carbon bond by the stannylation of a silylated phosphine–methylene compound [83]:

$$2Me_3SiCH{=}PMe_3 + Me_3SnCl \longrightarrow$$

$$[Me_3SiCH_2\text{-}PMe_3]^{\oplus} Cl^{\ominus} + Me_3SiC(SnMe_3){=}PMe_3.$$

Ferrocene can be lithiated with butyllithium and the product, with tributyltin chloride, gives tributylstannylferrocene [136].

Most of the organometallic reagents employed for the replacement of tin-bound halogen atoms by organic groups are those derived from magnesium or lithium. Aluminium alkyls can be used to alkylate stannic chloride but the presence of a Lewis base is necessary to obtain good yields of the fully alkylated product [66].

$$4R_3Al + 4R_3'N + 3SnCl_4 \longrightarrow 3R_4Sn + 4AlCl_3.NR_3'$$

Trimethyltin chloride is readily alkylated by diethylaluminium chloride [33] and an interesting recent development is the preparation, in quantitative yields, of tetraalkyltin compounds from organotin hydrides and aluminium alkyls [84].

$$R_{4-n}SnH_n + nR_3'Al \longrightarrow R_{4-n}SnR_n' + nR_2'AlH$$

There are occasional reports, in the current literature, of the use of organosodium [127, 135] and organopotassium [23] reagents for the preparation of organotin compounds.

Bromodichloromethylphenylmercury can be employed for the insertion of a dichloromethylene group into the Sn—Br bond of trimethyltin bromide [89].

$$Me_3SnBr + PhHgCCl_2Br \longrightarrow Me_3SnCCl_2Br + PhHgBr$$

The bromodichloromethyltin product can, in turn, be used as a dichlorocarbene insertion reagent. Trihalomethyltin compounds can also be made by the organolithium method [89,89a].

$$Me_3SnCl + LiCCl_3 \xrightarrow{-110°} Me_3SnCCl_3 + LiCl$$

Compounds containing β-carbonyl and β-ester substituents have been synthesised by heating organotin sulphides with the appropriate organomercury compounds [68].

$$(R_3Sn)_2S + Hg(CH_2COR')_2 \longrightarrow 2R_3SnCH_2COR' + HgS$$

Tetra(perfluorophenyl)tin is obtained in 60% yield when $(C_6F_5)_2Hg$ and tin powder are heated at 260° in a sealed tube for 9 days [11].

Iodomethyl and bromomethyl groups may be transferred from zinc to tin thus, for example, when iodomethylzinc iodide and trimethyltin chloride are heated together in tetrahydrofuran, trimethyliodomethyltin is produced in 86% yield [89a].

$$ICH_2ZnI + Me_3SnCl \longrightarrow Me_3SnCH_2I + ZnICl$$

In some cases good yields of organotin compounds can be obtained by reaction between an alkyl or aryl halide and alloys of tin with sodium or other metals. This is, potentially, an economic method for large-scale preparations. Luijten and van der Kerk[57] showed that the physical properties of tin–sodium alloys were such that it was impracticable to use an alloy containing more than one atomic proportion of sodium and, therefore, only one quarter of the tin could be utilised. The same authors proposed the use of a tin–magnesium alloy of composition Mg_2Sn where, in principle, all of the tin is available for conversion to an organotin compound. This alloy when treated with ethyl bromide in the presence of a mercuric chloride catalyst gave 54% of tetraethyltin together with a little triethyltin bromide. This method, however, is less successful when higher alkyl bromides are used.

In the Wurtz-type reaction for the preparation of tetrabutyltin:

$$4BuCl + 8Na + SnCl_4 \longrightarrow Bu_4Sn + 8NaCl$$

even under optimum conditions about 35% of the tin becomes unavailable due to a side reaction whereby the sodium reduces the stannic chloride to stannous chloride and tin[57]. The importance of controlling the size of the sodium particles and other factors which minimise side reactions have recently been discussed[52]. Since dibutyltin dichloride is not reduced by sodium the conversion of this compound to tetrabutyltin is a more efficient process[57]. The Wurtz reaction is sometimes used for the laboratory preparation of aryltin compounds, for example, a 45% yield of tetra-o-tolyltin was obtained from o-chlorotoluene, stannic chloride and sodium using a mercury catalyst[107].

A related method of preparation which can be used for the introduction of functionally substituted groups is the reaction between a compound containing tin bonded to an alkali metal and an alkyl or aryl halide. Thus acyltin compounds are formed from triphenyltin lithium and acid chlorides at $-70°$[71].

$$Ph_3SnLi + RCOCl \longrightarrow Ph_3SnCOR + LiCl$$

A similar reaction using an iodophenol:

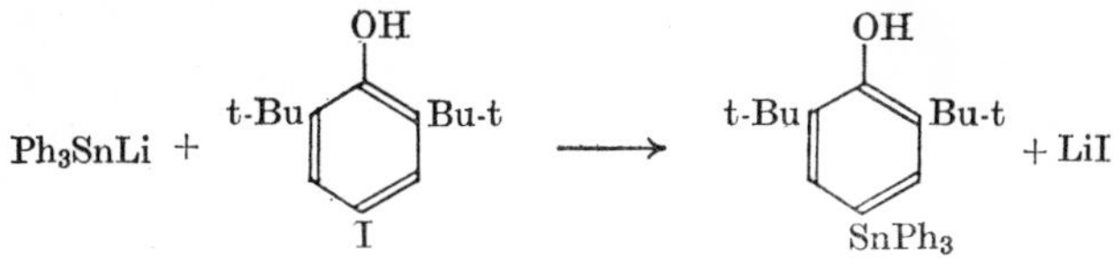

VI

is of particular interest since the product 2,6-di-t-butyl-4-triphenyl-stannylphenol (*VI*) can be dehydrogenated in benzene to give a green paramagnetic solution containing the radical *VII* in which the delocalisation of the odd electron was considered to extend to the tin atom [108].

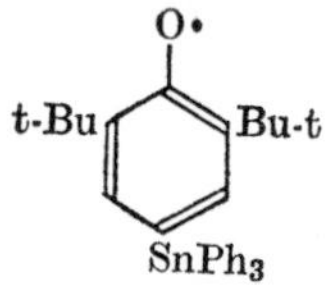

$$\text{t-Bu} \overset{\text{O·}}{\underset{\text{SnPh}_3}{\bigcirc}} \text{Bu-t}$$

VII

There is growing interest in the preparation of organometallic compounds by electrolytic methods. These processes are in an early stage of development but the preparation of alkyltin compounds by the electrolysis of solutions of alkyl halides in esters with zinc bromide as the electrolyte, and by the electrolysis of solutions of sodium tetra-alkylborate [128] or potassium tetraalkylaluminate [137] (in all cases a tin anode being used) are noted. With a few exceptions the preparative methods so far discussed are unsuitable for the introduction of organic groups carrying functional substituents onto tin. The preparation of *tetrakis*(2-cyanoethyl)tin in yields of up to 44% by the electrolytic reduction of acrylonitrile at a tin cathode [7, 111] is therefore of some interest.

An important method of forming tin–carbon bonds is by the hydrostannation of unsaturated compounds as in reactions of the following type.

$$R_3SnH + CH_2{=}CHR' \longrightarrow R_3SnCH_2CH_2R'$$
VIII

The R' moiety in the product *VIII* may contain a wide variety of groups and this is a general method (which is discussed in Chapter 7) for the preparation of functionally substituted organotin compounds.

A preparative method which is particularly attractive for commercial production is the direct reaction between metallic tin and organic halides. A detailed treatment of this method will be found in Chapter 4 since, in general, the major products are the organotin halides. However, tetra(perfluorophenyl)tin and di(perfluorobi-phenylene)tin (*IX*) are formed by heating, respectively, pentafluoro-iodobenzene and 3,4,5,6,3',4',5',6'-octafluoro-2,2'-diiodobiphenyl with tin at 240° in sealed tubes[16].

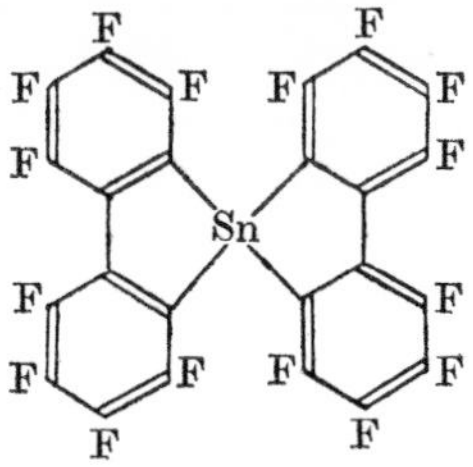

IX

This account of the preparation of alkyl- and aryl-tin compounds is not exhaustive but it includes all widely used methods and those less common procedures in which there is some current interest.

2.1.1. *Properties of alkyl- and aryl-tin compounds*

Compounds in this class are colourless with relatively high chemical and thermal stabilities. The symmetrical tetraaryltin compounds are highly crystalline solids which are sparingly soluble in organic solvents; the lower tetraalkyl compounds are liquids with higher solubilities and can be distilled at atmospheric pressure without decomposition.

TABLE II

Physical properties of some symmetrically substituted alkyl- and aryl-tin compounds

Compound	m.p.	b.p.	n_D^{20}	d^{20}
Me_4Sn	$-54°$	$78\cdot3°/740$ mm	$1\cdot4415$	$1\cdot2905\ (25°)$
Et_4Sn	$-136°$ to $-126°^a$	$181°$	$1\cdot4693\ (25°)$	$1\cdot1990$
Pr_4Sn	$-109°$	$221°-226°$	$1\cdot4748$	$1\cdot107$
Bu_4Sn	$-97°$	$127°/1\cdot7$ mm	$1\cdot4727$	$1\cdot0541$
$(C_6H_{11})_4Sn^b$	$261-263°$	—	—	—
$(PhCH_2)_4Sn$	$41\cdot5\!\!-\!\!43°$	—	—	—
Ph_4Sn	$228°$	—	—	$1\cdot521$
$(o\text{-}MeC_6H_4)_4Sn$	$217\cdot5-219\cdot5°$	—	—	—
$(m\text{-}MeC_6H_4)_4Sn$	$128\cdot4-129\cdot6°$	—	—	—
$(p\text{-}MeC_6H_4)_4Sn$	$238°$	—	—	—
$(1\text{-}C_{10}H_7)_4Sn$	$310-320°$	—	—	—
$(p\text{-}ClC_6H_4)_4Sn$	$199°$	—	—	—

a Several crystalline modifications.   *b* C_6H_{11} = cyclohexyl.

Chemical reactions which involve Sn—C bond fission are treated in Chapter 3 but mention should be made here of redistribution

TABLE III

Physical properties of some unsymmetrically substituted alkyl- and aryl-tin compounds

Compound	m.p.	b.p.	n_D^{20}	d^{20}
Me$_3$SnEt	—	106°/746 mm	1·4527	—
Me$_3$SnBu	—	149–150°/724 mm	1·4560	1·183
Me$_3$SnPh	—	205°	1·5330	—
Me$_3$SnCH$_2$Cl	—	44–48°/15 mm	1·4860 (25°)	1·556 (25°)
Et$_3$SnBu	—	73–75°/4 mm	1·4736	1·1457
Et$_3$SnPh	—	128·5°/12·5 mm	1·5349	—
Pr$_3$SnMe	—	94–96°/11 mm	—	1·125 (23°)
Pr$_3$SnPh	—	105°/15 mm	—	—
(C$_6$H$_{11}$)$_3$SnPh[a]	191°	—	—	—
Ph$_3$SnEt	56–58°	—	—	—
	(62–63°)			
Ph$_3$SnCH$_2$Ph	91–92°	—	—	—
Ph$_3$Sn(2-pyridyl)	178–179°	—	—	—
Ph$_3$SnC$_6$H$_4$Cl-p	141°	—	—	—
Me$_2$SnEt$_2$	—	131–132°	1·4650 (19°)	1·2319 (19°)
Me$_2$SnBu$_2$	—	70°/4·4 mm	1·4640 (25°)	1·124 (25°)
Me$_2$SnPh$_2$	60–61°	—	—	—
Et$_2$Sn(Pr-i)$_2$	—	74–75°/5 mm	1·4750	1·1513
Et$_2$SnPh$_2$	—	154–156°/4 mm	—	—
Me$_2$EtSnPr	—	153°/762 mm	—	—
Pr$_2$EtSnMe	—	183–184°/758 mm	—	—

[a] C$_6$H$_{11}$ = cyclohexyl.

reactions which may occur when unsymmetrical compounds are heated, for example:

$$2R_2SnR_2' \rightleftharpoons R_4Sn + R_4'Sn.$$

Equilibria of this type may be approached from either side and are catalysed by Lewis acids such as aluminium chloride[64, 75a].

A reaction which does not involve Sn—C bond cleavage is the insertion of dichlorocarbene into the C—H bonds of organotin compounds[95].

$$Me_3SnCH_2CH_2CH_3 + PhHgCCl_2Br \longrightarrow Me_3SnCH_2CH(CHCl_2)CH_3 + PhHgBr$$

Similarly, reactions at functional substituents in the organic groups can occur, for example[46]:

$$Et_3SnCH_2Cl + R_2NK \longrightarrow Et_3SnCH_2NR_2 + KCl.$$

The relative rates of halogen exchange for the reaction:

$$\text{Me}_3\text{MCH}_2\text{Cl} + \text{I}^\ominus \rightleftharpoons \text{Me}_3\text{MCH}_2\text{I} + \text{Cl}^\ominus$$

$$(\text{M} = \text{Si, Ge, Sn})$$

are Si 1·00, Ge 1·76, Sn 360 which is unexpected since the greater electron release associated with the Me_3Sn group would be expected to lower the rate. Substitution at carbon is, however, facilitated by initial coordination of the iodide ion to the metal(loid) atom and tin is the most powerful acceptor, also steric hindrance is minimised in the tin compound [6]. By alkaline hydrolysis of *tetrakis*(2-cyanoethyl)tin followed by acidification with the calculated quantity of hydrochloric acid at 0° the corresponding tetracarboxylic acid can be obtained [79].

$$\text{Sn}(\text{CH}_2\text{CH}_2\text{CN})_4 + 4\text{NaOH} + 4\text{H}_2\text{O} \longrightarrow \text{Sn}(\text{CH}_2\text{CH}_2\text{COONa})_4 + 4\text{NH}_3$$

$$\text{Sn}(\text{CH}_2\text{CH}_2\text{COONa})_4 + 4\text{HCl} \longrightarrow \text{Sn}(\text{CH}_2\text{CH}_2\text{COOH})_4 + 4\text{NaCl}$$

Treatment of the tetracarboxylic acid with diazomethane gives the corresponding tetramethyl ester $\text{Sn}(\text{CH}_2\text{CH}_2\text{COOMe})_4$ [135a].

A small but increasing number of publications deal with the exposure of organotin compounds to high energy radiation. For example, when tetramethyltin was exposed to X-rays, hexamethyl-distannane, *bis*(trimethylstannyl)methane, methane and ethane were obtained [30]. The electron spin resonance spectrum of γ-irradiated tetramethyltin indicated the presence of $\text{Me}_3\text{SnCH}_2\cdot$ radicals [53]. Studies have been made of the chemical effects created by the recoil atoms in nuclear reactions induced by neutron irradiation of tetra-phenyltin [14, 62]. When dilute solutions of tetraphenyltin in benzene were irradiated at 2537 Å the production of biphenyl was interpreted in terms of an energy transfer from excited benzene to the tin compound [72]. Tetraphenyltin dissolved in benzene experienced radio-sensitised decomposition when the solution was exposed to ^{60}Co γ-radiation [71a].

A number of authors have described gas chromatographic procedures for the separation of mixtures of tetraalkyltin and other organotin compounds [31, 76, 77, 113]. Gas chromatography and mass spectrometry were used to investigate the impurities present in a typical specimen of tetrabutyltin; the principal contaminants were Bu_3SnCl and $\text{Bu}_3\text{SnBu-}s$ [112].

A method has been published for the determination of tetrabutyl- and tetraethyl-tin in air [88].

2.2. ALKENYLTIN COMPOUNDS

In general the methods of attaching olefinic groups to tin, irrespective of the position of the double bond, are similar to those discussed above for alkyl- and aryl-tin compounds. Vinylmagnesium halides, made in tetrahydrofuran, give good yields of vinyltin compounds with stannic chloride or organotin chlorides [92].

$$R_{4-n}SnCl_n + nCH_2{=}CHMgX \longrightarrow R_{4-n}Sn(CH{=}CH_2)_n + nMgClX$$
$$(n = 1, 2, 3, 4)$$

Seyferth and his coworkers have shown that when propenyltin compounds are made by this method, for example:

$$Me_3SnBr + MeCH{=}CHMgBr \longrightarrow Me_3SnCH{=}CHMe + MgBr_2$$

the reaction proceeds with retention of configuration. The yield of trimethylprop-1-enyltin was 42% and this was more than doubled when prop-1-enyllithium was used instead of the Grignard reagent; in this case also the configuration at the double bond was retained [93]. Similar methods are used for the preparation of β-styrenyltin compounds such as $PhCH{=}CHSnMe_3$ [94].

An alternative route to vinyltin compounds, which is discussed in Chapter 7, is by the addition of organotin hydrides to acetylenes.

The preparation of alkenyltin compounds in which the double bond is in the allylic position presents little difficulty and the Grignard method is usually employed [86, 110]; p-styrenyltin compounds of the type $(p{-}CH_2{=}CHC_6H_4)_nSnR_{4-n}$ are made similarly [69, 114]. Cyclopentadienyl- and indenyl-tin compounds can be prepared from the hydrocarbon and the appropriate stannylamine [40] (see Chapter 6).

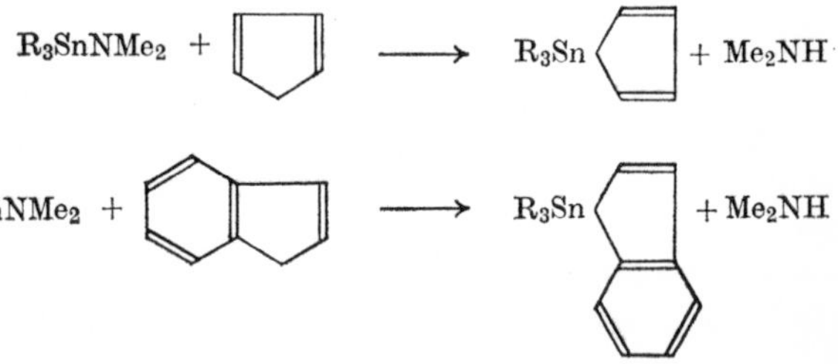

2.2.1. *Properties of alkenyltin compounds*

Although somewhat more reactive than the alkyl- and aryl-compounds the alkenyltin compounds are generally stable but some allyl and cyclopentadienyl derivatives polymerise on storage. Reactions in which the olefinic groups are cleaved from the tin atom are discussed in the general context of tin–carbon bond scission in Chapter 3. Here we consider those reactions which involve addition to the C=C bonds.

At room temperature addition of hydrogen polysulphide to allyltin compounds proceeded as follows:

$$R_3SnCH_2CH{=}CH_2 + H_2S_{5\cdot2} \longrightarrow [R_3SnCH_2CH(Me)]_2S_x.$$

The polysulphides formed contained 5–6 sulphur atoms and were decomposed on warming to give trialkyl(aryl)tin polysulphides $(R_3Sn)_2S_y$, sulphur and propylene[87]. Addition of thiocyanogen to the olefinic bond in p-trimethylstannylstyrene $(X, M = Sn)$ has been studied; the tin compound reacted more readily than the corresponding compounds X $(M = C, Ge$ and $Si)$[12]. In contrast the ease of hydrogenation, using a Raney nickel catalyst, of the same compounds (X) decreases in the order $M = Si > C > Ge > Sn$[13].

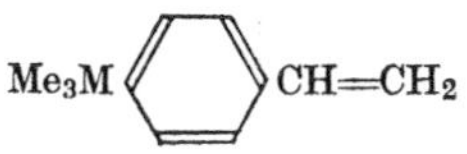

$$Me_3M \underset{X}{\left\langle \bigcirc \right\rangle} CH{=}CH_2$$

Under irradiation with ultraviolet light thiocarboxylic acids[133] and perfluoroalkyl iodides[132] add to alkenyltin compounds.

$$Bu_3SnCH{=}CH_2 + AcSH \xrightarrow{\text{U.V.}} Bu_3SnCH_2CH_2SAc$$

$$Cl_3SnCH{=}CH_2 + CF_3(CF_2)_8CF_2I \xrightarrow{\text{U.V.}} Cl_3SnCHICH_2C_{10}F_{21}$$

Partly because of the possibility of new technical applications the general topic of organotin polymers has attracted some attention and has been reviewed briefly[29, 34]. In other parts of this book descriptions are given of polymers produced by hydride additions (Chapter 7) and by the polymerisation of unsaturated organotin esters (Chapter 10); the discussion which follows is confined to the addition polymerisation of alkenyltin compounds.

In the presence of free-radical initiators, or on exposure to γ-rays, compounds containing vinyl or allyl groups directly attached to tin either do not polymerise or, at best, form low molecular weight polymers with great reluctance[28, 34, 63]. Although, on prolonged heating or storage, allyltin compounds polymerise, alkenyltin compounds, particularly the allyl derivatives, inhibit the polymerisation of methyl methacrylate, vinyl acetate and other vinyl monomers. The degree

3

TABLE IV

Physical properties of some alkenyltin compounds

Compound	m.p.	b.p.	n_D	d
$(CH_2{=}CH)_4Sn$	—	160–163°	1·4914 (25°) 1·4993 (25°)	1·257 (25°)
$(CH_2{=}CHCH_2)_4Sn$	—	87–88°/4 mm	1·5324 (32°)	1·243 (30·5°)
$CH_2{=}CHSnMe_3$	—	99–100°	1·4536 (25°)	1·265 (25°)
$CH_2{=}CHSnPh_3$	45°, 39–40°	—	—	—
$CH_2{=}CHCH_2SnMe_3$	—	128–130°	1·4734 (20°)	1·2547 (20°)
$CH_2{=}CHCH_2SnPh_3$	75·5–76°	—	—	—
$Ph_3SnC_6H_4CH{=}CH_2\text{-}p$	112–113°, 105·5–108° 102–105°	—	—.	—
$Ph_3Sn\diagdown$ 	130–131°	—	—	—
$(CH_2{=}CH)_2SnBu_2$	—	78–80°/2 mm	1·4749 (25°)	1·122 (25°)
$(CH_2{=}CH)_2SnPh_2$	—	153–154°/5 mm	1·5949 (25°)	1·334 (25°)

of inhibition depends upon the number of unsaturated groups attached to tin, tetraallyltin being the most effective [34]. However, copolymerisation of trialkylvinyltin compounds with styrene or methyl methacrylate, under free-radical conditions, is possible [63].

Triethylvinyltin was polymerised to give a vitreous solid and copolymerised with styrene to give a white powder, in both cases butyllithium was the catalyst [73, 140]. However, anionic polymerisations of this nature tend to be self-inhibiting due to inactivation of the catalyst by the formation of complexes such as $R_3SnCH{=}CH_2 . 2BuLi$ [1]. Some vinyl- and allyl-tin compounds give highly crystalline, isotactic polymers with Ziegler catalysts [130]. When 2-tributylstannyl-butadiene was polymerised the product had an all-*cis* configuration due to the large steric requirement of the Bu_3Sn group [3].

$$nCH_2{=}C(SnBu_3)CH{=}CH_2 \longrightarrow {+}CH_2C(SnBu_3){=}CHCH_2{+}_n$$

There are a number of accounts of the polymerisation of styrenyltin compounds of the type *XI* (R = H, Me; R′ = Me, Et, cyclohexyl, Ph)[29, 34, 70]. These compounds are usually polymerised under free-radical conditions to give colourless solids which form transparent films. With azo*bis*isobutyronitrile as initiator, trimethyl(*p*-vinylphenyl)tin (*XI*, R = H; R′ = Me) was copolymerised with butadiene giving a product of molecular weight 160,000 which could be vulcanised to give a rubber[65]. The same styrenyltin derivative (XI, R = H, R' = Me) is more readily polymerised than styrene[47]; these two monomers give block copolymers[82].

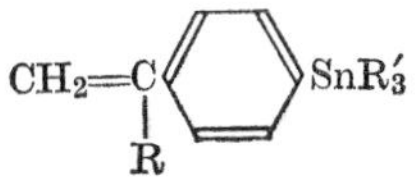

$$CH_2=C\underset{R}{\overset{|}{}}\hspace{-0.5em}\left\langle\hspace{-0.3em}\bigcirc\hspace{-0.3em}\right\rangle SnR_3'$$

XI

A novel approach to the production of polymers containing tin is the polymerisation of a solution of tetraphenyltin in styrene to give a product containing up to 11·1% of tin[5].

2.3. ALKYNYLTIN COMPOUNDS

This is a rapidly developing area of organotin chemistry and has been reviewed recently by Russian[58] and American[18] authors. The latter review covered the literature to the end of 1965 and some 120 alkynyltin compounds are listed.

Reactions between acetylenic Grignard reagents or Group 1 metal acetylides and stannic or organotin halides are the most common preparative procedures. The following equations for the preparation of stannic acetylide, stannic p-chlorophenylacetylide and trimethylsilyl(triphenylstannyl)acetylide illustrate these methods:

$$SnCl_4 + 4NaC\equiv CH \longrightarrow Sn(C\equiv CH)_4 + 4NaCl \qquad \text{(ref. 81)}$$

$$SnCl_4 + 4p\text{-}ClC_6H_4C\equiv CLi \longrightarrow Sn(C\equiv CC_6H_4Cl\text{-}p)_4 + 4LiCl \qquad \text{(ref. 26)}$$

$$Ph_3SnCl + Me_3SiC\equiv CMgBr \longrightarrow Me_3SiC\equiv CSnPh_3 + MgBrCl \qquad \text{(ref. 48)}$$

When sodium acetylide is treated with trialkyltin chlorides the product, $R_3SnC\equiv CH$, may be accompanied by considerable amounts of the distannylacetylenes $R_3SnC\equiv CSnR_3$[121]. Trialkyltin hydroxides react with sodium acetylide in the same manner as organotin chlorides giving a mixture of mono- and di-stannylacetylenes[42]. Sodioacetylenes have been used to make $Ph_3SnC\equiv CH$ and $Ph_3SnC\equiv CSnPh_3$[55] and to introduce the stannyl group into more highly substituted acetylenes[99].

$$CH_2=CHOCH_2CH_2OCH_2C\equiv CNa + Et_3SnCl \longrightarrow$$

$$CH_2=CHOCH_2CH_2OCH_2C\equiv CSnEt_3 + NaCl$$

Many stannyl derivatives of acetylene carboxylates evolve carbon dioxide when heated a few degrees above their melting points to give alkynyltin compounds in moderate to high yields[57a].

$$R_3SnOCOC\equiv CCOOSnR_3 \longrightarrow R_3SnC\equiv CSnR_3 + 2CO_2$$

When the disodium derivative of butadiyne, in liquid ammonia, was treated with trimethyltin chloride bis(trimethylstannyl)butadiyne (*XII*) was obtained[27].

$$2Me_3SnCl + NaC\equiv CC\equiv CNa \longrightarrow Me_3SnC\equiv CC\equiv CSnMe_3 + 2NaCl$$

XII

The trimethylstannyl groups in *XII* can be replaced by Ar_3Sn- or $(cycloC_6H_{11})_3Sn-$ groups by treatment with the appropriate organotin chloride[27].

$$2(p\text{-}MeC_6H_4)_3SnCl + Me_3SnC\equiv CC\equiv CSnMe_3 \longrightarrow$$
$$(p\text{-}MeC_6H_4)_3SnC\equiv CC\equiv CSn(C_6H_4Me\text{-}p)_3 + 2Me_3SnCl$$

Similar derivatives of butadiyne, such as $Ph_3SnC\equiv CC\equiv CSnPh_3$, have been prepared by the Grignard method[55] which has also been used to prepare $F_3CC\equiv CSnMe_3$[17]. Dialkynyltin compounds result from reaction between acetylenic Grignard reagents and organotin oxides[101].

$$2RC\equiv CMgBr + R'_2SnO \longrightarrow (RC\equiv C)_2SnR'_2 + MgBr_2 + MgO$$

Propargyltin compounds, made by the Grignard method, undergo partial isomerisation to the corresponding allene derivatives[56, 61, 106].

$$\gtrdot SnCHRC\equiv CR' \rightleftharpoons \gtrdot SnCR\!=\!C\!=\!CHR'$$

It is also possible to make alkynyltin compounds from haloacetylenes and organotin sodium compounds, for example[118]:

$$Et_3SnNa + BrC\equiv CC\equiv CEt \longrightarrow Et_3SnC\equiv CC\equiv CEt + NaBr.$$

When bromoacetylene itself is used the hydrogen atom is replaced first to give the stannylbromoacetylene which then reacts further[119, 120].

$$R_3SnNa + HC\equiv CBr \longrightarrow R_3SnC\equiv CBr + NaH$$
$$R_3SnNa + R_3SnC\equiv CBr \longrightarrow R_3SnC\equiv CSnR_3 + NaBr$$

In the presence of powdered potassium hydroxide, triethyltin chloride reacts with phenylacetylene to give $Et_3SnC\equiv CPh$ in 80% yield[103].

A different type of procedure for attaching acetylenic groups to tin is protolysis of an Sn—O bond by the acidic hydrogen of an acetylene [67, 123, 125, 139].

$$2RC{\equiv}CH + (R_3'Sn)_2O \longrightarrow 2RC{\equiv}CSnR_3' + H_2O$$

When the water is removed azeotropically or chemically a range of acetylenic groups, including some containing functional substituents such as —C≡CCN, can be introduced in yields of 80–90% [67]. Dehydration can also be effected by fractional distillation under reduced pressure as in the preparation of $Et_3SnC{\equiv}CPh$ [104] and the same method has been used recently to prepare $Et_3SnC{\equiv}CCH{=}CHOBu$, $Et_3SnC{\equiv}CCH{=}CHSBu$ and $Et_3SnC{\equiv}CCH{=}CHNEt_2$ [100]. The reaction takes a different course when the acetylenic moiety contains ester groups, for example, when propargyl acetate and *bis*(triethyltin)-oxide reacted the alkoxide (*XIII*) was formed:

$$HC{\equiv}CCH_2OAc + (Et_3Sn)_2O \longrightarrow HC{\equiv}CCH_2OSnEt_3 + AcOSnEt_3$$
$$XIII$$

Compound *XIII* was also formed in the corresponding reaction with propargyl alcohol [67]. When the reaction is not carried out under dehydrating conditions the compounds react in an equimolar ratio giving the organotin hydroxide as a by-product, for example [44]:

$$Me_3SiC{\equiv}CH + (Et_3Sn)_2O \longrightarrow Et_3SnOH + Me_3SiC{\equiv}CSnEt_3.$$

Shostakovskii and his coworkers [102] have shown that, in some cases better yields are obtained by using a trialkyltin alkoxide in place of the *bis*(trialkyltin) oxide and have used this modification of the method to prepare the compounds $R_3SnC{\equiv}CCH{=}CH_2$ [124, 129] and $R_3SnC{\equiv}CCH_2OCHMeOBu$ [105].

A similar route to alkynyltin compounds is protolysis of Sn—N bonds by acetylenes, for example [39]:

$$Et_3SnNMe_2 + PhC{\equiv}CH \longrightarrow Et_3SnC{\equiv}CPh + Me_2NH$$
$$2R_3SnNMe_2 + HC{\equiv}CH \longrightarrow R_3SnC{\equiv}CSnR_3 + 2Me_2NH.$$

A method for the preparation of organometallic acetylenes $R_3MC{\equiv}CM'R_3''$ in which M and M' are different Group IVb metals is by reactions which are thought to proceed as follows:

$$R_3MNR_2' + NaC{\equiv}CH \longrightarrow R_3MC{\equiv}CNa + R_2'NH$$
$$R_3MC{\equiv}CNa + R_3''M'Cl \longrightarrow R_3MC{\equiv}CM'R_3'' + NaCl.$$

Using this procedure the following compounds were prepared, $Ph_3SnC{\equiv}CSiPh_3$, $Ph_3SnC{\equiv}CGePh_3$, $Ph_3SnC{\equiv}CPbPh_3$, $Me_3SnC{\equiv}CSiMe_3$ and $Me_3SnC{\equiv}CGeMe_3$[22].

2.3.1. Properties of alkynyltin compounds

As would be expected the alkynyltin compounds are generally more reactive than the other types of compounds described in this chapter. The care needed in handling stannic acetylide is stressed; a specimen of this compound exploded violently during a vacuum transfer operation[81]. As before, discussion here is restricted to reactions not involving Sn—C bond cleavage.

Seyferth and his coworkers showed that *bis*(trimethyltin)acetylide and tetraphenylcyclopentadienone when heated together in boiling xylene for 138 hours gave a quantitative yield of 1,2-*bis*(trimethylstannyl)3,4,5,6-tetraphenylbenzene (*XIV*) and carbon monoxide, presumably by decomposition of an initially formed Diels-Alder adduct[91]:

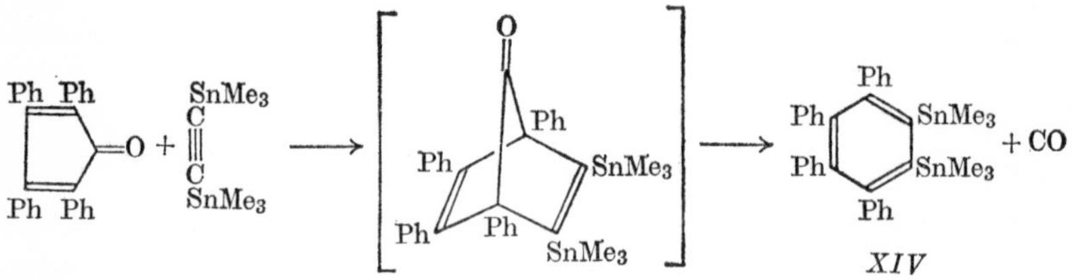

The monostannylacetylides $Me_3SnC{\equiv}CMe$ and $Me_3SnC{\equiv}CPh$ reacted similarly. When perchlorocyclopentadiene was used instead of the tetracyclone the bridged Diels–Alder adduct *XV* could be isolated[90].

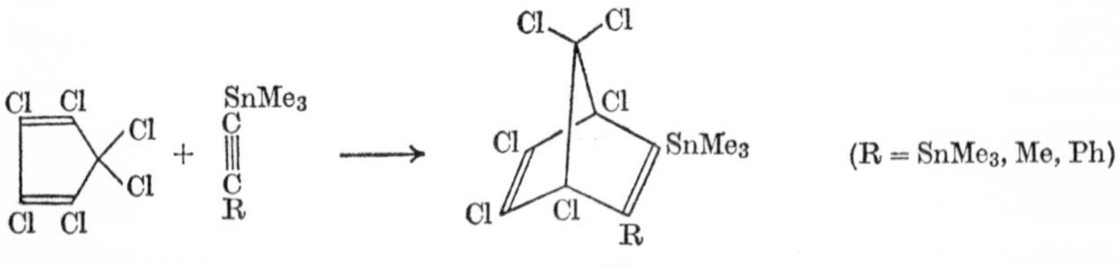

Triethylstannylacetylene reacts similarly[43]. The products (*XV*) were crystalline and reasonably stable but decomposed on extended storage.

When XV (R = SnMe$_3$ or Me) was irradiated with ultraviolet light the quadricyclane XVI (R = SnMe$_3$ or Me) was formed.

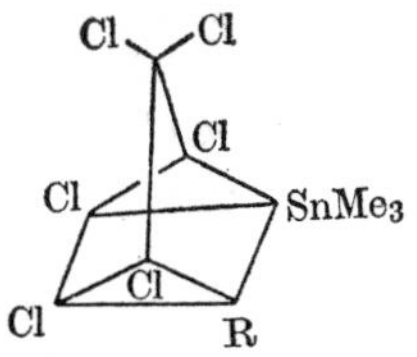

XVI

On replacing the perchlorocyclopentadiene by the closely related ketal XVII the Diels–Alder adduct again becomes unstable. The products isolated show that, in addition to the loss of bridgehead carbon, some Sn—C bond cleavage has occurred[90].

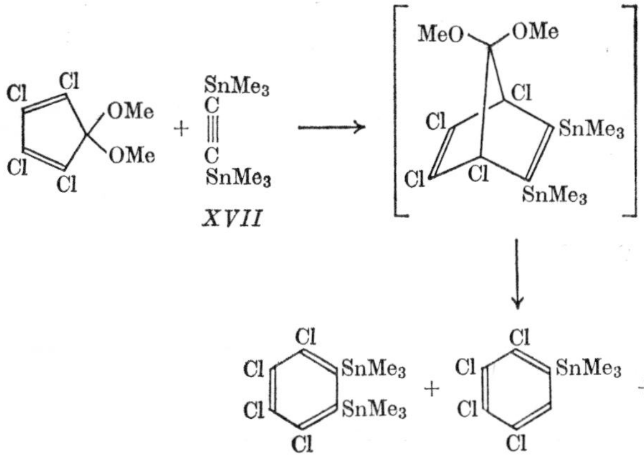

In an extension of this work it was shown that, by using α-pyrones (XVIII) as the dienes, functionally substituted aryltin compounds could be prepared[20].

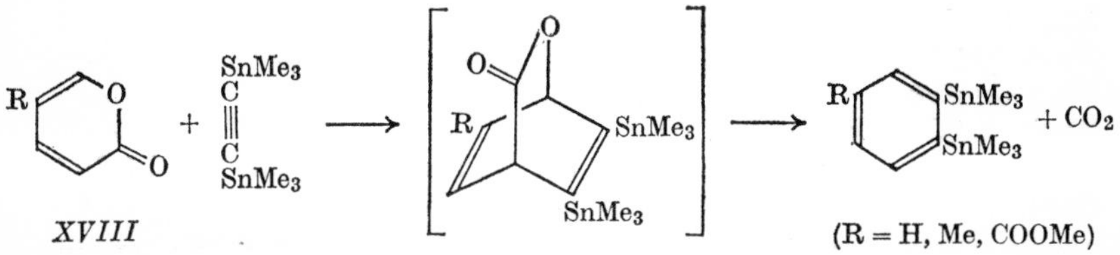

Due to unfavourable steric effects the dimethyl-α-pyrone XIX failed to react.

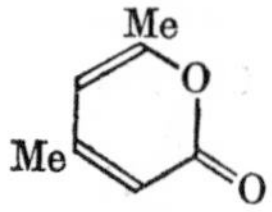

XIX

Triphenylstannyl phenyl acetylene undergoes a Diels–Alder reaction with 3,6-diphenyltetrazine yielding an unstable product which loses nitrogen to give a pyridazine [66a].

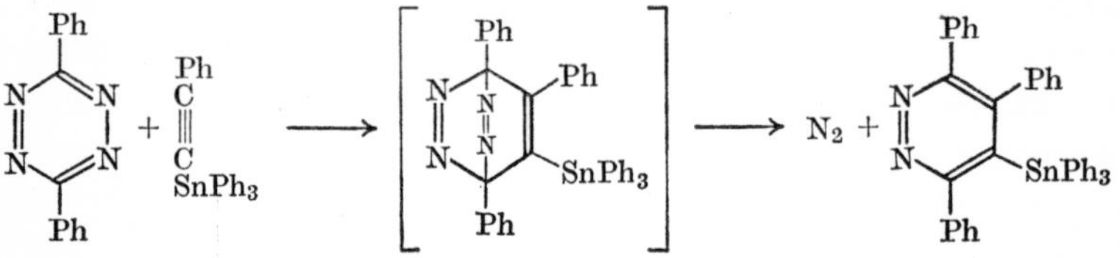

Diazomethane adds to the triple bond of triethylstannylacetylene to give a stannylpyrazole [97].

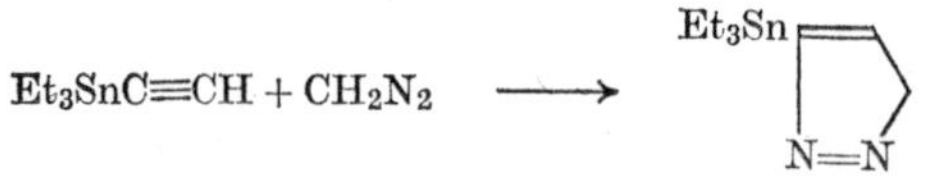

$$Et_3SnC{\equiv}CH + CH_2N_2 \longrightarrow$$

TABLE V

Physical properties of some alkynyltin compounds

Compound	m.p.	b.p.	n_D^{20}	d^{20}
$Me_3SnC{\equiv}CH$	—	86–92°	—	—
$Me_3SnC{\equiv}CPh$	67°	60°/0·2 mm	1·5720	1·3324
$Bu_3SnC{\equiv}CH$	—	85–90°/0·1 mm	—	—
$Bu_2Sn(C{\equiv}CPh)_2$	14°	—	1·588 (21°)	—
$BuSn(C{\equiv}CPh)_3$	70°	—	—	—
$Ph_3SnC{\equiv}CH$	34°	—	—	—
$Ph_3SnC{\equiv}CPh$	62°	—	—	—
$Ph_3SnC{\equiv}CC{\equiv}CSnPh_3$	68°	—	—	—
$Et_3SnC{\equiv}CCH_2OCHMeOBu$	—	105°/0·3 mm	1·4754	1·1474
$Et_3SnC{\equiv}CCHMeOSnEt_3$	—	111–112°/0·5 mm	1·5039	1·3313

Triethylstannylhaloacetylenes undergo Arbuzov reactions with triethylphosphite[117].

$$Et_3SnC\equiv CX + (EtO)_3P \longrightarrow Et_3SnC\equiv CP(O)(OEt)_2 + EtX$$
$$(X = Cl, Br)$$

Polymers containing $C\equiv C$ bonds and tin atoms in the backbone have been obtained by polymerisation of compounds of the types $R_2Sn(C\equiv CH)_2$ and $R_2Sn(CH_2C\equiv CH)_2$ (R = alkyl or aryl)[59].

References

1. Aleksandrov, A. Yu., Gol'danskii, V. I., Korytko, L. A., Mal'tsev, V. A. and Plate, N. A., *Vysokomol. Soedin., Ser. B*, 1968, **10**, 209 (*Chem. Abs.*, 1968, **69**, 3925).
2. Atwell, W. H. and Weyenberg, D. R., *J. Org. Chem.*, 1967, **32**, 885.
3. Aufdermarsh, C. A. and Pariser, R., *J. Polymer Sci., Pt. A*, 1964, **2**, 4727.
4. Bajer, F. J. and Post, H. W., *J. Organometallic Chem.*, 1968, **11**, 187.
5. Baroni, E. E., Kilin, S. F., Lebsadze, T. N., Rozman, I. M. and Shoniya, V. M., *At. Energ.* (*U.S.S.R.*), 1964, **17**, 497 (*Chem. Abs.*, 1965, **62**, 10604h).
6. Bott, R. W., Eaborn, C. and Swaddle, T. W., *J. Organometallic Chem.*, 1966, **5**, 233.
6a. Boué, S., Gielen, M. and Nasielski, J., *Tetrahedron Letters*, 1968, 1047.
7. Brago, I. N., Kaabak, L. V. and Tomilov, A. P., *Zh. Vses. Khim. Obshchest.*, 1967, **12**, 472 (*Chem. Abs.*, 1967, **67**, 104513u).
8. Bresadola, S., Rossetto, F. and Tagliavini, G., *Chem. Comm.*, 1966, 623.
9. Bresadola, S., Rossetto, F. and Tagliavini, G., *Ann. Chim.* (*Rome*), 1968, **58**, 597 (*Chem. Abs.*, 1968, **69**, 77385).
10. Bresadola, S., Rosetto, F. and Tagliavini, G., *Eur. Polym. J.*, 1968, **4**, 75.
11. Burdon, J., Coe, P. L. and Fulton, M., *J. Chem. Soc.*, 1965, 2094.
12. Chernyshev, E. A., Zelenetskaya, A. A. and Krasnova, T. L., *Izv. Akad. Nauk S.S.S.R. Ser. Khim.*, 1966, 1118.
13. Chernyshev, E. A., Zhukova, I. F., Krasnova, T. L. and Freidlin, L. K. H., *Zh. Obshch. Khim.*, 1968, **38**, 504 (*Chem. Abs.*, 1968, **69**, 51331).
14. Claridge, R. F. C., Merz, E. and Riedel, H. J., *Nukleonik*, 1965, **7**, 53 (*Chem. Abs.*, 1965, **62**, 16293d).
15. Cohen, S. C. and Massey, A. G., *J. Organometallic Chem.*, 1967, **10**, 471.
16. Cohen, S. C., Reddy, M. L. N. and Massey, A. G., *Chem. Comm.*, 1967, 451.
17. Cullen, W. R. and Waldman, M. C., *Inorg. Nucl. Chem. Lett.*, 1968, **4**, 205.
18. Davidsohn, W. E. and Henry M. C., *Chem. Rev.*, 1967, **67**, 73.
19. Eaborn, C., Hornfeld, H. L. and Walton, D. R. M., *J. Organometallic Chem.*, 1967, **10**, 529.
20. Evnin, A. B. and Seyferth, D., *J. Amer. Chem. Soc.*, 1967, **89**, 952.
21. Fenton, D. E. and Massey, A. G., *Tetrahedron*, 1965, **21**, 3009.
22. Findeiss, W., Davidsohn, W. E. and Henry, M. C., *J. Organometallic Chem.*, 1967, **9**, 435.
23. Foldesi, I. and Gomory, P., *Acta Chim. Acad. Sci. Hung.*, 1965, **45**, 231 (*Chem. Abs.*, 1965, **63**, 16377h).

23a. Gielen, M., Nasielski, J. and Topart, J., *Rec. Trav. Chim.*, 1968, **87**, 1051.
24. Gilman, H. and See-Yuen Sim, *J. Organometallic Chem.*, 1967, **7**, 249.
25. Gilman, H. and Zuech, E. A., *J. Amer. Chem. Soc.*, 1960, **82**, 2522.
26. Hartmann, H. and El A'ssar, M. K., *Naturwissenschaften*, 1965, **52**, 304.
27. Hartmann, H., Karbstein, B. and Reiss, W., *Naturwissenschaften*, 1965, **52**, 59.
28. Hayakawa, K., Kawase, K. and Matsuda, T., *Nature*, 1965, **206**, 1038.
29. Henry, M. C. and Davidson, W. E., *Ann. New York Academy of Sciences*, 1965, **125**, 172.
30. Hoeppner, K., *Proc. Tihany Symp. Radiat. Chem. 2nd Tihany. Hung.*, 1966, 33 (*Chem. Abs.*, 1967, **67**, 5948w).
31. Hoeppner, K., Proesch, U. and Zoepfl, H. J., *Abh. Deut. Akad. Wiss. Berlin, Kl. Chem., Geol. Biol.*, 1966, 393 (*Chem. Abs.*, 1967, **67**, 65827c).
32. Holmes, J. M., Peacock, R. D. and Tatlow, J. C., *J. Chem. Soc. (A)*, 1966, 150.
33. Horder, J. R. and Lappert, M. F., *J. Chem. Soc. (A)*, 1968, 1167.
34. Ingham, R. K. and Gilman, H., *Inorganic Polymers*. Ed. Stone, F. G. A. and Graham, W. A. G., Academic Press, New York and London, 1962, 368.
35. Ingham, R. K., Rosenberg, S. D. and Gilman, H., *Chem. Rev.*, 1960, **60**, 459.
36. Jaura, K. L., Churamani, L. K. and Sharma, K. K., *Indian J. Chem.*, 1966, **4**, 329.
37. Jaura, K. L., Hundal, H. S. and Handa, R. D., *Indian J. Chem.*, 1967, **5**, 211.
38. Johnson, W. A., *Tin and its Uses*, 1962, 5.
39. Jones, K. and Lappert, M. F., *Organometallic Chem. Revs.*, 1966, **1**, 67.
40. Jones, K. and Lappert, M. F., *J. Organometallic Chem.*, 1965, **3**, 295.
41. Kazankova, M. A., Belinka, M. A. and Lutsenko, I. F., *Zh. Obshch. Khim.*, 1967, **37**, 1710.
42. Komarov, N. V., Guseva, I. S. and L'vova, F. P., *Izv. Akad. Nauk S.S.S.R. Ser. Khim.*, 1966, 1479.
43. Komarov, N. V., Shostakovskii, M. F. and Burnashova, T. D., *Zh. Obshch. Khim.*, 1968, **38**, 1398 (*Chem. Abs.* 1968, **69**, 87132).
44. Komarov, N. V. and Yarosh, O. G., *Zh. Obshch. Khim.*, 1966, **36**, 101.
45. Korotaevskii, K. N., Lysenko, E. N., Smolyan, Z. S., Monastryskii, L. M. and Armenskaya, L. V., *Zh. Obshch. Khim.*, 1966, **36**, 167.
46. Kostyanovskii, R. G. and Prokof'ev, A. K., *Izv. Akad. Nauk S.S.S.R. Ser. Khim.*, 1965, 175.
47. Koton, M. M. and Dokukina, L. F., *Vysokomol. Soedin.*, 1964, **6**, 1791.
48. Kraihanzel, C. S. and Losee, M. L., *J. Organometallic Chem.*, 1967, **10**, 427.
49. Kuivila, H. G. and Beumel, O. F., *J. Amer. Chem. Soc.*, 1958, **80**, 3250.
50. Kupchik, E. J. and Perciaccante, V. A., *J. Organometallic Chem.*, 1967, **10**, 181.
51. Kupchik, E. J., Ursino, J. A. and Boudjouk, P. R., *J. Organometallic Chem.*, 1967, **10**, 269.
52. Kuschk, R., Kaltwasser, H. and Braun, W., *Chem. Tech. (Berlin)*, 1965, **17**, 749 (*Chem. Abs.*, 1966, **64**, 9761a).
53. Lassman, G. and Hoeppner, K., *Z. Naturforsch. A*, 1968, **23**, 622.
54. Leavitt, F. C., Manuel, T. A., Johnson, F., Matternas, L. U. and Lehmann, D. S., *J. Amer. Chem. Soc.*, 1960, 82, 5099.

55. Le Quan, M. and Cadiot, P., *Bull. Soc. Chem. France*, 1965, 35.
56. Le Quan, M. and Cadiot, P., *Bull. Soc. Chem. France*, 1965, 45.
57. Luijten, J. G. A. and van der Kerk, G. J. M., *Investigations in the Field of Organotin Chemistry*, Tin Research Institute, Greenford, 1955.
57a.Luijten, J. G. A. and van der Kerk, G. J. M., *Rec. Trav. Chim. Pays-Bas*, 1964, **83**, 295.
58. Luneva, L. K., *Usp. Khim.*, 1967, **36**, 1140.
59. Luneva, L. K., Sladkov, A. M. and Korshak. V. V., *Vysokomol. Soedin.*, 1965, **7**, 427 (*Chem. Abs.*, 1965, **63**, 1879e).
60. Maire, J. C., *Ann. Chim. Paris*, 1961, **6**, 969.
61. Masson, J. C., Le Quan, M. and Cadiot, P., *Bull. Soc. Chim. France*, 1967, 777.
62. Merz, E. and Riedel, H. J., *Chem. Effects Nucl. Transformations, Proc. Symp., Vienna*, 1964, **2**, 179 (*Chem. Abs.*, 1965, **63**, 9314c).
63. Minoura, Y., Suzuki, Y., Sakanaka, Y. and Doi, H., *J. Polymer Sci. Pt. A*, 1966, **4**, 2757.
64. Moedritzer, K., *Organometallic Chem. Revs.*, 1966, **1**, 179.
65. Nagibina, T. D., Yasenkova, L. S., Alikberova, G. I., Petrov, A. D., Chernyshev, E. A. and Krasnova, T. L., *Kauch. Rezina*, 1966, **26**, 2 (*Chem. Abs.*, 1967, **66**, 38678c).
66. Neumann, W. P., *Angew. Chem. Intern. Ed. Engl.*, 1963, **2**, 165.
66a.Neumann, W. P. and Kleiner, F. G., *Annalen*, 1968, **716**, 29.
67. Neumann, W. P., and Kleiner, F. G., *Tetrahedron Letters*, 1964, 3779.
68. Nguyen, D. H., Fainberg, V. S., Baukov, Yu. I. and Lutsenko, I. F., *Zh. Obshch. Khim.*, 1968, **38**, 191 (*Chem. Abs.*, 1968, **69**, 19274).
69. Noltes, J. G., Budding, H. A. and van der Kerk, G. J. M., *Rec. Trav. Chim.*, 1960, **79**, 408.
70. Noltes, J. G., Budding, H. A. and van der Kerk, G. J. M., *Rec. Trav. Chim.*, 1960, **79**, 1076.
71. Peddle, G. J. D., *J. Organometallic Chem.*, 1968, **14**, 139.
71a. Peterson, D. B., Arakawa, T., Walmsley, D. A. G. and Burton, M., *J. Phys. Chem.*, 1965, **69**, 2880.
72. Peterson, D. B., Walmsley, D. A. G., Povinelli, R. J. and Burton, M., *J. Phys. Chem.*, 1967, **71**, 4506.
73. Plate, N. A., Mal'tsev, V. V., Davydova, S. L. and Kargin, V. A., *Vysokomol. Soedin.*, 1966, **8**, 1890 (*Chem. Abs.*, 1967, **66**, 18886r).
74. Pohlmann, J. L. W., Brinckman, F. E., Tesi, G. and Donadio, R. E., *Z. Naturforsch.*, 1965, **206**, 1.
75. Pollard, F. H., Nickless, G. and Nolan, D. N., *Chem. and Ind.*, 1965, 1027.
75a. Pollard, F. H., Nickless, G. and Uden, P. C., *J. Chromatog.*, 1965, **19**, 28.
76. Putnam, R. C. and Pu, H., *J. Gas Chromatog.*, 1965, **3**, 160.
77. Putnam, R. C. and Pu, H., *J. Gas Chromatog.*, 1965, **3**, 289.
78. Rausch, M. D. and Klemann, L. P., *J. Amer. Chem. Soc.*, 1967, **89**, 5732.
79. Reifenberg, G. H. and Considine, W. J., *J. Organometallic Chem.*, 1967, **9**, 495.
80. Rubinchik, G. F. and Manulkin, Z. M., *Zh. Obshch. Khim.*, 1966, **36**, 261; 748; 1301.
81. Sacher, R. E., Lemmon, D. H. and Miller, F. A., *Spectrochim. Acta*, 1967, **23A**, 1169.

34   THE CHEMISTRY OF ORGANOTIN COMPOUNDS

82. Sandler, S. R., Dannin, J. and Tsou, K. C., *J. Polymer Sci. Pt. A*, 1965, **3**, 3199.
83. Schmidbauer, H. and Tronich, W., *Chem. Ber.*, 1967, **100**, 1032.
84. Schneider, B. and Neumann, W. P., *Annalen*, 1967, **707**, 7.
85. Schroeder, H., Papetti, S., Alexander, R. P., Sieckhaus, J. F. and Heying, T. L., *U.S. Clearinghouse Fed. Sci. Tech. Inform.*, *A.D.* 652379 (*Chem. Abs.*, 1968, **68**, 134463); *Inorg. Chem.*, 1969, **8**, 2444.
86. Schwartz, W. T. and Post, H. W., *J. Organometallic Chem.*, 1964, **2**, 357.
87. Schwartz, W. T. and Post, H. W., *J. Organometallic Chem.*, 1964, **2**, 425.
88. Selivokhin, P. I., *Gigiena i sanit.*, 1966, **31**, 68 (*Chem. Abs.*, 1966, **65**, 15973g).
89. Seyferth, D., Armbrecht, F. M., Prokai, B. and Ross, R. J. C., *J. Organometallic Chem.*, 1966, **6**, 573.
89a. Seyferth, D. and Andrews, S. B., *J. Organometallic Chem.*, 1969, **18**, P21.
89b. Seyferth, D. and Armbrecht, F. A., *J. Organometallic Chem.*, 1969, **16**, 249.
90. Seyferth, D. and Evnin, A. B., *J. Amer. Chem. Soc.*, 1967, **89**, 1468.
91. Seyferth, D., Sarafidis, G. and Evnin, A. B., *J. Organometallic Chem.*, 1964, **2**, 417; 437.
92. Seyferth, D. and Stone, F. G. A., *J. Amer. Chem. Soc.*, 1957, **79**, 515.
93. Seyferth, D. and Vaughan, L. G., *J. Organometallic Chem.*, 1963, **1**, 138.
94. Seyferth, D., Vaughan, L. G. and Suzuki, R., *J. Organometallic Chem.*, 1964, **1**, 437.
95. Seyferth, D. and Washburne, S. S., *J. Organometallic Chem.*, 1966, **5**, 389.
96. Seyferth, D. and Welch, D. E., *J. Organometallic Chem.*, 1964, **2**, 1.
97. Sharanina, L. G., Zavgorodnii, V. S. and Petrov, A. A., *Zh. Obshch. Khim.*, 1968, **38**, 1146 (*Chem. Abs.*, 1968, **69**, 77374).
98. Shiina, K., Brennan, T. and Gilman, H., *J. Organometallic Chem.*, 1968, **11**, 471.
99. Shostakovskii, M. F., Atavin, A. S., Vyalykh, E. P., Trofimov, B. A. and Yakubov, R. D., *Izv. Akad. Nauk S.S.S.R. Ser. Khim.*, 1967, 2118.
100. Shostakovskii, M. F., Komarov, N. V., Burnashova, T. D. and Akchurina, I. S., *Izv. Akad. Nauk S.S.S.R. Ser. Khim.*, 1968, 625 (*Chem. Abs.*, 1968, **69**, 67500).
101. Shostakovskii, M. F., Komarov, N. V., Misyunas, V. K. and Sklyanova, A. M., *Dokl. Akad. Nauk S.S.S.R.*, 1965, **161**, 370.
102. Shostakovskii, M. F., Vlasov, V. M. and Mirskov, R. G., *Dokl. Akad. Nauk S.S.S.R.*, 1964, **159**, 689.
103. Shostakovskii, M. F., Vlasov, V. M. and Mirskov, R. G., *Zh. Obshch. Khim.*, 1965, **35**, 750.
104. Shostakovskii, M. F., Vlasov, V. M., Mirskov, R. G. and Koroteva, I. M., *Zh. Obshch. Khim.*, 1965, **35**, 401.
105. Shostakovskii, M. F., Vlasov, V. M., Mirskov, R. G. and Loginova, I. E., *Zh. Obshch. Khim.*, 1964, **34**, 3178.
106. Sladkov, A. M. and Luneva, L. K., *Zh. Obshch. Khim.*, 1966, **36**, 553.
107. Srivastava, T. N. and Bhattacharya, S. N., *Z. Anorg. Allgem. Chem.*, 1966, **344**, 102.
108. Stegmann, H. B. and Scheffler, K., *Tetrahedron Letters*, 1964, 3387.
109. Tamborski, C., Soloski, E. J. and Dec, S. M., *J. Organometallic Chem.*, 1965, **4**, 446.

110. Tillyaev, K. S. and Manulkin, Z. M., *Dokl. Akad. Nauk Uz. S.S.R.*, 1965, **22**, 45 (*Chem. Abs.*, 1965, **63**, 5668c).
111. Tomilov, A. P., Smirnov, Yu. D. and Varshavskii, S. L., *Zh. Obshch. Khim.*, 1965, **35**, 391.
112. Umilin, V. A. and Tsinovoi, Yu. N., *Izv. Akad. Nauk. S.S.S.R., Ser. Khim.*, 1968, 1409 (*Chem. Abs.*, 1968, **69**, 87138).
113. Wowk, A. and Dogiovanni, S., *Anal. Chem.*, 1966, **38**, 742.
114. Yakubova, F. A., Rashkes, A. M., Kuchkarev, A. B. and Manulkin, Z. M., *Zh. Obshch. Khim.*, 1965, **35**, 387.
115. Zakharin, L. I., Bregadze, V. I. and Okhlobystin, O. Yu., *J. Organometallic Chem.*, 1965, **4**, 211.
116. Zakharin, L. I., Kalinin, V. N. and Podvisotskaya, L. S., *Izv. Akad. Nauk S.S.S.R., Ser. Khim.*, 1968, 679 (*Chem. Abs.*, 1968, **69**, 87147).
117. Zavgorodnii, V. S., Ionin, B. I. and Petrov, A. A., *Zh. Obshch. Khim.*, 1967, **37**, 949.
118. Zavgorodnii, V. S. and Petrov, A. A., *Zh. Obshch. Khim.*, 1965, **35**, 760.
119. Zavgorodnii, V. S. and Petrov, A. A., *Zh. Obshch. Khim.*, 1965, **35**, 931.
120. Zavgorodnii, V. S. and Petrov, A. A., *Zh. Obshch. Khim.*, 1966, **36**, 1480.
121. Zavgorodnii, V. S., Sharanina, L. G. and Petrov, A. A., *Zh. Obshch. Khim.*, 1967, **37**, 1548.
122. Zimmer, H., Blewett, C. W. and Brakas, A., *Tetrahedron Letters*, 1968, 1615.
123. British Patent 1,084,522. Sept. 27, 1967 (*Chem. Abs.*, 1967, **67**, 108763y).
124. British Patent 1,092,036, Nov. 22, 1967 (*Chem. Abs.*, 1968, **68**, 22059).
125. French Patent 1,415,111, Oct. 22, 1965 (*Chem. Abs.*, 1966, **64**, 6694c).
126. French Patent 1,449,872, Aug. 19, 1966 (*Chem. Abs.*, 1967, **66**, 95200q).
127. French Patent 1,467,549, Jan. 27, 1967 (*Chem. Abs.*, 1968, **68**, 49769).
128. German Patent 1,212,085, March 10, 1966 (*Chem. Abs.*, 1966, **64**, 19675a).
129. German Patent 1,235,917, March 9, 1967 (*Chem. Abs.*, 1967, **67**, 22014t).
130. Italian Patent 589,299, March 4, 1959 (*Chem. Abs.*, 1961, **55**, 5034c).
131. Netherlands Patent 6,507,716, Dec. 17, 1965 (*Chem. Abs.*, 1966, **64**, 17640a).
132. Netherlands Patent 6,509,546, Jan. 31, 1966 (*Chem. Abs.*, 1966, **65**, 750c).
133. U.S. Patent 3,206,489, Sept. 14, 1965 (*Chem. Abs.*, 1965, **63**, 18152f).
134. U.S. Patent 3,234,239, Feb. 8, 1966 (*Chem. Abs.*, 1966, **64**, 11251c).
135. U.S. Patent 3,240,795, March 15, 1966 (*Chem. Abs.*, 1966, **64**, 14220d).
135a. U.S. Patent 3,332,970, July 25, 1967 (*Chem. Abs.*, 1968, **68**, 13179).
136. U.S. Patent 3,350,434, Oct. 31, 1967 (*Chem. Abs.*, 1968, **68**, 49789).
137. U.S. Patent 3,372,097, March 5, 1968 (*Chem. Abs.*, 1968, **68**, 101279).
138. U.S.S.R. Patent 172,785, July 7, 1965 (*Chem. Abs.*, 1966, **64**, 1662d).
139. U.S.S.R. Patent 173,758, Aug. 6, 1965 (*Chem. Abs.*, 1966, **64**, 3603d).
140. U.S.S.R. Patent 176,408, Nov. 2, 1965 (*Chem. Abs.*, 1966, **64**, 9838e).

CLEAVAGE OF TIN–CARBON BONDS

Reactions of the general type:

$$\geq Sn-C\leq \; + \; A-B \;\longrightarrow\; \geq Sn-A \; + \; \geq C-B$$

are of the utmost theoretical and practical importance in organotin chemistry. Although the reactivity of tin–carbon bonds depends upon molecular environment, they are susceptible to attack by a wide variety of reagents so that $A - B$ in the above equation may be halogen, mineral acid, carboxylic acid, thiol, phenol, alcohol, metallic or non-metallic halide, alkali, alkali metal and so on.

3.1. HALOGEN CLEAVAGE

This reaction is used particularly for the preparation of organotin mono- and di-halides.

$$R_4Sn + X_2 \;\longrightarrow\; R_3SnX + RX$$
$$R_3SnX + X_2 \;\longrightarrow\; R_2SnX_2 + RX \qquad (X = Cl, Br, I)$$

Where the halogen is iodine the reaction is carried out by heating in a solvent or by mixing the finely powered reactants. Bromine is used as a solution, which is usually added dropwise at $0°$ to avoid uncontrolled cleavage reactions. Because the reagent is inconvenient to handle much less work has been done with chlorine.

When unsymmetrical organotin compounds are treated with one molecular proportion of halogen it is useful to be able to predict which group will be preferentially removed. From a number of studies of the cleavage reactions of unsymmetrical compounds, such as the R_3SnR' or R_2SnR_2' types, sequences have been obtained showing the relative ease with which groups are cleaved from tin, for example [48], o-tolyl > p-tolyl > phenyl > benzyl > vinyl > methyl > ethyl > propyl > isobutyl > butyl > isoamyl > amyl > hexyl > heptyl > octyl. (This series can readily be extended; thus alkynyl groups are more readily removed than phenyl groups [47].) Such sequences are of undoubted use in synthetic chemistry and will usually predict correctly the product of cleavage of one group from a compound R_3SnR'. Thus when the ditin compound $Ph_3Sn(CH_2)_4SnPh_3$ is treated with two molecular proportions of iodine one phenyl group is removed from each tin atom and the Sn—C (alkyl) bonds survive [91].

$$Ph_3Sn(CH_2)_4SnPh_3 + 2I_2 \longrightarrow Ph_2ISn(CH_2)_4SnPh_2I + 2PhI$$

Predictions are less reliable, however, when dealing with the $R_2SnR'_2$ type of compound. Moreover it has been pointed out that these sequences have been compiled from reactions using differing cleavage reagents under a variety of conditions and careful examination of the literature reveals a number of contradictions [41].

This situation has been clarified in a series of papers by Gielen and Nasielski [17, 41, 42] who have studied the kinetics of bromo- and iodo-demetallation of a number of tetraalkyltin compounds in a range of solvents. In polar, nucleophilic, solvents such as methanol or acetic acid the tin compound is solvated and electrophilic attack on a carbon attached to tin is thought to occur by an S_E2 process.

$$R_4Sn + solvent \rightleftharpoons solvent \rightarrow SnR_4$$

$$solvent \rightarrow \underset{R \quad R}{\overset{R}{\underset{|}{Sn}}}\!\!-\!\!R + X_2 \rightleftharpoons [solvent \rightarrow \underset{R \quad R}{\overset{R}{\underset{|}{Sn}}}\!\!\cdots\!\!R\!\cdots\!\overset{\delta\oplus}{X}\!\cdots\!\overset{\delta\ominus}{X}] \rightleftharpoons$$

$$R_3SnX.solvent + RX \quad (X = Br, I)$$

Nucleophilic assistance is rendered by coordination of the solvent to the tin atom thereby increasing the polarity of the Sn—C bonds. The reactivity sequence observed for cleavage of the R group in Me_3SnR is R = Me > Et > Bu > Pr > i-Pr > t-Bu; this steric rather than inductive sequence is unexpected and there is no entirely satisfactory explanation for it. The spread of rate constants in such a series is considerable so that there is a high degree of selectivity in halodemetallation reactions in polar solvents. The reactivity of the group R' in R_3SnR' is profoundly affected by the nature of the leaving group R_3Sn, thus the relative rates of cleavage, by iodine in methanol, of a methyl group from Me_4Sn and Me_3SnBu-t are $1.77:0.01$[17].

In less polar, weakly nucleophilic, solvents such as chlorobenzene the necessary nucleophilic assistance to the electrophilic substitution is rendered by the halogen. There is an initial pre-rate-determining coordination of a halogen molecule to tin.

$$R_4Sn + X_2 \rightleftharpoons X_2 {\rightarrow} SnR_4 \quad (X = Br, I)$$

In the rate-determining step electrophilic attack on carbon may involve (a) the same, or (b) a second, halogen molecule.

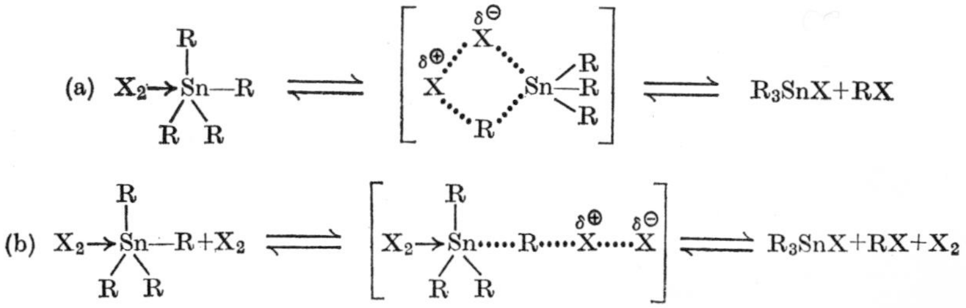

The observed reactivity sequence for halodemetallation of Me_3SnR in chlorobenzene is $R = t\text{-}Bu > Me = i\text{-}Pr > Pr = Bu$ with the rate constants lying quite close together so that halogen in solvents of low polarity does not discriminate well between different alkyl groups[17]. A claim that bromine in carbon tetrachloride reacts selectively with either Bu_3SnEt or Bu_3SnPr to produce the alkyl bromide with the lowest molecular weight[40] was not confirmed[21]. The high degree of selectivity of halodemetallations in methanol has been exploited in the first practicable synthesis of a fully unsymmetrically substituted tetraalkylstannane[20, 42a]:

$$Me_3SnC_6H_{11} + Br_2 \xrightarrow{0°, MeOH} Me_2C_6H_{11}SnBr + MeBr$$

$$Me_2C_6H_{11}SnBr + i\text{-}PrMgBr \longrightarrow Me_2C_6H_{11}SnPr\text{-}i + MgBr_2$$

$$Me_2C_6H_{11}SnPr\text{-}i + Br_2 \xrightarrow{0°, MeOH} MeC_6H_{11}i\text{-}PrSnBr + MeBr$$

$$MeC_6H_{11}i\text{-}PrSnBr + EtMgBr \longrightarrow MeC_6H_{11}i\text{-}PrSnEt + MgBr_2.$$

$$(C_6H_{11} = cyclohexyl)$$

The realisation of the importance of the nature of the solvent has been a significant step in the elucidation of the halodemetallation reaction[33]. The mechanisms of these reactions have also been discussed in the wider context of electrophilic substitution at the carbon atom of a metal–carbon bond[1]. Under the influence of ultraviolet light, bromodemetallation of tetraalkyltin compounds occurs by a radical mechanism[18].

It has recently been demonstrated that the relative rates of cleavage

4

by bromine of the groups R and R′ in R_3SnR' are substantially affected if a second tetralkylstannane $R_4''Sn$ is present. This effect which is observed in the ionic, and to a lesser extent in the radical, reaction is ascribed to the formation of a complex, $Br_2{\rightarrow}SnR_4''$, which acts as a modified halogen towards a second tetraalkyltin compound [19].

Iododemetallation of the optically active cyclopropyltin compound I in carbon tetrachloride proceeds with racemisation. A radical

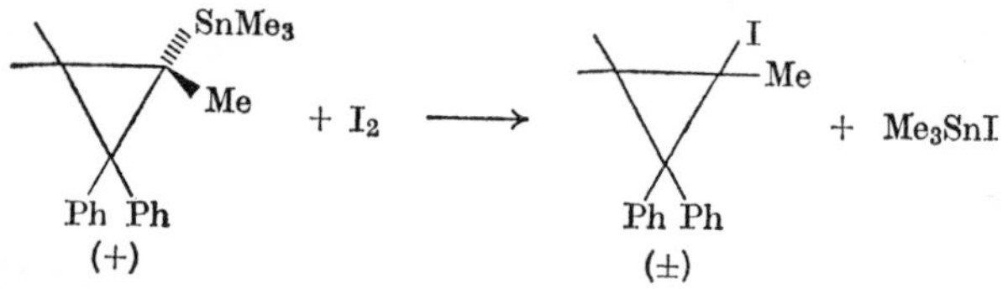

$$\begin{array}{ccc} \text{(+)} & & \text{(±)} \\ \text{Ph Ph} & & \text{Ph Ph} \\ I & & \end{array}$$

mechanism was proposed for this reaction[84]. In contrast, retention of configuration was observed in the iodo- and bromo-demetallation of 1-methyl-2-trimethylstannylcyclopropane, II, in a number of solvents i.e. *trans-II* gave only *trans*-1-methyl-2-halogenocyclopropane and *cis-II* gave only the *cis*-halogenocompound[10]. Two transition states were proposed, depending upon solvent polarity and it was concluded that a radical mechanism had been avoided by using very pure solvents in the absence of light.

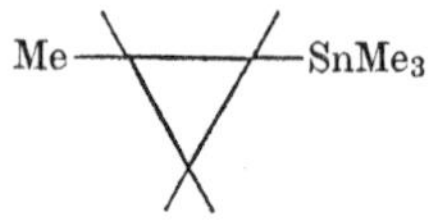

$$II$$

In a kinetic and stereochemical investigation of the mechanism of cleavage of vinyl groups from tin by iodine in methanol it was found that, for derivatives of $Me_3SnCH{=}CH_2$, the rate varied according to the substituents in the vinyl group in the sequence:

$$Me_2C{=}CH \gg cis\text{-}MeCH{=}CMe \sim trans\text{-}MeCH{=}CMe \gg CH_2{=}CH$$
$$\sim CH_2{=}CEt.$$

For the series $R_3SnCH{=}CH_2$ the rate depended upon the nature of R in the order R = Me $\sim$ Et > Bu > i-Pr > $CH_2{=}CH$. To account for these sequences and for the fact that, in all cases, the configuration at the double bond was retained, the transition state III was proposed[9].

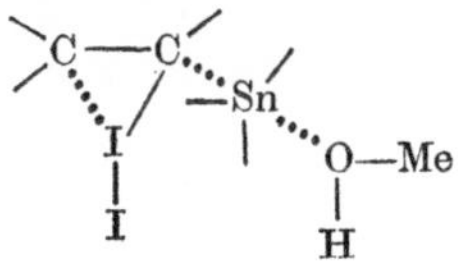

III

Cleavage of the aryl group in trialkylaryltin compounds, R_3SnAr, by iodine in carbon tetrachloride is a third-order reaction which is thought to occur by participation, in the rate determining step, of a π-complex between iodine and the arylstannane together with a second iodine molecule[16]. On the other hand the iododemetallation of Me_3SnPh in methanol was found to be a simple aromatic substitution[23]. The rate of substitution is sensitive to the size of the leaving group so that when the methyl radicals in $PhSnMe_3$ were replaced by larger alkyl groups there was a marked drop in the rate of cleavage[25]. Since the ratio of the second-order rate constants for the iododemetallation of $PhSnMe_3$ and $PhSn(Pr-i)_3$ is the same in methanol or propanol it was concluded that the rate determining step in these substitutions is electrophilic attack by iodine.

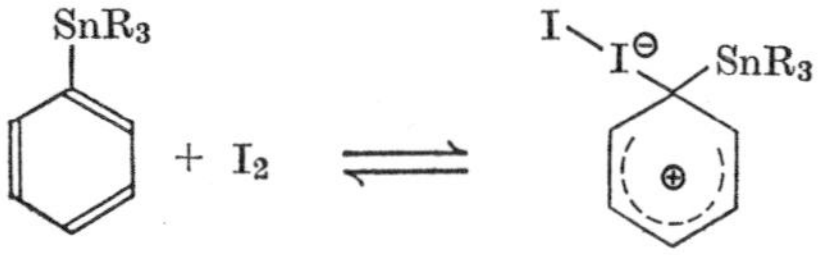

The nucleophilic assistance by the solvent occurring in a later, kinetically non-significant step[57].

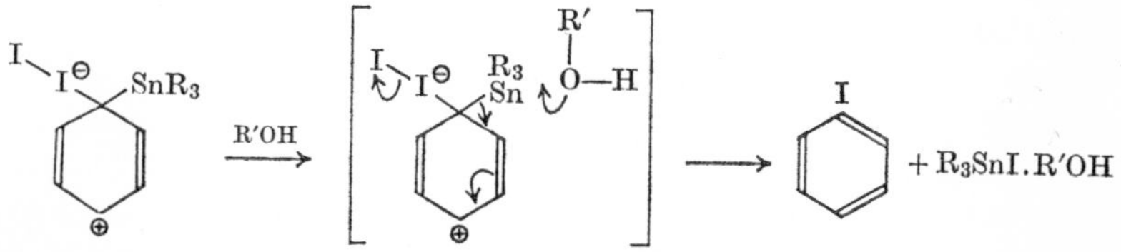

For the iododemetallation of trimethylaryltin compounds in methanol the relative rates are $PhSnMe_3$ 1·00; $2,4\text{-}Me_2C_6H_3SnMe_3$ 15·9; $2,6\text{-}Me_2C_6H_3SnMe_3$ 9·6 whereas, in the tripropyltin series, the diortho-substituted phenyl group was cleaved much more readily, the relative rates being $PhSnPr_3$ 1·00; $2,4\text{-}Me_2C_6H_3SnPr_3$ 16·7; $2,6\text{-}Me_2C_6H_3SnPr_3$ 25·8. It is suggested[58] that this is an example of steric acceleration, the interference between the $SnPr_3$ group and the ortho-substituents on the aryl group being relieved on the change from sp^2 to sp^3 hybridisation in the transition state.

3.2. CLEAVAGE BY ACIDS AND BASES

A variety of acids will effect Sn—C bond scission and, for example, hydrochloric acid is sometimes used to prepare organotin chlorides.

$$R_4Sn + HCl \longrightarrow R_3SnCl + RH$$

In 100% sulphuric acid tetramethyltin is converted to trimethyltin hydrogen sulphate:

$$Me_4Sn + H_2SO_4 \longrightarrow Me_3SnHSO_4 + CH_4$$

whereas all the Sn—C bonds of tetraphenyltin are broken giving benzenesulphonic acid and $H_2Sn(HSO_4)_6$ [43]. For unsymmetrical tetraalkyltin compounds the overall reactivity sequence quoted on p. 37 may be applied but here also the sequence can only be regarded as an approximate guide to reactivity and many exceptions will be found.

Acid cleavage of an organic group from tin is, essentially, an electrophilic substitution at carbon [24, 57a] and a phenyl group carrying electron releasing substituents is more readily attacked than unsubstituted phenyl, i.e. p-$MeOC_6H_4SnR_3 > PhSnR_3$ [38]. The rate of acid cleavage of the aryl group in $C_6F_5SnMe_3$ was, however, greater than would be predicted from the inductive effects of the 5 fluorine atoms and steric acceleration of the reaction is thought to occur [36].

The relative ease with which a phenyl group can be removed from a Group IVb element M in the compounds $PhMEt_3$ by aqueous ethanolic perchloric acid is strongly dependent on the nature of M as shown by the relative rates Si = 1·0, Ge = 36, Sn = 3·5 × 10^5. This is strong evidence for a change of mechanism on passing from the germanium to the tin compound [35]. In terms of the concentration of acid needed to effect cleavage, the ease of removal of the phenylethynyl group in the compounds $PhC{\equiv}CMR_3$ shows a similar dependence upon the nature of the Group IVb element M with Si < Ge ≤ Sn < Pb; in fact the group can be cleaved from the tin compound with neutral aqueous methanol [15, 55].

Tin tetracarboxylates have been made by reaction between carboxylic acids and tetravinyltin.

$$Sn(CH{=}CH_2)_4 + 4RCOOH \longrightarrow Sn(OCOR)_4 + 4CH_2{=}CH_2$$

The corresponding reaction with tetraphenyltin did not go to completion[46], however a 7% yield of $Sn(OCOCF_3)_4$ was obtained from trifluoroacetic acid and tetraphenyltin [67].

Phenyltin compounds undergo Sn—C bond fission when heated with 8-hydroxyquinoline (=HOx) as in the following examples [53, 54, 64].

$$Ph_2SnCl_2 + 2HOx \longrightarrow Cl_2SnOx_2 + 2PhH$$

$$Ph_2SnOx_2 + 2HOx \longrightarrow SnOx_4 + 2PhH$$

β-Diketones and β-ketoesters effect Sn—C bond cleavage in stannylacetones [62].

$$R_3SnCH_2COCH_3 + MeCOCH_2COR' \longrightarrow R_3SnOCMe{=}CHCOR' + CH_3COCH_3$$

$$(R' = \text{alkyl or alkoxyl})$$

The relative rates for the reactions:

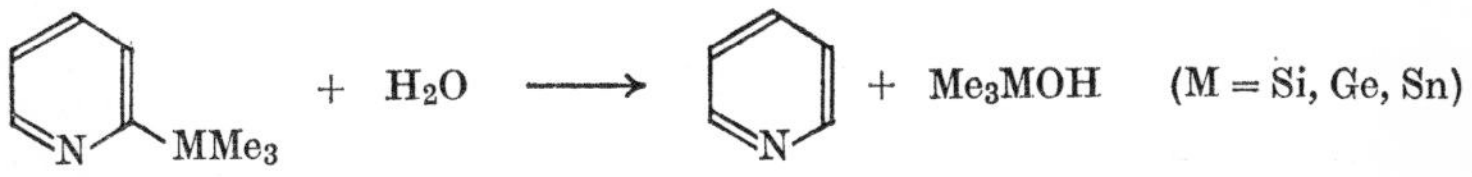

are $Si = 1$, $Ge \sim 10^{-4}$, $Sn = 22$; these figures, and other experimental data, are shown to be consistent with a mechanism involving a rate-determining nucleophilic attack by the oxygen of H_2O on the metal(loid) atom [8].

An example of Sn—C bond cleavage by a base is the formation of $K_2Sn(NH_2)_6$ and benzene from tetraphenyltin and potassium amide in liquid ammonia at 0°; these reactions probably occur by nucleophilic attack at the tin atom [68]. The cleavage of a $PhCH_2CH{=}CH—$ group from tin by aqueous ethanolic alkali was considered to involve a rate-determining S_N2 attack by hydroxide ion at the tin atom with expulsion of a carbanion[65]. However, the effect of *para*-substituents on the rate of cleavage of an aryl group from a series of compounds p-$XC_6H_4SnMe_3$ by aqueous methanolic alkali, was abnormal since, both when X was electron withdrawing, and electron releasing, there was an increase in rate relative to $X = H$ [34]. The ready cleavage of the $PhC{\equiv}C$ group in $PhC{\equiv}CSnR_3$ by aqueous methanol has already been referred to; this reaction is susceptible to base catalysis [37]. Fission of the Sn—C bonds in stannylcarboranes by ethanolic KOH occurs readily [22].

3.3. ORGANOTIN COMPOUNDS AS ALKYLATING (AND ARYLATING) REAGENTS

The transfer of an organic group from tin to another element is a cleavage reaction which can be used to prepare a wide range of compounds. In the general equation:

$$R_4Sn + M - X \longrightarrow R_3SnX + RM$$

M may be an alkali metal, boron, mercury, germanium, silicon, phosphorus etc. and, while X is usually halogen, M − X may be a lithium alkyl or even lithium metal.

Seyferth and Weiner[80] showed that vinyllithium could be conveniently prepared by treating tetravinyltin with phenyllithium in ether or petroleum ether.

$$(CH_2{=}CH)_4Sn + 4PhLi \longrightarrow 4CH_2{=}CHLi + Ph_4Sn$$

Tetrapropenyltin, $(MeCH{=}CH)_4Sn$, is cleaved by metallic lithium in diethyl ether to give propenyllithium and the configuration at the double bond is not retained. Pure *cis-* or *trans*-propenyltrimethyltin, when treated with lithium in diethyl ether, is isomerised to an equilibrium mixture of the *cis-* and *trans*-compounds; with lithium in tetrahydrofuran complete cleavage of the propenyl group and partial cleavage of the methyl groups occurs[77]. Similarly addition of a small amount of an organolithium compound to *cis-* or *trans*-crotyltrimethyltin, $Me_3SnCH_2CH{=}CHMe$, caused isomerisation to a *cis/trans* mixture. Hence, in the cleavage reaction:

$$Me_3SnCH_2CH{=}CHMe + BuLi \longrightarrow Me_3SnBu + MeCH{=}CHCH_2Li$$

the configuration at the double bond was not retained[74]. In contrast, the following reactions occurred with retention of configuration[79].

$$trans\text{-}Ph_3SnCH{=}CHPh + PhLi \longrightarrow Ph_4Sn + trans\text{-}PhCH{=}CHLi$$
$$trans\text{-}Me_3SnCH{=}CHPh + MeLi \longrightarrow Me_4Sn + trans\text{-}PhCH{=}CHLi$$

Treatment of *bis*(trimethylstannyl)dichloromethane with butyllithium results, at least partly, in the reaction:

$$Me_3SnCCl_2SnMe_3 + BuLi \longrightarrow Me_3SnCCl_2Li + BuSnMe_3$$

but further reaction of the lithium compound with reagents such as Me_3SiCl gives a mixture of products[69]. It has been suggested[86] that these transmetallation reactions occur *via* ate complexes.

$$R_4Sn + R'Li \rightleftarrows [R_4SnR']^{\ominus}Li^{\oplus} \rightleftarrows R_3SnR' + RLi$$

Studies of Sn—C bond scission by alkali metals are mainly confined to lithium but, for example, treatment of (4-biphenylyl)trimethyltin with sodium led to cleavage of the aryl-tin bond [31].

A recent review [59] on the synthesis of organohaloboranes summarises the literature on the use of organotin compounds for the introduction of organic groups onto boron. A typical reaction is:

$$R_4Sn + 2BCl_3 \longrightarrow R_2SnCl_2 + 2RBCl_2$$

and, in general, only two of the organic groups attached to tin can be utilised although, by use of excess tin compound, the R_2BCl and even R_3B species can be obtained. Tetravinyltin is exceptional in that all four vinyl groups can be utilised and the preparation of vinyldifluoroborane by this method has been described [92]. Alkoxyhaloboranes undergo similar reactions, for example [82]:

$$Et_3SnC\equiv CCH=CH_2 + (BuO)_2BCl \longrightarrow Et_3SnCl + (BuO)_2BC\equiv CCH=CH_2.$$

Treatment of trichloromethyltrimethyltin with boron trichloride at 25° resulted in preferential cleavage of a methyl group.

$$Me_3SnCCl_3 + BCl_3 \longrightarrow Me_2Sn(Cl)CCl_3 + MeBCl_2$$

At higher temperatures the Cl_3C group was removed but the product was unstable [29]. Tetra(perfluorophenyl)tin is too unreactive to be used as a perfluorophenyl transfer reagent and the best method for preparing perfluorophenyl- [27] and perchlorophenyl- [29] haloboranes is by the following reaction.

$$Me_3SnC_6X_5 + 2BCl_3 \longrightarrow Me_2SnCl_2 + MeBCl_2 + C_6X_5BCl_2$$
$$(X = F \text{ or } Cl)$$

A similar type of reaction has been used to prepare a heterocyclic boron compound [50].

A convenient method for the preparation of pure phenylmercury chloride and bromide is by reaction between the mercuric halide and tetraphenyltin [71].

$$HgX_2 + Ph_4Sn \longrightarrow Ph_3SnX + PhHgX \qquad (X = Cl, Br)$$

The mechanism of corresponding reactions between mercuric iodide and tetraalkyltin compounds has been discussed in detail [2]. The transfer of propenyl groups from $(MeCH=CH)_4Sn$ to mercury occurs with retention of configuration and this method has been used to prepare pure *cis-* and pure *trans*-MeCH=CHHgBr [78]. The relative rates of cleavage of substituted phenyl groups from a series of compounds $XC_6H_4SnEt_3$, by mercuric acetate, have been studied and possible mechanisms for these reactions discussed[45]. It has been suggested [65a] that, when alkenyl groups are cleaved from tin by mercuric halides, significant π-interactions occur between the C=C bonds and the tin atoms in the transition state.

By treating a germanium tetrahalide with varying proportions of a methyl trialkylstannylacetate one to four of the halogen atoms can be replaced [3, 51].

$$GeX_4 + nR_3SnCH_2COOMe \longrightarrow X_{4-n}Ge(CH_2COOMe)_n + nR_3SnX$$
$$(X = Cl, Br, I; R = Et, Bu; n = 1, 2, 3, 4)$$

From the same tin compounds and chlorogermanium trimethoxide, $(MeO)_3GeCH_2COOMe$ can be prepared [51]. Similarly, transfer of alkynyl groups from $Et_3SnC{\equiv}CR$ (R = H, Me) gave $Cl_3GeC{\equiv}CR$ and $Cl_3SiC{\equiv}CR$ [82, 90].

Transfer of the CH_2COOMe group to silicon can proceed normally as in the reaction between silicon tetrachloride and methyl tripropylstannylacetate at 70° during 1·5 hours.

$$SiCl_4 + Pr_3SnCH_2COOMe \longrightarrow Cl_3SiCH_2COOMe + Pr_3SnCl$$

Under milder conditions (40°, 1 hour) however, the product is the *O*-silylketene acetal $CH_2{=}C(OMe)OSiCl_3$; similar reactions occur when $MeSiCl_3$ or Me_2SiCl_2 are used instead of $SiCl_4$ [26, 52]. When the CH_2COCH_3 group migrates from tin to silicon the sole product is the *O*-silyl derivative of the enol form [11, 12, 13].

$$Pr_3SnCH_2COCH_3 + Me_3SiCl \longrightarrow Me_3SiOC(Me){=}CH_2 + Pr_3SnCl$$

Organotin compounds can be used to phenylate and methylate olefins in the presence of molar quantities of palladium compounds, thus methyl cinnamate (PhCH=CHCOOMe) is produced from reaction between tetraphenyltin, methyl acrylate and Li_2PdCl_4 in methanol [45a].

Alkyltetrafluorophosphoranes have been prepared by alkyl transfer from tin to phosphorus [87].

$$R_4Sn + 2PF_5 \longrightarrow RPF_4 + R_3SnPF_6$$

The reaction was allowed to take place for several days at room temperature and the yields of the alkylfluorophosphorane were ~60% for R = Me, Et, Pr and 26% for R = vinyl. Although the reaction was not successful when tetraphenyltin was used there is evidence for the formation of Ph_3SnPF_6 when Ph_4Sn and PF_5 are heated [83]. In a similar manner the CF_3 group can be transferred from tin to phosphorus by heating together Me_3SnCF_3 and PF_5 [28]. The reactions between stannylacetones and a number of substituted phosphorus halides have been studied [60]; examples are as follows.

$$Bu_3SnCH_2COCH_3 + (EtO)_2PCl \longrightarrow (EtO)_2POC(Me){=}CH_2 + Bu_3SnCl$$
$$Et_3SnCH_2COCH_3 + (Et_2N)_2PCl + S \longrightarrow (Et_2N)_2P(S)OC(Me){=}CH_2 + Et_3SnCl$$

In each of these cases P—O rather than P—C bonds are formed. Phenylacetylene groups can be transferred from tin to phosphorus or arsenic [44].

$$3R_3SnC{\equiv}CPh + MCl_3 \longrightarrow M(C{\equiv}CPh)_3 + 3R_3SnCl \quad (M = P, As)$$

When methyl triethylstannylacetate is heated at 90–100° with antimony chloride Sb—C bonds are formed [14].

$$3Et_3SnCH_2COOMe + SbCl_3 \longrightarrow Sb(CH_2COOMe)_3 + 3Et_3SnCl$$

Allyltin compounds will add across the C=O bond of aldehydes in a manner analogous to that of a Grignard reagent [49].

$$Et_3SnCH_2CH{=}CH_2 + RCHO \longrightarrow RCH(OSnEt_3)CH_2CH{=}CH_2$$

A similar reaction occurs when triethylstannylphenylacetylene is heated with chloral in a sealed tube [56,58a].

$$Et_3SnC{\equiv}CPh + Cl_3CCHO \longrightarrow Cl_3CCH(OSnEt_3)C{\equiv}CPh$$

3.4. MISCELLANEOUS CLEAVAGE REACTIONS

The treatment of trifluoromethyltrimethyltin with sodium iodide in 1,2-dimethoxyethane at 80° in the presence of an olefin gives good yields of the *gem*-difluorocyclopropane [72, 81].

$$Me_3SnCF_3 + NaI + RR'C{=}CR''R''' \longrightarrow \begin{array}{c} RR'C{-}{-}{-}CR''R''' \\ \diagdown \diagup \\ CF_2 \end{array} + Me_3SnI + NaF$$

The reaction is thought to involve nucleophilic displacement, by $I^{\ominus}$, of $CF_3^{\ominus}$ which then loses fluoride ion to give difluorocarbene [72]. When Me_3SnCCl_2Br was used as a dihalomethylene transfer reagent most of the product was derived from $:CCl_2$ insertion but a small amount of : CClBr insertion also occurs [70]. Other methods of promoting dihalomethylene transfer reactions need more vigorous conditions, for example, the decarboxylation of organotin halogenoacetates:

$$Ph_3SnOCOCCl_3 \longrightarrow Ph_3SnCl + CO_2 + : CCl_2$$

occurs at 140° and was unsuccessful with the corresponding trifluoro compound. Another method of generating difluorocarbene is by the thermal decomposition of trifluoromethyltrimethyltin [30].

$$Me_3SnCF_3 \xrightarrow{150°} Me_3SnF + :CF_2$$

When the cyclopropane derivative IV is treated with zinc chloride, trimethyltin chloride is eliminated and chloroprene is formed [75].

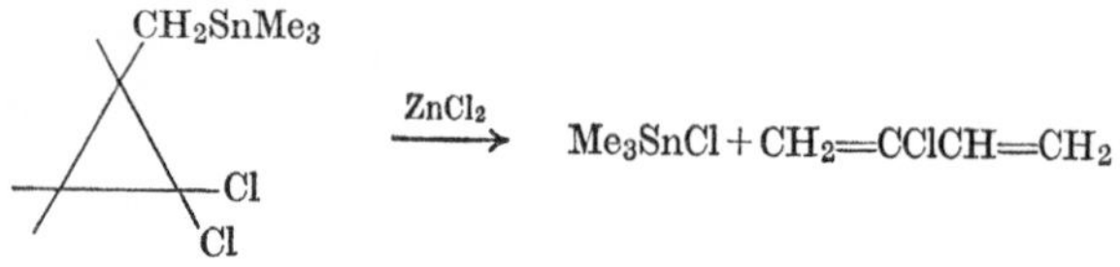

$$\xrightarrow{ZnCl_2} Me_3SnCl + CH_2{=}CClCH{=}CH_2$$

Providing suitable catalysts are present many organic halides can cause Sn—C bond fission. Thus the acetonyl residue in tributyl-stannylacetone is removed by allyl bromide and by benzyl bromide, at somewhat similar rates, as indicated by the competitive reaction [61]:

$$MeCOCH_2SnBu_3 + CH_2{=}CHCH_2Br + PhCH_2Br \xrightarrow[140°, 48 \, hr.]{(Me_2N)_3PO}$$
$$Bu_3SnBr + MeCOCH_2CH_2CH{=}CH_2(34\%) + MeCOCH_2CH_2Ph\,(48\%).$$

Similarly acetyl chloride [66] and isoamyl bromide [32], both in the presence of aluminium bromide, and carbon tetrachloride in the presence of $BiCl_3$ [63] have all been shown to effect Sn—C bond scission. When compound V was heated self condensation occurred by a reaction which involved Sn—C bond cleavage by an aryl iodide [76].

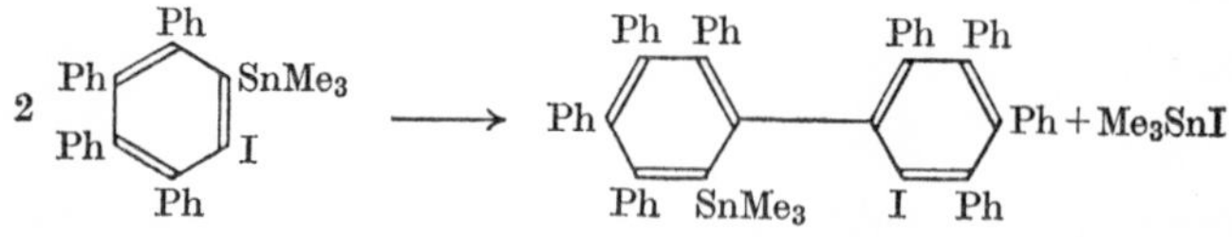

Other highly substituted phenyl groups can be removed from tin by heating the compound with an aryl iodide [73].

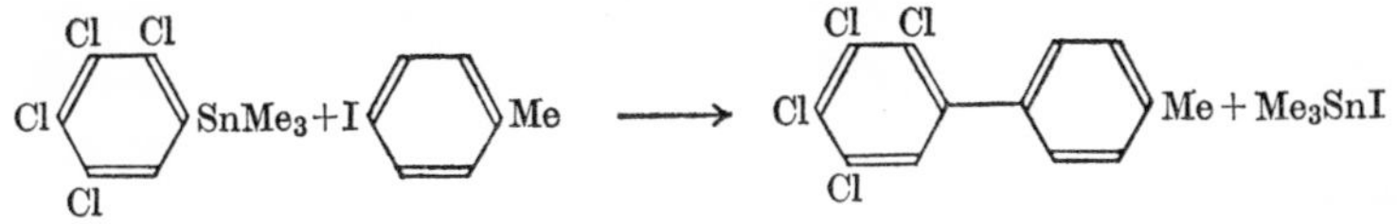

$$Cl_3C_6H_2(Cl)SnMe_3 + I\text{-}C_6H_4\text{-}Me \longrightarrow Cl_3C_6H_2(Cl)\text{-}C_6H_4\text{-}Me + Me_3SnI$$

However, the reaction is less successful when the phenyl group carries fewer substituents and only 12% of 2-fluorobiphenyl was obtained in the reaction [76]:

$$2\text{-}FC_6H_4SnMe_3 + PhI \longrightarrow 2\text{-}FC_6H_4Ph + Me_3SnI.$$

Similarly, although some self-condensation took place when trimethyl(o-iodophenyl)tin was heated a redistribution reaction also occurred [39].

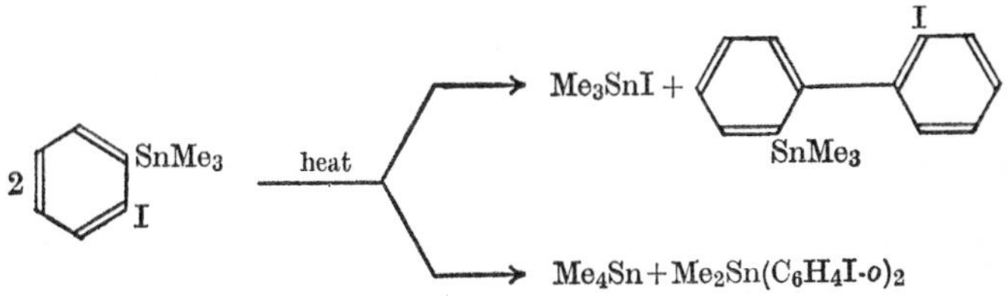

$$2\ (o\text{-}I\text{-}C_6H_4)SnMe_3 \xrightarrow{\text{heat}} \begin{cases} Me_3SnI + (I)(C_6H_4)\text{-}(C_6H_4)(SnMe_3) \\ Me_4Sn + Me_2Sn(C_6H_4I\text{-}o)_2 \end{cases}$$

Although tetraphenylsilane can be hydrogenated under pressure, using a nickel catalyst, to give tetracyclohexylsilane, when the same reaction was attempted with tetraphenylstannane Sn—C bond cleavage occurred and no cyclohexyltin compounds were formed [85].

Ozone removes ethyl groups from tetraethyltin; at −68° an oligomeric compound containing the Et_2Sn unit is formed whereas at 70° Et_2SnO, Et_3SnOH and other compounds are produced [6, 7]. A number of oxidation reactions of tetraalkyltin compounds brought about by irradiation with ultraviolet light [4, 5] or by treatment with peroxides [89] lead to Sn—C bond fission. When $Me_3SiCH_2CH_2COOOCMe_3$ was decomposed in the presence of tetraethyltin the products obtained indicated that homolysis of Sn—C and C—H bonds had occurred [88].

References

1. Abraham, M. H. and Hill, J. A., *J. Organometallic Chem.*, 1967, **7**, 11.
2. Abraham, M. H. and Spalding, T. R., *J. Chem. Soc.* (*A*), 1968, 2530; 1969, 399; 784.
3. Adveeva, V. I., Burlachenko, G. S., Baukov, Yu. I. and Lutsenko, I. F., *Zh. Obshch. Khim.*, 1966, **36**, 1679.
4. Aleksandrov, Yu. A., Glushakova, V. N. and Radbil, B. A., *Tr. Khim. Khim. Tekhnol.*, 1967, 69 (*Chem. Abs.*, 1968, **69**, 27506).
5. Aleksandrov, Yu. A. and Radbil, B. A., *Zh. Obshch. Khim.*, 1967, **37**, 2345.

6. Aleksandrov, Yu. A. and Sheyanov, N. G., *Zh. Obshch. Khim.*, 1966, **36**, 953.
7. Aleksandrov, Yu. A., Sheyanov, N. G. and Shushunov, V. A., *Zh. Obshch. Khim.*, 1968, **38**, 1352 (*Chem. Abs.*, 1968, **69**, 77388).
8. Anderson, D. G. and Webster, D. E., *J. Chem. Soc.* (*B*), 1968, 765.
9. Baekelmans, P., Gielen, M., Malfroid, P. and Nasielski, J., *Bull. Soc. Chim. Belges*, 1968, **77**, 85.
10. Baekelmans, P., Gielen, M. and Nasielski, J., *Tetrahedron Letters*, 1967, 1149.
11. Baukov, Yu. I., Burlachenko, G. S., Belavin, I. Yu. and Lutsenko, I. F., *Zh. Obshch. Khim.*, 1966, **36**, 153.
12. Baukov, Yu. I., Burlachenko, G. S. and Lutsenko, I. F., *J. Organometallic Chem.*, 1965, **3**, 478.
13. Baukov, Yu. I., Burlachenko, G. S. and Lutsenko, I. F., *Zh. Obshch. Khim.*, 1965, **35**, 757.
14. Besolova, E. A., Foss, V. L. and Lutsenko, I. F., *Zh. Obshch. Khim.*, 1968, **38**, 1574 (*Chem. Abs.*, 1968, **69**, 106849).
15. Bott, R. W., Eaborn, C. and Walton, D. R. M., *J. Organometallic Chem.*, 1964, **1**, 420.
16. Bott, R. W., Eaborn, C. and Waters, J. A., *J. Chem. Soc.*, 1963, 681.
17. Boué, S., Gielen, M. and Nasielski, J., *J. Organometallic Chem.*, 1967, **9**, 443.
18. Boué, S., Gielen, M. and Nasielski, J., *J. Organometallic Chem.*, 1967, **9**, 461.
19. Boué, S., Gielen, M. and Nasielski, J., *J. Organometallic Chem.*, 1967, **9**, 481.
20. Boué, S., Gielen, M. and Nasielski, J., *Tetrahedron Letters*, 1968, 1047.
21. Boué S., Gielen M., Nasielski, J., Autin, J. and Limbourg, M., *J. Organometallic Chem.*, 1968, **15**, 267.
22. Bregadze, V. I. and Okhlobystin, O. Yu., *Izv. Akad. Nauk. S.S.S.R.*, *Ser. Khim.*, 1967, 2084.
23. Buchman, O., Grosjean, M. and Nasielski, J., *Helv. Chim. Acta*, 1964, **47**, 1679.
24. Buchman, O., Grosjean, M. and Nasielski, J., *Helv. Chim. Acta*, 1964, **47**, 1695.
25. Buchman, O., Grosjean, M., Nasielski, J. and Wilmet-Devos, B., *Helv. Chim. Acta*, 1964, **47**, 1688.
26. Burlachenko, G. S., Khasapov, B. N., Petrovskaya, L. I., Baukov, Yu. I. and Lutsenko, I. F., *Zh. Obshch. Khim.*, 1966, **36**, 512.
27. Chambers, R. D. and Chivers, T., *J. Chem. Soc.*, 1965, 3933.
28. Chan, S. S. and Willis, C. J., *Can J. Chem.*, 1968, **46**, 1237.
29. Chivers, T. and David, P., *J. Organometallic Chem.*, 1967, **10**, P35.
30. Clark, H. C. and Willis, C. J., *J. Amer. Chem. Soc.*, 1960, **82**, 1888.
31. Curtis, M. D. and Allred, A. L., *J. Amer. Chem. Soc.*, 1965, **87**, 2554.
32. Dao-Huy-Giao, *Compt. Rend.*, 1965, **260** (Groupe 8), 6937.
33. Dessy, R. E. and Kitching, W., *Advances in Organometallic Chemistry*, Academic Press, New York and London, 1966, **4**, 267.
34. Eaborn, C., Hornfield, H. L. and Walton, D. R. M., *J. Chem. Soc.* (*B*), 1967, 1036.
35. Eaborn, C. and Pande, K. C., *J. Chem. Soc.*, 1960, 1566.
36. Eaborn, C., Treverton, J. A. and Walton, D. R. M., *J. Organometallic Chem.*, 1967, **9**, 259.
37. Eaborn, C. and Walton, D. R. M., *J. Organometallic Chem.*, 1965, **4**, 217.

38. Eaborn, C. and Waters, J. A., *J. Chem. Soc.*, 1961, 542.
39. Evnin, A. B. and Seyferth, D., *J. Amer. Chem. Soc.*, 1967, **89**, 952.
40. Faleschini, S. and Tagliavini, G., *Gazz. Chim. Ital.*, 1967, **97**, 1401 (*Chem. Abs.*, 1968, **68**, 95053).
41. Gielen, M. and Nasielski, J., *J. Organometallic Chem.*, 1963, **1**, 173.
42. Gielen, M. and Nasielski, J., *J. Organometallic Chem.*, 1967, **7**, 273.
42a. Gielen, M., Nasielski, J. and Topart, J., *Rec. Trav. Chim.*, 1968, **87**, 1051.
43. Gillespie, R. J., Kapoor, R. and Robinson, E. A., *Can J.. Chem.*, 1966, **44**, 1197.
44. Hartmann, H., *Annalen*, 1968, **714**, 1.
45. Hashimoto, H. and Morimoto, Y., *J. Organometallic Chem.*, 1967, **8**, 271.
45a. Heck, R. F., *J. Amer. Chem. Soc.*, 1968, **90**, 5531.
46. Henderson, A. and Holliday, A. K., *J. Organometallic Chem.*, 1965, **4**, 377.
47. Ibekwe, S. and Newlands, M. J., *J. Chem. Soc.*, 1965, 4608.
48. Ingham, R. K., Rosenberg, S. D. and Gilman, H., *Chem. Rev.*, 1960, **60**, 479.
49. Koenig, K. and Neumann, W. P., *Tetrahedron Letters*, 1967, 495.
50. Leusink, A. J., Drenth, W., Noltes, J. G. and van der Kerk, G. J. M., *Tetrahedron Letters*, 1967, 1263.
51. Lutsenko, I. F., Baukov, Yu. I. and Burlachenko, G. S., *J. Organometallic Chem.*, 1966, **6**, 496.
52. Lutsenko, I. F., Baukov, Yu. I., Burlachenko, G. S. and Khasapov, B. N., *J. Organometallic Chem.*, 1966, **5**, 20.
53. Martin, D. F., Maybury, P. C. and Walton, R. D., *J. Organometallic Chem.*, 1967, **7**, 362.
54. Martin, D. F. and Walton, R. D., *J. Organometallic Chem.*, 1966, **5**, 57.
55. Masson, J. C., Le Quan Minh, and Cadiot, P., *Bull. Soc. Chim. France*, 1968, 1085.
56. Mirskov, R. G. and Vlasov, V. M., *Zh. Obshch. Khim.*, 1966, **36**, 562.
57. Nasielski, J., Buchman, O., Grosjean, M. and Hannecart, E., *Bull. Soc. Chim. Belges*, 1968, **77**, 15.
57a. Nasielski, J., Buchman, O., Grosjean, M. and Jauquet, M., *J. Organometallic Chem.*, 1969, **19**, 353.
58. Nasielski, J., Buchman, O., Grosjean, M., Shurter, J. J. and Vandendunghen, G., *Bull. Soc. Chim. Belges*, 1968, **77**, 349.
58a. Neumann, W. P. and Kleiner, F. G., *Annalen*, 1968, **716**, 29.
59. Niedenzu, K., *Organometallic Chem. Rev.*, 1966, **1**, 305.
60. Novikova, Z. S., Proskurnina, M. V., Petrovskaya, L. I., Bogdanova, I. V., Galitskova, N. P. and Lutsenko, I. F., *Zh. Obshch. Khim.*, 1967, **37**, 2080.
61. Pereyre, M., Colin, G. and Valade, J., *Compt. Rend.*, 1967, **264**, 1204.
62. Ponamarev, S. V., Machigin, E. V. and Lutsenko, I. F., *Zh. Obshch. Khim.*, 1966, **36**, 548.
63. Puchinyan, E. A. and Manulkin, Z. M., *Tr. Tashkent Farmatsevt. Inst.*, 1966, **4**, 354 (*Chem. Abs.*, 1968, **68**, 78382).
64. Ramaiah, K. and Martin, D. F., *Chem. Comm.*, 1965, 130.
65. Roberts, R. M. G. and El Kaissi, F., *J. Organometallic Chem.*, 1968, **12**, 79.
65a. Roberts, R. M. G., *J. Organometallic Chem.*, 1969, **18**, 307.
66. Sakurai, H., Tominaga, K., Watanabe, T. and Kumada, M., *Tetrahedron Letters*, 1966, 5493.

67. Sartori, P. and Weidenbruch, M., *Chem. Ber.*, 1967, **100**, 2049.
68. Schmitz-Dumont, O., Mueller, G. and Schaal, W., *Z. Anorg. Allgem. Chem.*, 1964, **332**, 263.
69. Seyferth, D., Armbrecht, F. M. and Hanson, E. M., *J. Organometallic Chem.*, 1967, **10**, P25.
70. Seyferth, D., Armbrecht, F. M., Prokai, B. and Cross, R. J., *J. Organometallic Chem.*, 1966, **6**, 573.
71. Seyferth, D. and Burlitch, J. M., *J. Organometallic Chem.*, 1965, **4**, 127.
72. Seyferth, D., Dertouzos, H., Suzuki, R. and Yick-Pui Mui, J., *J. Org. Chem.*, 1967, **32**, 2980.
73. Seyferth, D. and Evnin, A. B., *J. Amer. Chem. Soc.*, 1967, **89**, 1468.
74. Seyferth, D. and Jula, T. F., *J. Organometallic Chem.*, 1967, **8**, P13.
75. Seyferth, D. and Jula, T. F., *J. Amer. Chem. Soc.*, 1968, **90**, 2938.
76. Seyferth, D., Sarafidis, C. and Evnin, A. B., *J. Organometallic Chem.*, 1964, **2**, 417; 437.
77. Seyferth, D., Suzuki, R. and Vaughan, L. G., *J. Amer. Chem. Soc.*, 1966, **88**, 286.
78. Seyferth, D. and Vaughan, L. G., *J. Organometallic Chem.*, 1966, **5**, 580.
79. Seyferth, D., Vaughan, L. G. and Suzuki, R., *J. Organometallic Chem.*, 1964, **1**, 437.
80. Seyferth, D. and Weiner, M. A., *Chem. and Ind.*, 1959, 402; *J. Amer. Chem. Soc.*, 1961, **83**, 3583.
81. Seyferth, D., Yick-Pui Mui, J., Gordon, M. E. and Burlitch, J. M., *J. Amer. Chem. Soc.*, 1965, **87**, 681.
82. Sharanina, L. G., Zavgorodnii, V. S. and Petrov, A. A., *Zh. Obshch. Khim.*, 1966, **36**, 1154.
83. Sharp, D. W. A. and Winfield, J. M., *J. Chem. Soc.*, 1965, 2278.
84. Sisido, K., Kozima, S. and Takizawa, K., *Tetrahedron Letters*, 1967, 33.
85. Spialter, L., Buell, G. and Harris, C. W., *J. Org. Chem.*, 1965, **30**, 375.
√86. Tochterman, W., *Angew. Chem. Intern. Edit. Engl.*, 1966, **5**, 351.
87. Treichel, P. M. and Goodrich, R. A., *Inorg. Chem.*, 1965, **4**, 1424.
88. Vyazankin, N. S., Razuvaev, G. A. and Brevnova, T. N., *Dokl. Akad. Nauk S.S.S.R.*, 1965, **163**, 1389.
89. Vyazankin, N. S., Razuvaev, G. A. and Brevnova, T. N., *Zh. Obshch. Khim.*, 1965, **35**, 2033.
90. Zavgorodnii, V. S., Sharanina, L. G. and Petrov, A. A., *Zh. Obshch. Khim.*, 1968, **38**, 1150 (*Chem. Abs.*, 1968, **69**, 59355).
91. Zimmer, H. and Miller, J. J., *Naturwissenschaften*, 1966, **53**, 38.
92. British Patent 973,636, Oct. 28, 1964 (*Chem. Abs.*, 1965, **62**, 9173b).

4

ORGANOTIN HALIDES AND PSEUDOHALIDES

The organotin halides are key intermediates in the industrial and laboratory preparation of other classes of organotin compounds. The principal commercial application for organotin compounds is in the stabilisation of poly(vinyl chloride) (see Chapter 15). For this purpose, large quantities of dialkyltin dichlorides are needed as intermediates, and a considerable amount of research work is directed at producing these compounds more cheaply. In the laboratory almost every class of organotin compound can be made by replacement of halogen in the corresponding halide and although there is now an increasing use of organotin-oxides, -hydrides and -amines in synthesis these classes are themselves made from the halides. The chemistry of the organotin halides has been recently reviewed [56].

The main emphasis in this chapter is on the synthesis of the halides and, although a summary of reactions is included, the preparative use of these compounds is discussed in more detail in the appropriate sections of Chapters 5–10. Organotin isocyanates, isothiocyanates, cyanides and azides are also discussed here as pseudohalides.

4.1. PREPARATION

The principal synthetic routes to the organotin halides are (a) cleavage reactions, using halogen or hydrogen halide, which have already been discussed in Chapter 3, (b) redistribution reactions and (c) direct synthesis by reaction between an alkyl halide and metallic tin. All of these methods tend to give equilibrium mixtures containing mono-, di- and tri-halides so that, although they are adequate for most purposes, they are not suitable for the preparation of compounds of high purity [75]. Since the physical properties of the three types of halide often do not permit clean separation by traditional purification procedures it may be necessary to resort to chromatography (p. 60), differentiation by complex formation (Chapter 11) or other chemical methods (p. 60).

4.1.1. *Redistribution reactions*

The treatment of a tetraalkyl(aryl)tin compound with the requisite proportion of a stannic halide is a widely used procedure for the preparation of organotin halides and has the advantage, over the halogen cleavage method, that none of the organic groups is lost. The reactions are usually formulated as follows.

$$3R_4Sn + SnX_4 \longrightarrow 4R_3SnX$$
$$R_4Sn + SnX_4 \longrightarrow 2R_2SnX_2$$
$$R_4Sn + 3SnX_4 \longrightarrow 4RSnX_3$$

(where R = alkyl or aryl and X = Cl or Br)

The chlorides are formed more readily than the bromides, stannic iodide usually reacts so slowly that the method is rarely used for the preparation of iodides [73]. The two compounds are heated at temperatures in the region of 200° in the absence of a solvent. The reactions may be facilitated by the presence of a Lewis-acid catalyst [102] or by heating under pressure [98].

The general field of redistribution reactions at tin has been reviewed [42, 43]; studies of the reactions have been made using gas chromatography [46] and nmr spectroscopy [19, 73] and much useful information obtained concerning the manner in which the redistributions occur. At temperatures of 0–50° there is one dominant spontaneous reaction.

$$R_4Sn + SnX_4 \longrightarrow RSnX_3 + R_3SnX$$

This reaction is complete over a wide range of concentrations and molar ratios before other reactions are observed. The activation energy for this reaction when R = Me and X = Cl is estimated at 3·5–5 kcal. mole^{-1} [19]. Subsequent reactions occur at higher temperatures and depend upon the initial molar ratio of the reactants. With equimolar proportions i.e. $R_4Sn : SnX_4 = 1$ the major process is:

$$R_3SnX + RSnX_3 \longrightarrow 2R_2SnX_2.$$

When $R_4Sn : SnX_4 > 1$ the subsequent reactions are:

$$R_4Sn + RSnX_3 \longrightarrow R_3SnX + R_2SnX_2$$
$$R_4Sn + R_2SnX_2 \longrightarrow 2R_3SnX.$$

Finally, when the molar ratio is $R_4Sn : SnX_4 < 1$ the major product, the trihalide, is formed by the following reactions.

$$R_3SnX + SnX_4 \longrightarrow R_2SnX_2 + RSnX_3$$
$$R_2SnX_2 + SnX_4 \longrightarrow 2RSnX_3$$

The ease with which the initial spontaneous reaction occurs has been exploited in a preparation of butyltin trichloride and tributyltin chloride, the products being separated by distillation [85]. For the commercial production of triphenyltin chloride a process based on a similar reaction between tetraphenyltin and stannic chloride has been proposed, the phenyltin trichloride is separated and reconverted to tetraphenyltin by a Wurtz reaction [80].

$$Ph_4Sn + SnCl_4 \longrightarrow Ph_3SnCl + PhSnCl_3$$
$$PhSnCl_3 + 6Na + 3PhCl \longrightarrow Ph_4Sn + 6NaCl$$

It has not been found possible to produce pure methyltin trichloride by a simple redistribution reaction but when dimethyltin dichloride and stannic chloride were heated in dimethylsulphoxide (= DMSO) the adduct of the trichloride was obtained in 98% yield [27].

$$Me_2SnCl_2 + SnCl_4 + 4DMSO \longrightarrow 2[MeSnCl_3.2DMSO]$$

Organotin dichlorides having two different organic groups, $RR'SnCl_2$, have been prepared by redistribution reactions between $RSnCl_3$ and R'_4Sn [26]. A method has been proposed for the preparation of dichlorides, R_2SnCl_2, by the following redistribution reaction.

$$Me_2SnR_2 + SnCl_4 \longrightarrow Me_2SnCl_2 + R_2SnCl_2 \qquad (R \neq Me)$$

The products are separated and the dimethyltin dichloride converted to more Me_2SnR_2 [39].

4.1.2. Direct Synthesis

Although the details were not published until later it is clear that Frankland prepared the first organotin compound in 1849 by the direct reaction between ethyl iodide and metallic tin at 160° [15].

$$2EtI + Sn \longrightarrow Et_2SnI_2$$

Today there is considerable interest in the preparation of dialkyltin dihalides by the direct method since this is, potentially, far more economical than the synthesis of a tetraalkyltin compound followed by a redistribution reaction with a stannic halide. In the general reaction:

$$2RX + Sn \longrightarrow R_2SnX_2$$

the reactivity of the alkyl halide is $RI > RBr > RCl$ and a catalyst is usually necessary. Benzyl chloride, however, is exceptionally reactive and gives high yields of dibenzyltin dichloride or tribenzyltin chloride depending upon the reaction conditions. In solvents of low polarity,

5

such as benzene, the predominant product is the dichloride but if the reaction is carried out in polar media such as butanol or water the monochloride is formed preferentially [62].

Examples of the preparation of dialkyltin dihalides from metallic tin and organic halides in the presence of catalysts are given in Table I. It can be seen that, in most cases, a Lewis base which is usually an ether or an alcohol must be present and that most of the reactions require the presence of an iodide either as a reactant or as a catalyst component. Despite this it has been suggested that an iodine source is not essential for the direct synthesis of organotin dibromides [63]. Good yields of organotin halides were obtained when alkyl- or aryl-halides were heated with a suspension of activated tin particles in the presence of metallic iodides[83]. The reaction between butyl iodide and an activated mixture of tin and zinc is an effective method for the preparation of butyltin compounds [49]. Chloromethyl methyl ether reacts with tin but the exact composition of the product, formulated as $(MeOCH_2)_nSnCl_{4-n}$, was not determined [59]. For the preparation of dibutyltin dibromide the optimum proportions of reagents and catalyst are 2 moles BuBr, 0·2 mole BuI, 1 g.atom Sn, 0·05 g.atom Li and 0·15 mole $EtOCH_2CH_2OH$; a detailed mechanism for this reaction has been proposed [51]. If, in the direct synthesis of alkyltin chlorides, the proportion of tin and triethylamine catalyst is increased then the yield of R_3SnCl is increased at the expense of the R_2SnCl_2 product; this is due to a conversion of the initially formed dichloride to the monochloride [61].

Alkyl bromides react with tin when irradiated with γ-rays [3, 11, 12, 13, 41, 103], the main product is the dialkyltin dibromide but the proportion of trialkyltin bromide is influenced by the size of the tin particles and by the temperature [10]; certain compounds, e.g. alcohols, function as activators for this process [104]. In the presence of catalysts alkyl halides react directly with stannous halides to give alkyltin trihalides.

$$RX + SnX_2 \longrightarrow RSnX_3 \quad (X = Cl, Br, I)$$

An equimolar mixture of the reactants is heated in the region of 160–200° and the reactivity sequences are $RI > RBr > RCl$ and $SnI_2 > SnBr_2 > SnCl_2$ [79]. Simple primary amines have been used as catalysts [95] but usually two- or three-component catalyst systems are used such as a Lewis base with a metallic salt [79], tetrabutylammonium iodide with diethyleneglycol diethyl ether [100], an organic disulphide or

TABLE I

Examples of the preparation of organotin dihalides by the direct method [a]

Organic halide	Catalyst	Principal product [b]	References
MeCl	Et$_3$N, MeI	Me$_2$SnCl$_2$	81
CH$_2$=CHCH$_2$Br	HgCl$_2$, Et$_3$N	(CH$_2$=CHCH$_2$)$_2$SnBr$_2$	63a
BuCl	BuNH$_2$, BuI	Bu$_2$SnCl$_2$	87
BuCl	C$_{12}$H$_{25}$SH, Mg, I$_2$	Bu$_2$SnCl$_2$	92
CH$_3$(CH$_2$)$_7$Cl	SbI$_3$, CH$_3$(CH$_2$)$_7$I	[CH$_3$(CH$_2$)$_7$]$_2$SnCl$_2$	96
CH$_3$COCH$_2$CH$_2$Cl	MgO, BuI, BuOH	(CH$_3$COCH$_2$CH$_2$)$_2$SnCl$_2$	36
BuBr	SnCl$_2$, [Bu$_4$N]Br	Bu$_2$SnBr$_2$	97
BuBr	BuI, Li, EtOCH$_2$CH$_2$OH	Bu$_2$SnBr$_2$	51
BuBr	Bu$_2$SnBr$_2$, LiBr, (MeOCH$_2$CH$_2$)$_2$O	Bu$_2$SnBr$_2$	99
BuI	Li, EtOCH$_2$CH$_2$OH	Bu$_2$SnI$_2$	50
BuI	HgCl$_2$, Et$_3$N, (MeOCH$_2$CH$_2$)$_2$O	Bu$_2$SnI$_2$	88
BrCH$_2$CH$_2$COOBu	Mg, BuOH	Br$_2$Sn(CH$_2$CH$_2$COOBu)$_2$	37
BrCH$_2$CH$_2$COOMe	Mg, BuOH, BuI	Br$_2$Sn(CH$_2$CH$_2$COOMe)$_2$	34
Br(CH$_2$)$_3$COOMe	Mg, THF, BuI	Br$_2$Sn[(CH$_2$)$_3$COOMe]$_2$	47
BrCH$_2$CH$_2$CONH$_2$	Mg, THF, BuI, HOCH$_2$CH$_2$OH	Br$_2$Sn(CH$_2$CH$_2$CONH$_2$)$_2$	23
BrCH$_2$CHMeCONMe$_2$	Mg, THF, BuI, HOCH$_2$CH$_2$OH	Br$_2$Sn(CH$_2$CHMeCONMe$_2$)$_2$	20
EtOCOCHBrCH$_2$COOEt	Mg, THF, BuI	Br$_2$Sn[CH(CH$_2$COOEt)COOEt]$_2$	38, 52, 53
ICH$_2$CH$_2$CN	Mg or THF	I$_2$Sn(CH$_2$CH$_2$CN)$_2$	35
ICH$_2$CHMeCOOMe	Mg, THF, BuI	I$_2$Sn(CH$_2$CHMeCOOMe)$_2$	48
ICH$_2$CH$_2$COEt	Mg, BuOH, BuI	I$_2$Sn(CH$_2$CH$_2$COEt)$_2$	34
I(CH$_2$)$_3$OPh	Mg, THF, BuI	I$_2$Sn[(CH$_2$)$_3$OPh]$_2$	21
ClCH$_2$CH$_2$CONHCH$_2$COOEt	Mg, THF	Cl$_2$Sn(CH$_2$CH$_2$CONHCH$_2$COOEt)$_2$	22
ClCH$_2$CH=CClMe	Cu, H$_2$O	Cl$_2$Sn(CH$_2$CH=CClMe)$_2$	18

[a] Reaction conditions vary considerably, the reactants are heated for periods of 1–50 hr. at temperatures 100–180°.
[b] In general the yields are in the range 65–95%.

thiol with magnesium and iodine[93]. Dibutyltin dibromide can be prepared by the anodic dissolution of tin, using a magnesium cathode, in a mixture of butyl bromide, butyl acetate and bromine[2]. When methyl chloride was passed into a suspension of liquid tin in a mixture of molten KCl and $SnCl_2$ at 300° a 54% yield of trimethyltin chloride was formed [68, 86].

4.1.3. *Other methods of preparation*

The organotin oxides and hydroxides are normally prepared from the chlorides (Chapter 5) and are not usually considered as starting materials even though they are readily converted to organotin halides by treatment with the hydrogen halide. In the laboratory, however, it is sometimes convenient to convert a commercially available organotin chloride to the bromide or iodide by the sequence:

$$2R_3SnCl + 2NaOH \longrightarrow R_3SnOSnR_3 + 2NaCl + H_2O$$
$$R_3SnOSnR_3 + 2HX \longrightarrow 2R_3SnX + H_2O \qquad (X = Br, I).$$

When an organotin fluoride is formed in an equilibrium reaction its low solubility in most solvents ensures a high, often quantitative, yield. Many organotin species such as halides, oxides, esters, etc. when dissolved in an organic solvent and shaken with aqueous sodium fluoride give an immediate precipitate of the organotin fluoride. Other methods for the preparation of fluorides include reaction between an organotin oxide and H_2SiF_6[89] and also between a dialkylmercury compound and stannous fluoride[33].

Partial alkylation of a stannic halide by means of a Grignard reagent is usually unsatisfactory since attempts to carry out a reaction such as:

$$2RMgBr + SnCl_4 \longrightarrow R_2SnCl_2 + MgBr_2 + MgCl_2$$

give mixtures in which the $RSnCl_3$ and R_3SnCl species are also present. However, if the organic groups are bulky partial substitution using a Grignard reagent is feasible, for example, good yields of tri(cyclohexyl)tin chloride were obtained by the reaction[94]:

$$C_6H_{11}MgCl + (C_6H_{11})_2SnCl_2 \longrightarrow (C_6H_{11})_3SnCl + MgCl_2.$$

When stannic chloride reacted with a Grignard reagent containing the 2,2-dimethyl-2-phenylethyl group the organotin monochloride was the principal product[55]. Partial alkylation of stannic chloride is practicable with other reagents and trihexyltin chloride was made using trihexylaluminium[82] and methylation with hexamethyldisiloxane gave methyltin trichloride[77].

4.2. PROPERTIES OF ORGANOTIN HALIDES

The physical properties of some representative compounds are shown in Table II from which it can be seen that the fluorides have high melting points. This is, at least partly, a consequence of their highly associated structures (Chapter 12).

TABLE II

Physical properties of some organotin halides

Compound	m.p.	b.p.	n_D^{20}	d
Me₃SnF	375° decomp.	—	—	—
Me₃SnCl	37·5–39·5°	154–156°	—	—
Me₃SnBr	26–27°	163–165°	—	—
Me₃SnI	−5 to −3°	160–170°	—	—
Me₂SnCl₂	107–108°	185–190°	—	—
MeSnCl₃	45–46°	—	—	—
Bu₃SnF	218–219°	—	—	—
Bu₃SnCl	—	152–156°/14 mm	1·4930	1·2105 (20°)
Bu₂SnCl₂	41–42°	140–143°/10 mm	—	—
Bu₂SnBr₂	21–22°	90·5–92°/0·3 mm	1·5400 (25°)	1·3913 (25°)
Bu₂SnI₂	—	145°/6 mm	1·6042 (25°)	1·996 (25°)
BuSnCl₃	—	102–103°/12 mm	1·5233	—
Ph₃SnF	357° decomp.	—	—	—
Ph₃SnCl	106°	—	—	—
Ph₂SnCl₂	42–44°	180–185°/5 mm	—	—
PhSnCl₃	—	142–143°/25 mm	1·5871	—

The reason for the wide use of organotin halides in preparative work is the ease with which the halogens can be replaced by a variety of nucleophilic reagents as indicated in the following chart.

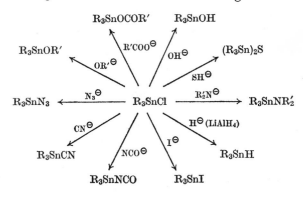

Further details of these reactions are given in the appropriate sections of this book.

The conversion of a monohalide, R_3SnX, to the corresponding hydroxide R_3SnOH is usually carried out by shaking an ethereal solution of the halide with aqueous alkali. Under the same conditions the dihalide and trihalide are converted to the oxide and stannonic acid respectively (see Chapter 5).

$$R_2SnX_2 + 2NaOH \longrightarrow R_2SnO + 2NaX + H_2O$$
$$RSnX_3 + 3NaOH \longrightarrow RSnO_2H + 3NaX + H_2O$$

In favourable cases these reactions can be used for the purification of an organotin halide. If, for example, a specimen of R_2SnCl_2 which is contaminated with R_3SnCl and $RSnCl_3$ is dissolved in ether and shaken with aqueous alkali the R_3SnOH formed is ether-soluble, the $RSnO_2H$ dissolves in the alkali and the insoluble R_2SnO can be filtered off and reconverted to R_2SnCl_2 by treatment with concentrated hydrochloric acid. The lower alkyltin halides are completely dissociated in dilute aqueous solution and their polarographic behaviour has been studied [9a, 39a].

There is an interesting report that organotin bromides will add across the double bonds of allyl bromide and 3-bromocyclohexene though the products were not fully characterised, for example[40]:

$$Ph_2SnBr_2 + CH_2{=}CHCH_2Br \xrightarrow{\text{pyridine}}$$
$$Ph_2BrSnCH(CH_2Br)_2.2py \quad or \quad Ph_2BrSnCH_2CHBrCH_2Br.2py.$$

By the use of deuterium labelling, and other methods, evidence was obtained that tetraphenylcyclobutadiene is formed as an unstable intermediate in the pyrolysis of the dibromide I [58].

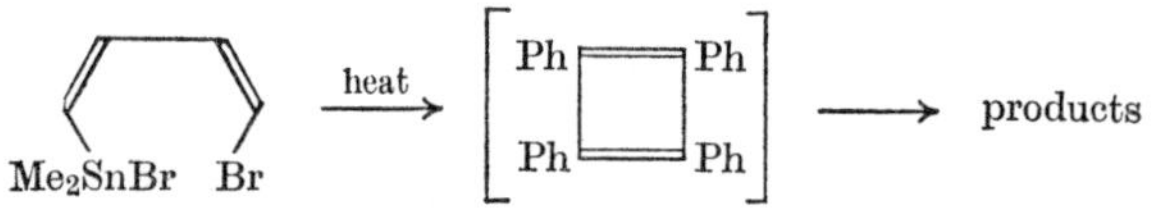

When triethyltin chloride in hexane was treated with oxygen containing a small amount of ozone the final product obtained was formulated as $EtSnOCl.Et_3SnCl$ [1] (it should be noted that the melting point of this product is very close to that of $ClEt_2SnOSnEt_2Cl$).

The separation of organotin halides by gas- [16, 17, 46, 67, 72], thin-layer- [7, 8, 25], and paper- [76] chromatography and their estimation by polarographic techniques [6, 14, 24] has been studied by a number of workers. Details of the amperometric titration of organotin chlorides have been given [54] as well as a method for the estimation of small quantities of these substances in air [60].

4.3. ORGANOTIN PSEUDOHALIDES

Two recent review articles include accounts of the organotin pseudohalides [28, 70] and in a review of organometallic azides [69], 7 organotin examples were mentioned.

Organotin azides are commonly prepared by treating an organotin halide with an alkali metal or alkaline earth metal azide [64, 101].

$$R_3SnCl + NaN_3 \longrightarrow R_3SnN_3 + NaCl$$

Alternatively azides are produced, in near quantitative yields, by reaction between a stannyldialkylamine and hydrazoic acid in dry ether [31].

$$R_3SnNR'_2 + HN_3 \longrightarrow R_3SnN_3 + R'_2NH$$
$$R_2Sn(NR'_2)_2 + 2HN_3 \longrightarrow R_2Sn(N_3)_2 + 2R'_2NH$$

Many organotin azides react with phosphines to give N-stannylphosphinimines [31]:

$$R_3SnN_3 + PR'_3 \longrightarrow R_3SnN{=}PR'_3 + N_2$$
$$(R = alkyl; R' = Bu, Ph, NMe_2 \text{ etc.})$$

The ability of azides to undergo this reaction has been proposed as the basis of a method of classification [69] but the validity of this method has been questioned [29]. When trimethyltin azide reacts with a phosphine no $Me_3SnN{=}PR_3$ is formed, instead the following reactions occur [58a].

$$2Me_3SnN_3 \longrightarrow Me_4Sn + Me_2Sn(N_3)_2$$
$$Me_2Sn(N_3)_2 + PR_3 \longrightarrow Me_2(N_3)SnN{=}PR_3 + N_2$$

Electron spin resonance measurements showed that photolysis of triphenyltin azide at $-160°$ gave the corresponding nitrene [44].

$$Ph_3SnN_3 \xrightarrow{h\nu} Ph_3SnN + N_2$$

Organotin azides can function as Lewis bases forming adducts with stannic chloride, boron tribromide and other electron-pair acceptors [71].

The organotin isocyanates, isothiocyanates, selenocyanates and cyanides are all conveniently prepared by reaction between an organotin halide and an appropriate salt as in the following examples.

$$Et_3SnCl + KCN \longrightarrow Et_3SnCN + KCl \qquad \text{(ref. 32)}$$

$$Ph_3SnCl + AgNCO \longrightarrow Ph_3SnNCO + AgCl \qquad \text{(ref. 45)}$$

$$2(PhCH_2)_3SnI + Pb(NCO)_2 \longrightarrow 2(PhCH_2)_3SnNCO + PbI_2 \qquad \text{(ref. 66)}$$

$$Me_2SnCl_2 + 2NaNCS \longrightarrow Me_2Sn(NCS)_2 + 2NaCl \qquad \text{(ref. 74)}$$

$$(p\text{-}MeC_6H_4)_3SnCl + AgNCS \longrightarrow (p\text{-}MeC_6H_4)_3SnNCS + AgCl \qquad \text{(ref. 65)}$$

$$(ClBu_2Sn)_2O + 2NH_4NCS \longrightarrow (SCNBu_2Sn)_2O + 2NH_4Cl \qquad \text{(ref. 90)}$$

$$Ph_3SnCl + NaSeCN \longrightarrow Ph_3SnNCSe + NaCl \qquad \text{(ref. 5)}$$

In addition the cyanides can be prepared from the stannylamines and hydrogen cyanide [30]:

$$R_3SnNEt_2 + HCN \xrightarrow{-30^\circ} R_3SnCN + Et_2NH$$

and also by reaction between alkyl isocyanides and organotin hydrides in the presence of a peroxide [57].

$$PhCH_2NC + Bu_3SnH \xrightarrow{\text{t-BuOOBu-t}} Bu_3SnCN + PhMe$$

The isocyanates and isothiocyanates are formed when (thio)urea or urethans are heated with organotin oxides or alkoxides.

$$(Bu_3Sn)_2O + 2NH_2CONH_2 \xrightarrow{140^\circ} 2Bu_3SnNCO + 2NH_3 + H_2O$$
$$\text{(refs. 84, 91)}$$

$$(Bu_3Sn)_2O + 2NH_2CSNH_2 \xrightarrow{170^\circ} 2Bu_3SnNCS + 2NH_3 + H_2O \qquad \text{(ref. 84)}$$

$$Et_3SnOMe + NH_2COOEt \xrightarrow{60-70^\circ} Et_3SnNCO + EtOH + MeOH \qquad \text{(ref. 9)}$$

It is possible that, in the reaction with urea, the reactive species is cyanic acid since this substance readily converts an organotin oxide to the isocyanate [66a].

$$(R_3Sn)_2O + 2HNCO \longrightarrow 2R_3SnNCO + H_2O$$

When triphenyltin chloride reacted with sodium urethan the product was triphenyltin isocyanate, presumably formed by spontaneous decomposition of the initially formed stannylurethan [45].

$$Ph_3SnCl + NaNHCOOEt \longrightarrow [Ph_3SnNHCOOEt] + NaCl$$
$$\downarrow$$
$$Ph_3SnNCO + EtOH$$

(The isomeric fulminates can be made from reaction between an organotin halide and silver fulminate [4]:

$$Ph_3SnCl + AgCNO \longrightarrow Ph_3SnCNO \text{ (m.p. 146-8°)} + AgCl.)$$

The general properties of these compounds are typical of metallic psueudohalides and are unlike those of their organic counterparts, for example treatment with water results in displacement of the pseudohalogen group and not addition to it.

$$R_3SnCN + H_2O \longrightarrow R_3SnOH + HCN \qquad \text{(refs. 32, 78)}$$
$$R_2Sn(NCO)_2 + H_2O \longrightarrow R_2SnO + 2HNCO \qquad \text{(ref. 45)}$$

The diisocyanates are very susceptible to hydrolysis and, although $Bu_2Sn(NCO)_2$ was isolated, only partially hydrolysed specimens of $Ph_2Sn(NCO)_2$ could be obtained[45]. (The hydrolysis of the diisocyanates is more involved than is indicated in the above equation; various intermediate products which are isolated are discussed in Chapter 5.)

Certain of the isocyanates undergo exchange reactions with organosilicon halides, for example[9]:

$$Pr_3SnNCO + Me_3SiCl \longrightarrow Me_3SiNCO + Pr_3SnCl.$$

TABLE III

Physical properties of some organotin pseudohalides

Compound	m.p.	b.p.	n_D^{20}	d^{20}
Me_3SnN_3	119·5–121·5°	—	—	—
$(PhCH_2)_3SnN_3$	119°	—	—	—
Ph_3SnN_3	115–116°	—	—	—
Et_3SnNCO	53–54°	113–114°/8 mm	—	—
Bu_3SnNCO	—	144–147°/1·3 mm	1·4885	—
$Bu_2Sn(NCO)_2$	44–48°	—	—	—
Ph_3SnNCO	101–103°	—	—	—
Et_3SnNCS	33°	130°/1 mm	—	—
Bu_3SnNCS	—	146°/2 mm	1·5432	1·235
$Bu_2Sn(NCS)_2$	145°	—	—	—
Ph_3SnNCS	172–173°	—	—	—
Me_3SnCN	183–184·5°	—	—	—
Ph_3SnCN	255–256°	—	—	—

References

1. Aleksandrov, Yu. A. and Sheyanov, N. G., *Zh. Obshch. Khim.*, 1967, **37**, 2136 (*Chem. Abs.*, 1968, **68**, 29822).
2. Armenskaya, L. V., Monastyrskii, L. M., Smolyan, Z. S. and Lysenko, E. N., *Khim. Prom.* (*Moscow*), 1968, **44**, 665 (*Chem. Abs.*, 1968, **69**, 102461).

3. Averbukh, B. S., Abramova, L. V., Breger, A. K. H., Vainshtein, B. I., Gol'din, V. A., Kocheshkov, K. A., Syrkus, N. P., Shalyapin, N. K. and Sheverdina, N. I., *Zh. Fiz. Khim.*, 1964, **38**, 2445.
4. Beck, W. and Schuierer, E., *Chem. Ber.*, 1964, **97**, 3517.
5. Boehland, H. and Niemann, E., *Z. Chem.*, 1968, **8**, 191.
6. Bork, V. A. and Selivokhin, P. I., *Plast. Massy*, 1968, **56** (*Chem. Abs.*, 1968, **69**, 10519).
7. Braun, D. and Heimes, H. T., *Z. Anal. Chem.*, 1968, **239**, 6.
8. Burger, K., *Z. Anal. Chem.*, 1963, **192**, 280.
9. Chauzov, V. A., Litvinova, O. V. and Baukov, Yu. I., *Zh. Obshch. Khim.*, 1966, **36**, 952.
9a. Devaud, M. and Laviron, E., *Rev. Chim. Miner.*, 1968, **5**, 427 (*Chem. Abs.*, 1968, **69**, 112852).
10. Fentiman, A. F., Wyant, R. E., Jeffrey, D. A. and Kircher, J. F., *J. Organometallic Chem.*, 1965, 4, 302.
11. Fentiman, A. F., Wyant, R. E., McFarling, J. L., Kahler, J. E. and Kircher, J. F., A. E. C. Accession No. 10870, Rept. No. BMI-1753 (*Chem. Abs.*, 1966, **65**, 16293c).
12. Fentiman, A. F., Wyant, R. E., McFarling, J. L. and Kircher, J. F., *J. Organometallic Chem.*, 1966, **6**, 645.
13. Fentiman, A. F., Wyant, R. E., Steinmeyer, R. D., Jeffrey, D. A., Kircher, J. F. and Kahler, E. J., *U.S. At. Energy Comm. BMI*-1713, 1965, (*Chem. Abs.*, 1965, **63**, 2549f).
14. Flerov, V. N. and Tyurin, Yu. M., *Zh. Obshch. Khim.*, 1968, **38**, 1669 (*Chem. Abs.*, 1968, **69**, 102484).
15. Frankland, E., *Annalen*, 1849, **71**, 171.
16. Geissler, H. and Kriegsmann, H., *Z. Chem.*, 1965, **5**, 423.
17. Gerrard, W., Mooney, E. F. and Rees, R. G., *J. Chem. Soc.*, 1964, 740.
18. Gevorkyan, A. A. and Sarksyan, Zh. G., *Arm. Khim. Zh.*, 1968, **21**, 269 (*Chem. Abs.*, 1968, **69**, 106832).
19. Grant, D. and van Wazer, J. R., *J. Organometallic Chem.*, 1965, 4, 229.
20. Hayashi, T., Kikkawa, S. and Matsuda, S., *Kogyo Kagaku Zasshi*, 1967, **70**, 1389 (*Chem. Abs.*, 1968, **68**, 59672).
21. Hayashi, T., Kikkawa, S. and Matsuda, S., *Kogyo Kagaku Zasshi*, 1968, **71**, 710 (*Chem. Abs.*, 1968, **69**, 87127).
22. Hayashi, T., Kikkawa, S., Matsuda, S. and Fujita, K., *Kokyo Kagaku Zasshi*, 1967, **70**, 2298 (*Chem. Abs.*, 1968, **68**, 87363).
23. Hayashi, T., Uchimura, J., Matsuda, S. and Kikkawa, S., *Kogyo Kagaku Zasshi*, 1967, **70**, 714 (*Chem. Abs.*, 1968, **68**, 13109).
24. Issleib, K., Matschiner, H. and Naumann, S., *Talanta*, 1968, **15**, 379 (*Chem. Abs.*, 1968, **69**, 8314).
25. Jitsu, Y., Kudo, N. and Sugiyama, T., *Noyaku Seisan Gijutsu*, 1967, **17**, 17 (*Chem. Abs.*, 1968, **68**, 56448).
26. Kuivila, H. G., Sommer, R. and Green, D. C., *J. Org. Chem.*, 1968, **33**, 1119.
27. Langer, H. G., *Tetrahedron Letters*, 1967, 43.
28. Lappert, M. F. and Pyszora, H., *Advances in Inorganic and Radiochemistry*, eds. Emeléus, H. J. and Sharpe, A. G., Academic Press, New York and London, 1966, **9**, 133.

29. Lehn, W. L., *Inorg. Chem.*, 1967, **6**, 1061.
30. Lorberth, J., *Chem. Ber.*, 1965, **98**, 1201.
31. Lorberth J., Krapf, H. and Noeth, H., *Chem. Ber.*, 1967, **100**, 3511.
32. Luijten, J. G. A. and van der Kerk, G. J. M., *Investigations in the Field of Organotin Chemistry*, The Tin Research Institute, Greenford, 1955, 102.
33. Malnar, M., *Acta Pharm. Jugoslav.*, 1968, **18**, 65 (*Chem. Abs.*, 1968, **69**, 106847).
34. Matsuda, S., *Asahi Garasu Kogyo Gijutsu Shorei-Kai Kenkyu Hokoku*, 1966, **12**, 329 (*Chem. Abs.*, 1968, **68**, 39756).
35. Matsuda, S., Kikkawa, S. and Hayashi, T., *Kogyo Kagaku Zasshi*, 1966, **69**, 256 (*Chem. Abs.*, 1966, **65**, 5482h).
36. Matsuda, S., Kikkawa, S. and Kashiwa, N., *Kogyo Kagaku Zasshi*, 1966, **69**, 1036 (*Chem. Abs.*, 1966, **65**, 20160g).
37. Matsuda, S., Kikkawa, S. and Nomura, M., *Kogyo Kagaku Zasshi*, 1966, **69**, 649 (*Chem. Abs.*, 1966, **65**, 18612h).
38. Matsuda, S., Kikkawa, S. and Omae, I., *Kokyo Kagaku Zasshi*, 1966, **69**, 646 (*Chem. Abs.*, 1966, **65**, 18612f).
39. Matsuda, S., Matsuda, H., Iwamoto, N. and Matsumoto, A., *Kogyo Kagaku Zasshi*, 1967, **70**, 1747 (*Chem. Abs.*, 1968, **68**, 37373).
39a. Mehner, H., Jehring, H. and Kriegsmann, H., *J. Organometallic Chem.*, 1968, **15**, 97.
40. Meyer, H., *J. Organometallic Chem.*, 1968, **11**, 525.
41. Miretskii, V. Yu., Vereshchinskii, I. V., Ivanov, A. Yu., Kocheshkov, K. A., Pozdeev, V. V. and Sheverdina, N. I., *Khim. Prom.*, 1968, **44**, 258 (*Chem. Abs.*, 1968, **69**, 77397).
42. Moedritzer, K., *Advances in Organometallic Chemistry*, Academic Press, New York and London, 1968, **6**, 171.
43. Moedritzer, K., *Organometallic Chem. Rev.*, 1966, **1**, 179.
44. Moriarty, R. M., Rahman, M. and King, G. J., *J. Amer. Chem. Soc.*, 1966, **88**, 842.
45. Mufti, A. S. and Poller, R. C., *J. Chem. Soc.*, 1965, 5055.
46. Neumann, W. P. and Burkhardt, G., *Annalen*, 1963, **663**, 11.
47. Nomura, M., Ando, S. and Matsuda, S., *Kogyo Kagaku Zasshi*, 1968, **71**, 394 (*Chem. Abs.*, 1968, **69**, 77384).
48. Nomura, M. Matsuda, S. and Kikkawa, S., *Kogyo Kagaku Zasshi*, 1967, **70**, 710 (*Chem. Abs.*, 1968, **68**, 13108).
49. Nosek, J., *Collection Czech. Chem. Commun.*, 1964, **29**, 3173 (*Chem. Abs.*, 1965, **62**, 4046e).
50. Oakes, V. and Hutton, R. E., *J. Organometallic Chem.*, 1965, **3**, 472.
51. Oakes, V. and Hutton, R. E., *J. Organometallic Chem.*, 1966, **6**, 133.
52. Omae, I., Matsuda, S. and Kikkawa, S., *Kogyo Kagaku Zasshi*, 1967, **70**, 1759 (*Chem. Abs.*, 1968, **68**, 78383).
53. Omae, I., Matsuda, S., Kikkawa, S. and Sato, R., *Kogyo Kagaku Zasshi*, 1967, **70**, 705 (*Chem. Abs.*, 1968, **68**, 13107).
54. Plazzogna, G. and Pilloni, G., *Anal. Chim. Acta*, 1967, **37**, 260.
55. Reichle, W. T., *Inorg. Chem.*, 1966, **5**, 87.
56. Ruidisch, I., Schmidbauer, H. and Schumann, H., *Halogen Chemistry*, ed. Gutman, V., Academic Press, New York and London, 1967, **2**, 269.

66 THE CHEMISTRY OF ORGANOTIN COMPOUNDS

57. Saegusa, T., Kobayashi, S., Ito, Y. and Yasuda, N., *Amer. J. Chem. Soc,*. 1968, **90**, 4182.
58. Sandel, V. R. and Freedman, H. H., *J. Amer. Chem. Soc.*, 1968, **90**, 2059.
58a. Schmidbauer, H. and Wolfsberger, W., *Chem. Ber.*, 1968, **101**, 1664.
59. Schollkopk, U. and Traenckner, H. J., *J. Organometallic Chem.*, 1966, **5**, 300.
60. Selivokhin, P. I., *Vestn. Tekhn. i Ekon. Inform. Nauchn.—Issled. Gos. Kom. Khim. Prom. pri Gosplane S.S.S.R.*, 1964, **8**, 27 (*Chem. Abs.*, 1965, **63**, 17155d).
61. Sisido, K. and Kozima, S., *J. Organometallic Chem.*, 1968, **11**, 503.
62. Sisido, K., Kozima, S. and Hanada, T., *J. Organometallic Chem.*, 1967, **9**, 99.
63. Sisido, K., Kozima, S. and Tuzi, T., *J. Organometallic Chem.*, 1967, **9**, 109.
63a. Sisido, K. and Takeda, Y., *J. Org. Chem.*, 1961, **26**, 2301.
64. Srivastava, T. N. and Bhattacharya, S. N., *J. Inorg. Nucl. Chem.*, 1966, **28**, 1480.
65. Srivastava, T. N. and Bhattacharya, S. N., *J. Inorg. Nucl. Chem.*, 1966, **28**, 2445.
66. Srivastava, T. N. and Bhattacharya, S. N., *J. Inorg. Nucl. Chem.*, 1967, **29**, 1873.
66a. Stamm, W., *J. Org. Chem.*, 1965, **20**, 693.
67. Steinmeyer, R. D., Fentiman, A. F. and Kahler, E. J., *Anal. Chem.*, 1965, **37**, 520.
68. Sundermeyer, W. and Verbeek, W., *Angew. Chem. Intern. Ed. Engl.*, 1966, **5**, 1.
69. Thayer, J. S., *Organometallic Chem. Rev.*, 1966, **1**, 157.
70. Thayer, J. S. and West, R., *Advances in Organometallic Chemistry*, Academic Press, New York and London, 1967, **5**, 169.
71. Thayer, J. S. and West, R., *Inorg. Chem.*, 1965, **4**, 114.
72. Tonge, B. L., *J. Chromatog.*, 1965, **19**, 182.
73. Van den Berghe, E. V. and van der Kelen, G. P., *J. Organometallic Chem.*, 1966, **6**, 522.
74. Wada, M., Nishino, M. and Okawara, R., *J. Organometallic Chem.*, 1965, **3**, 70.
75. Weiss, R. W., Editor, *Organometallic Compounds*, 2nd Edn., Springer-Verlag, New York, 1967, **2**, p. 268.
76. Williams, D. J. and Price, J. W., *Analyst*, 1964, **89**, 220.
77. Yastrebov, V. V. and Chernyshev, A. I., *Zh. Obshch. Khim.*, 1967, **37**, 2140.
78. Zimmer, H. and Luebke, K., *Chem. Ber.*, 1952, **85**, 1119.
79. British Patent, 1,064,178, April 5, 1967, (*Chem. Abs.*, 1967, **67**, 43924e).
80. British Patent 1,070,942, June 7, 1967 (*Chem. Abs.*, 1967, **67**, 43925f).
81. French Patent 1,393,779, March 26, 1965 (*Chem. Abs.*, 1965, **63**, 9985f).
82. French Patent 1,399,552, May 21, 1965 (*Chem. Abs.*, 1965, **63**, 9985e).
83. French Patent 1,456,268, Oct. 21, 1966 (*Chem. Abs.*, 1967, **66**, 115806x).
84. German Patent 1,206,901, Dec. 16, 1965 (*Chem. Abs.*, 1966, **64**, 9767a).
85. German Patent, 1,222,503, Aug. 11, 1966 (*Chem. Abs.*, 1966, **65**, 13763f).
86. German Patent 1,239,687, May 3, 1967 (*Chem. Abs.*, 1968, **68**, 2989).
87. Japanese Patent 24,958, Nov. 6, 1965 (*Chem. Abs.*, 1965, **62**, 14726g).

88. Japanese Patent 1615, Feb. 7, 1966 (*Chem. Abs.*, 1966, **64**, 12724b).
89. Japanese Patent 21,021, Dec. 8, 1966 (*Chem. Abs.*, 1967, **66**, 46486d).
90. Japanese Patent 8015, March 31, 1967 (*Chem. Abs.*, 1967, **67**, 108759b).
91. Netherlands Patent 6,411,266, April 5, 1965 (*Chem. Abs.*, 1965, **63**, 11614a).
92. Netherlands Patent 6,415,330, July 1, 1965 (*Chem. Abs.*, 1966, **64**, 2129a).
93. Netherlands Patent 6,504,226, Oct. 7, 1965 (*Chem. Abs.*, 1966, **64**, 8240b).
94. Netherlands Patent 6,505,767, Nov. 8, 1965 (*Chem. Abs.*, 1966, **64**, 12723g).
95. Netherlands Patent 6,506,444, Nov. 22, 1965 (*Chem. Abs.*, 1966, **64**, 11251b).
96. Netherlands Patent 6,511,702, March 14, 1966 (*Chem. Abs.*, 1966, **65**, 5490a).
97. Netherlands Patent 6,512,145, March 21, 1966 (*Chem. Abs.*, 1966, **65**, 8962g).
98. Netherlands Patent 6,513,659, April 25, 1966 (*Chem. Abs.*, 1966, **65**, 7218c).
99. Netherlands Patent 6,515,697, June 3, 1966 (*Chem. Abs.*, 1966, **65**, 13763g).
100. Netherlands Patent 6,614,326, April 12, 1967 (*Chem. Abs.*, 1967, **67**, 100247n).
101. U.S. Patent 3,232,958, Feb. 1, 1966 (*Chem. Abs.*, 1966, **64**, 17640c).
102. U.S. Patent 3,297,732, Jan. 10, 1967 (*Chem. Abs.*, 1967, **66**, 115807y).
103. U.S. Patent, 3,397,131, Aug. 13, 1968 (*Chem. Abs.*, 1968, **69**, 77503).
104. U.S.S.R. Patent 181,104, April 15, 1966 (*Chem. Abs.*, 1966, **65**, 8962f).

ORGANOTIN OXIDES, HYDROXIDES AND OTHER COMPOUNDS CONTAINING Sn—O BONDS

Compounds containing the grouping $\rangle$Sn—O—X will be considered here i.e. the hydroxides (X = H), oxides (X = Sn), alkoxides and phenoxides (X = C), peroxides (X = O) and oxides of other elements (X = a metal or metalloid). Other types of compounds containing tin–oxygen bonds are discussed in Chapter 10.

5.1. TRIALKYL(ARYL)TIN OXIDES AND HYDROXIDES

Alkaline hydrolysis of the monohalides gives the organotin hydroxides.

$$R_3SnX + NaOH \longrightarrow R_3SnOH + NaX \quad (X = Cl, Br, I)$$

Very few unsymmetrical hydroxides, $R_2R'SnOH$, have been completely characterised and dibutylvinyltin hydroxide disproportionates at room temperature [25].

$$2Bu_2(CH_2=CH)SnOH \longrightarrow Bu_2SnO + Bu_2Sn(CH=CH_2)_2 + H_2O$$

In most cases the hydroxides are readily dehydrated to the oxides (organodistannoxanes).

$$2R_3SnOH \longrightarrow R_3SnOSnR_3 + H_2O$$

The ease with which the latter reaction occurs varies considerably and, whereas triphenyltin hydroxide loses water on storage [67], trimethyltin hydroxide is often purified by sublimation though it can be converted to $Me_3SnOSnMe_3$ by treatment with sodium in benzene. Most hydroxides, however, are readily dehydrated by heating under either atmospheric or reduced pressure and, with the exception of $Me_3SnOSnMe_3$ (which readily rehydrates), the oxides are stable to moisture. Many of the alkyltin oxides and hydroxides absorb atmospheric carbon dioxide to give trialkyltin carbonates.

There is considerable current interest in the addition reactions of organotin oxides and, for example, Bloodworth and Davies [14, 15, 16, 17] have shown that the addition products obtained with isocyanates undergo some useful reactions. The initial product from *bis*(tri-butyltin) oxide and ethyl isocyanate is tributylstannyl N-ethyl-N-tributylstannylcarbamate (I) resulting from addition to the C=N bond. By heating compound I or, more conveniently, by heating the

$$(Bu_3Sn)_2O + EtNCO \longrightarrow Bu_3SnN(Et)COOSnBu_3$$
$$I$$

oxide and isocyanate in a 1:2 molar ratio *bis*(tributylstannyl)ureas can be formed [15, 95].

$$(Bu_3Sn)_2O + 2RNCO \longrightarrow Bu_3SnNRCONRSnBu_3 + CO_2$$

If, on the other hand, the product I is treated with more isocyanate at room temperature a slow reaction occurs giving in 1–25 days the alkyl isocyanurate. This product apparently results from repeated addition reactions.

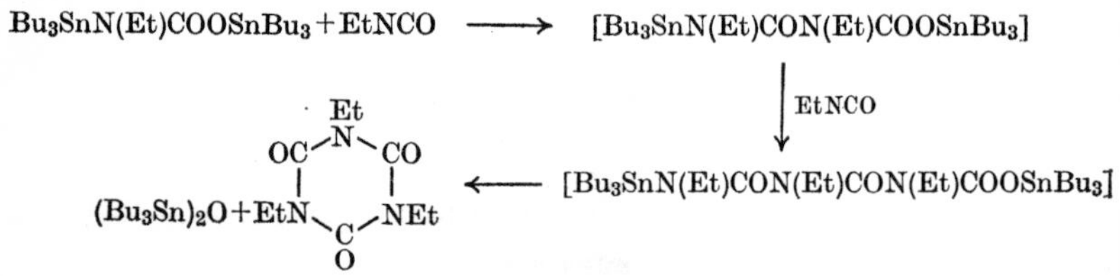

$$Bu_3SnN(Et)COOSnBu_3 + EtNCO \longrightarrow [Bu_3SnN(Et)CON(Et)COOSnBu_3]$$

These reactions have been carried out with a number of isocyanates and the last reaction has been modified to yield 'mixed' isocyanurates, for example with one phenyl and two ethyl substituents [16]. Similar isocyanate addition reactions have been used to prepare 'mixed' biurets of the type RNHCONR'CONHR" where R,R' and R" are different alkyl or aryl groups.

The *bis*(trialkyltin) oxides are believed to add to the C=S bond of isothiocyanates to give an unstable product which breaks down to the organotin sulphide and the isocyanate [19].

$$(Bu_3Sn)_2O + PhNCS \longrightarrow [PhN=C(OSnBu_3)SSnBu_3]$$
$$\downarrow$$
$$(Bu_3Sn)S + PhNCO$$

The latter then reacts with more oxide so that the carbamate is isolated as the final product and the overall equation is:

$$2(Bu_3Sn)_2O + PhNCS \longrightarrow Bu_3SnN(Ph)COOSnBu_3 + (Bu_3Sn)_2S.$$

Other examples of addition reactions are as follows [19]:

$$(Bu_3Sn)_2O + CO_2 \longrightarrow Bu_3SnOCOOSnBu_3$$

$$(Bu_3Sn)_2O + SO_2 \longrightarrow Bu_3SnOSOOSnBu_3$$

$$(Bu_3Sn)_2O + RN{=}C{=}NR \longrightarrow Bu_3SnNRC(OSnBu_3){=}NR.$$

Oxides will add to the cyanide and carbonyl groups only when these are directly attached to carbon atoms bearing electronegative substituents.

$$(Bu_3Sn)_2O + Cl_3CCN \longrightarrow Bu_3SnN{=}C(CCl_3)OSnBu_3 \qquad \text{(ref. 19)}$$

$$(Bu_3Sn)_2O + Cl_3CCHO \longrightarrow Bu_3SnOCH(CCl_3)OSnBu_3 \qquad \text{(ref. 35)}$$

(Similar addition reactions occur with organotin alkoxides, see p. 77.)

A large number of protic species will attack the hydroxides and

TABLE I

Physical properties of some trialkyl(aryl)tin hydroxides and oxides

Compound	m.p.	b.p.	n_D^{20}	d^{20}
Me$_3$SnOH	118°	—	—	—
(Me$_3$Sn)$_2$O	—	84°/22 mm	—	—
Et$_3$SnOH	49–50°	—	—	—
(Et$_3$Sn)$_2$O	—	272°	1·4975 (1·529, 1·4635)	1·3774 (1·319)
Bu$_3$SnOH	—	186–190°/5 mm[a]	—	—
(Bu$_3$Sn)$_2$O	—	210–214°/10 mm	1·4880	1·1715
(Bu$_2$MeSn)$_2$O	—	135–137°/1 mm	1·4868	1·251
(PhCH$_2$)$_3$SnOH	122–124°	—	—	—
[(PhCH$_2$)$_3$Sn]$_2$O	120°	—	—	—
Ph$_3$SnOH	124°	—	—	—
(Ph$_3$Sn)$_2$O	123–124°	—	—	—

[a] The significance of this figure is doubtful as Bu$_3$SnOH is converted to (Bu$_3$Sn)$_2$O on distillation under reduced pressure.

oxides in equilibrium reactions to give an organotin product and water. The equilibrium can be made to favour the products by using excess reagent, for example in the conversion to chlorides.

$$(R_3Sn)_2O + 2HCl \longrightarrow 2R_3SnCl + H_2O$$

Or, in the following examples, the water formed is usually removed by azeotropic distillation with benzene or toluene.

$$(R_3Sn)_2O + 2R'COOH \longrightarrow 2R_3SnOCOR' + H_2O$$
$$(R_3Sn)_2O + 2R'OH \longrightarrow 2R_3SnOR' + H_2O$$
$$(R_3Sn)_2O + R'SO_3H \longrightarrow 2R_3SnO_3SR' + H_2O$$

As was noted in Chapter 2, a similar reaction is used for the preparation of alkynyltin compounds[54].

$$(R_3Sn)_2O + 2R'C\equiv CH \longrightarrow 2R_3SnC\equiv CR' + H_2O$$

By the same method oximes can be stannylated[105].

$$(Bu_3Sn)_2O + 2Me_2C\text{=}NOH \longrightarrow 2Me_2C\text{=}NOSnBu_3 + H_2O$$

Methyllithium reacts with hexamethyldistannoxane in ether at room temperature to give lithium trimethylstannoxide, Me_3SnOLi, a useful reagent for the preparation of other metallostannoxanes[78] (p. 84). Irradiation of a solution of hexaethyldistannoxane in nonane with ultraviolet light produced $Et\cdot$ radicals and the final products included Et_2SnO, Et_4Sn, ethylene and ethane[2]. A number of anions were extracted from aqueous solutions into an organic phase which contained triphenyltin hydroxide[80].

5.2. DIALKYL(ARYL)TIN OXIDES AND RELATED COMPOUNDS

Hydrolysis of organotin dihalides or dipseudohalides with excess alkali gives the polymeric oxides.

$$xR_2SnX_2 + 2xNaOH \longrightarrow (R_2SnO)_x + 2xNaCl + xH_2O$$

The diisocyanates $R_2Sn(NCO)_2$ hydrolyse spontaneously in moist air and it was possible to follow the course of the reaction and show that it proceeds through at least three intermediates i.e. compounds *II*, *III* and *IV* (X = NCO) in the following sequence[52].

$$R_2SnX_2 \longrightarrow \underset{II}{R_2Sn(X)OH} \longrightarrow \underset{III}{R_2SnXOSnR_2X} \longrightarrow$$

$$\underset{IV}{R_2SnXOSn(OH)R_2} \longrightarrow \underset{V}{R_2Sn(OH)OSn(OH)R_2} \longrightarrow R_2SnO$$

(These intermediates are formulated, for simplicity, as monomers though their structures are generally more complex—see Chapter 12.) The hydroxyhalides *II* are labile and rarely encountered but the intermediates *III* and *IV* are relatively stable and are obtained when a dihalide is treated with a deficiency of alkali or with a weak base[8].

The symmetrical distannoxanes *III* can be made by hydrolysis in the presence of a base such as triethylamine (these compounds can arise inadvertently e.g. when the complex Bu_2SnCl_2. 2pyridine is crystallised from moist ethanol only $Bu_2SnClOSnBu_2Cl$ is isolated) or by heating an equimolar mixture of the oxide and dihalide in a solvent such as toluene [9].

$$R_2SnO + R_2SnX_2 \longrightarrow R_2SnXOSnR_2X$$
$$III$$

There is now a wide range of compounds with formula *III* including examples where X = halogen, carboxylate [48, 57], camphorsulphonate [9], phenoxide [24, 74, 88, 103], alkyl peroxide [18, 27], and the trimethylsiloxy group [58]. Direct substitution of the X groups in the distannoxanes has been reported [91].

$$Bu_2SnClOSnBu_2Cl + 2NaNCS \longrightarrow Bu_2Sn(NCS)OSn(NCS)Bu_2$$

A few 'mixed' compounds such as $Bu_2Sn(OAc)OSn(OMe)Bu_2$ are known [92].

Controlled partial hydrolysis of the distannoxanes, *III*, gives the hydroxydistannoxanes, *IV*, which can also be synthesised by heating the oxide and dihalide in a moist solvent [8, 60, 100].

$$3R_2SnO + R_2SnX_2 + H_2O \longrightarrow 2R_2SnXOSnR_2OH$$
$$IV$$

Crystallisation of compounds of type *IV* from an alcohol results in alcoholysis, a reaction which is reversed when the alkoxide is crystallised from a moist solvent or exposed for some days to the atmosphere [8, 60].

$$R_2SnXOSnR_2OH \underset{H_2O}{\overset{R'OH}{\rightleftarrows}} R_2SnXOSnR_2OR'$$

The first clearly defined dihydroxydistannoxane $R_2(OH)Sn-OSn(OH)R_2$ (*V*) to be isolated was $R = o\text{-}PhOC_6H_4$ [66] and it was assumed that steric hindrance by the bulky aryl groups inhibited further condensation. More recently however the compound *V*, $R = CH_2CH_2CN$ was isolated and it appears that the size of the R groups may not be the only factor in determining the existence of these compounds. In special cases the dihydroxides $R_2Sn(OH)_2$ may occur as unstable intermediates in the hydrolysis of dihalides; there is some evidence for the transient existence of these compounds in the earlier literature [86] and an impure specimen of $(o\text{-}PhOC_6H_4)_2Sn(OH)_2$ has been isolated [66].

When dibutyltin chloride and the corresponding oxide are heated together, in various molar ratios, a range of oligomeric stannoxanes can be obtained. Compounds corresponding to x = 1–6, 9 and 12

$$Bu_2SnCl_2 + xBu_2SnO \longrightarrow ClBu_2Sn(OSnBu_2)_xCl$$

were isolated [32]. With an oxide and a trichloride the unsymmetrical chlorodistannoxane is obtained; these compounds are hydrolysed in

$$R_2SnO + R'SnCl_3 \longrightarrow ClR_2SnOSnR'Cl_2$$

the atmosphere. In a similar reaction with monochlorides the products were obtained as thick greases [30].

$$R_2SnO + R'_3SnCl \longrightarrow ClR_2SnOSnR'_3$$

Unsymmetrical organotin monohalides have been prepared by reaction between oxides R_2SnO and Grignard reagents $R'MgBr$ followed by treatment of the products with hydrochloric acid to give $R_2R'SnCl$ [42]. The dialkyl(aryl)tin oxides are polymeric compounds which melt at high, non-characteristic, temperatures often with decomposition.

TABLE II

Melting points of some organodistannoxanes

Compound	m.p.
$Me_2ClSnOSnClMe_2$	300° decomp.
$Et_2ClSnOSnClEt_2$	175°
$Et_2BrSnOSnBrEt_2$	172–173°
$Bu_2ClSnOSnClBu_2$	112°
$Bu_2(NCS)SnOSn(NCS)Bu_2$	84°
$Bu_2(OAc)SnOSn(OAc)Bu_2$	58–60°
$Bu_2(t\text{-}BuOO)SnOSn(OOBu\text{-}t)Bu_2$	78–79°
$Et_2ClSnOSn(OH)Et_2$	236°
$Bu_2ClSnOSn(OEt)Bu_2$	85–140°

5.3. ALKYL(ARYL)STANNONIC ACIDS

When an organotin trihalide is treated with excess aqueous alkali the stannonic acid salt is formed.

$$RSnCl_3 + 4NaOH \longrightarrow RSnO_2Na + 3NaCl + 2H_2O$$

The alkali metal stannonate is obtained only when excess alkali is used and the stannonic acid can be precipitated by passing carbon dioxide into the alkaline solution. When lower concentrations of alkali are used the free RSnOOH is obtained directly [40]. The effect of variations in pH on the hydrolysis products of ethyltin trichloride has been examined [38]. Partial hydrolysis of the organotin trichlorides can be effected, the products formed being dependent not only on the reaction conditions but also on the nature of the organic group [42a,49]. Thus octyltin trichloride when mixed with water gives $C_8H_{17}Sn(OH)_2Cl$; the corresponding ethyl and butyl compounds are best obtained by treating the trichlorides with two equivalents of sodium hydroxide.

$$RSnCl_3 + 2NaOH \longrightarrow RSn(OH)_2Cl + 2NaCl \qquad (R = Et, Bu)$$

With three equivalents of sodium hydroxide BuSnOOH is obtained from $BuSnCl_3$ but $C_8H_{17}SnCl_3$ gives $C_8H_{17}Sn(OH)_2OSn(OH)ClC_8H_{17}$. No intermediate hydrolysis products could be isolated from the aryltin trihalides [49]. Ethylstannonic acid when treated with hydrochloric acid gave $EtSn(OH)_2Cl$, and this compound reacted with ethyltin trichloride to give $EtSn(OH)Cl_2$ [39]. Alkylstannonic acids can also be prepared from alkali metal stannates and alkyl halides.

$$RCl + Na_2SnO_2 \longrightarrow RSnO_2Na + NaCl$$

The stannonic acids are, generally, infusible polymeric amorphous powders which, in some cases, can be obtained in the dehydrated form with the empirical formula $R_2Sn_2O_3$ [22].

5.4. ORGANOTIN ALKOXIDES AND PHENOXIDES

Alkoxides can be prepared by reaction between an alcohol and an organotin halide in the presence of a base [81] or by a metathetical reaction from an alkali-metal alkoxide [11, 77].

$$R_3SnCl + R'ONa \longrightarrow R_3SnOR' + NaCl$$

It is often more convenient, however, to allow the organotin oxide and alcohol to react by heating them in boiling benzene or toluene, for example [99]:

$$(Bu_3Sn)_2O + 2BuOCH_2CH_2OH \longrightarrow 2Bu_3SnOCH_2CH_2OBu + H_2O.$$

The water formed can be collected in a Dean and Stark separator and good yields are obtained after 1 hour's boiling under reflux [34]. Triethyltin oxide reacts with acetylenic alcohols at room temperature.

$$(Et_3Sn)_2O + 2HOCR_2C{\equiv}CH \longrightarrow 2Et_3SnOCR_2C{\equiv}CH + H_2O$$
$$(R = H \text{ or alkyl})$$

If the mixture is heated and water is removed azeotropically the product reacts further [43, 85].

$$(Et_3Sn)_2O + 2Et_3SnOCR_2C\equiv CH \longrightarrow 2Et_3SnOCR_2C\equiv CSnEt_3 + H_2O$$

Trialkyltin alkoxides can also be obtained, in good yields, by heating together an oxide and a dialkyl carbonate [34].

$$(R_3Sn)_2O + (R'O)_2CO \longrightarrow 2R_3SnOR' + CO_2$$

Suitable dialkyltin compounds with glycols give cyclic alkoxides. (Mercaptoalcohols such as $HSCH_2CH_2OH$ react similarly [87].)

$$Bu_2Sn(OEt)_2 + HO(CH_2)_nOH \longrightarrow Bu_2Sn \overset{\displaystyle O}{\underset{\displaystyle O}{\diagup}}(CH_2)_n + 2EtOH$$

$(n = 4, 5, 6)$

(ref. 50)

$$2R_2SnO + 2HOCH_2CH_2OH \longrightarrow R_2Sn \overset{\displaystyle OCH_2CH_2O}{\underset{\displaystyle OCH_2CH_2O}{\diagup\diagdown}} SnR_2 + 2H_2O$$

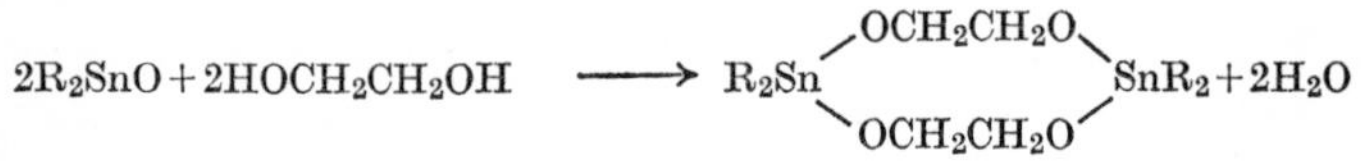

VI   (refs. 23, 90)

Although compound VI, R = Bu has a molecular weight [23, 50] corresponding to the dimeric structure shown a number of 1,2-glycols can give monomeric cyclic alkoxides, for example, compounds VII, VIII and IX [23].

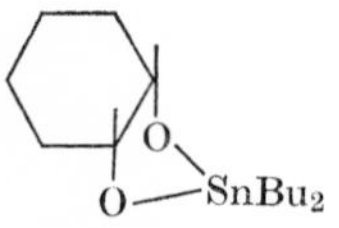

VII 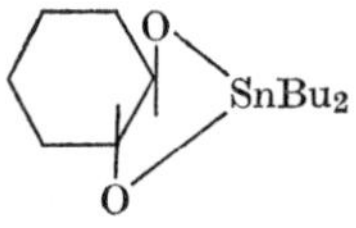   VIII 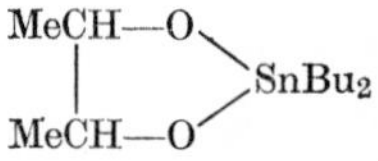   IX

Other workers have suggested that, in solution, these cyclic compounds exist as an equilibrium mixture of monomer and dimer [50].

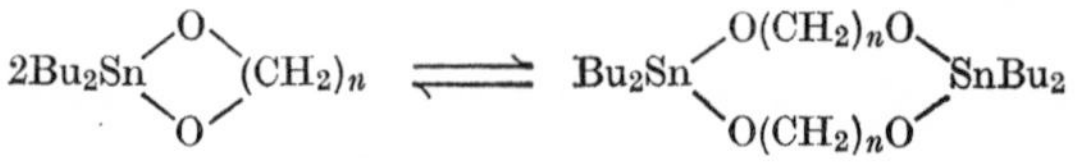

$$2Bu_2Sn \overset{\displaystyle O}{\underset{\displaystyle O}{\diagup}}(CH_2)_n \rightleftharpoons Bu_2Sn \overset{\displaystyle O(CH_2)_nO}{\underset{\displaystyle O(CH_2)_nO}{\diagup\diagdown}} SnBu_2$$

A number of α-substituted organotin compounds add to the carbonyl groups of aldehydes and ketones to give substituted alkoxides, for example [56]:

$$Bu_3SnCH_2CN + PhCHO \longrightarrow Bu_3SnOCHPhCH_2CN.$$

Derivatives of the enol forms of β-ketoesters and β-diketones can be made by protolysis of reactive tin–carbon bonds[72].

$$Et_3SnCH_2COMe + MeCOCH_2COOEt \longrightarrow Et_3SnOC(Me){=}CHCOOEt + Me_2CO$$

The alkoxides must be protected from moisture which causes rapid hydrolysis and the readiness with which these compounds participate in both substitution and addition reactions is being increasingly exploited. It was shown in Section 5.1 that the organotin oxides, $(R_3Sn)_2O$, will add across many unsaturated linkages; very similar reactions are shown by the alkoxides as exemplified in the following equations.

$Et_3SnOEt + ArNCO$	$\longrightarrow$	$Et_3SnNArCOOEt$	(refs. 14, 16)
$Bu_3SnOMe + PhNCS$	$\longrightarrow$	$Bu_3SnSC(OMe){=}NPh$	(ref. 19)
$Bu_3SnOMe + CO_2$	$\longrightarrow$	$Bu_3SnOCOOMe$	(ref. 19)
$Bu_3SnOMe + SO_2$	$\longrightarrow$	$Bu_3SnOSOOMe$	(ref. 19)
$Bu_3SnOMe + ArN{=}C{=}NAr$	$\longrightarrow$	$Bu_3SnN(Ar)C(OMe){=}NAr$	(ref. 19)
$Bu_3SnOMe + Cl_3CCN$	$\longrightarrow$	$Bu_3SnN{=}C(OMe)CCl_3$	(ref. 106)
$Bu_3SnOMe + Cl_3CCHO$	$\longrightarrow$	$Bu_3SnOCH(OMe)CCl_3$	(ref. 35)
$Et_3SnOC{\equiv}CMe + Cl_3CCHO$	$\longrightarrow$	$Et_3SnOCH(CCl_3)OC{\equiv}CMe$	(ref. 111)
$Et_3SnOMe + CH_2{=}CO$	$\longrightarrow$	Et_3SnCH_2COOMe	(ref. 47)

The product from addition to the isocyanate will undergo protolysis with an alcohol to give the urethan and the regenerated organotin alkoxide.

$$Et_3SnNArCOOEt + EtOH \longrightarrow ArNHCOOEt + Et_3SnOEt$$

It can be thus seen that a small amount of an organotin compound can catalyse the reaction between alcohols and isocyanates and this explains the known efficacy of organotin catalysts for the production of polyurethans[14]. Addition to carbonyl compounds occurs with a number of aldehydes and with reactive ketones such as hexachloroacetone. Since the products are themselves alkoxides further additions may occur and, in the case of chloral, a polymeric product was obtained[35, 108].

$$Bu_3SnOMe + xCl_3CCHO \longrightarrow Bu_3Sn(OCHCCl_3)_xOMe$$

The product from addition of tributyltin methoxide to hexachloroacetone is unstable and decomposes to give the trichloromethyltin derivative[36].

$$Bu_3SnOMe + Cl_3CCOCCl_3 \longrightarrow Bu_3SnOCOMe \overset{\displaystyle CCl_3}{\underset{\displaystyle CCl_3}{|}} \searrow$$
$$Bu_3SnCCl_3 + Cl_3CCOOMe$$

Dialkoxides undergo similar addition reactions to those already discussed for the monoalkoxides and, in most cases, 1:1 and 1:2 adducts can be isolated [29].

$$Bu_2Sn(OMe)_2 + A{=}B \longrightarrow Bu_2Sn(OMe)ABOMe$$
$$\downarrow A{=}B$$
$$Bu_2Sn(ABOMe)_2$$

(A=B = isocyanate, isothiocyanate, aldehyde, sulphur dioxide, carbon dioxide, cyanide, carbodiimide.) Dibutyltin dimethoxide undergoes a redistribution reaction with a dibutyltin dihalide and the resulting halomethoxide also participates in addition reactions [28, 29].

$$Bu_2Sn(OMe)_2 + Bu_2SnCl_2 \longrightarrow 2Bu_2Sn(Cl)OMe$$
$$Bu_2Sn(Cl)OMe + A{=}B \longrightarrow Bu_2Sn(Cl)ABOMe$$

Turning next to substitution reactions the alkoxides are sensitive to a wide range of protic species; reaction with water has already been mentioned and acetylenes readily displace alkoxyl groups [94].

$$Et_3SnOMe + HC{\equiv}CCH{=}CH_2 \longrightarrow Et_3SnC{\equiv}CCH{=}CH_2 + MeOH$$

N-Stannylamides are conveniently prepared from alkoxides.

$$Et_3SnOMe + PhNHCOMe \longrightarrow Et_3SnNPhCOMe + MeOH$$

However, when the corresponding thioamides are used the products have Sn—S rather than Sn—N bonds [55].

$$Et_3SnOMe + PhNHCSMe \longrightarrow Et_3SnSC(Me){=}NPh$$

Alkyl halides and alkoxides react to give ethers and organotin halides.

$$Bu_3SnOMe + CH_2{=}CHCH_2Br \longrightarrow Bu_3SnBr + CH_2{=}CHCH_2OMe$$
(ref. 70)

$$Et_3SnOCH_2CH_2C{\equiv}CH + MeOCH_2Cl \longrightarrow Et_3SnCl + HC{\equiv}CCH_2CH_2OCH_2OMe$$
(ref. 39)

$$R_3SnOCH_2C{\equiv}CSnR_3 + MeI \longrightarrow R_3SnI + R_3SnC{\equiv}CCH_2OMe$$
(refs. 83, 84, 110)

Irradiation with ultraviolet light and the addition of Lewis acids did not increase the yields in these reactions [70]. In a similar manner alkoxyl groups can be transferred to silicon by the use of silyl halides.

$$Bu_3SnOMe + Et_3SiCl \longrightarrow Bu_3SnCl + Et_3SiOMe \qquad \text{(ref. 69)}$$

$$R_3SnOCH_2C\equiv CSnR_3 + R_3'SiCl \longrightarrow R_3SnCl + R_3'SiOCH_2C\equiv CSnR_3$$
$$\text{(refs. 83, 109)}$$

$$3Bu_3SnOCH\!\!=\!\!CMe_2 + MeSiCl_3 \longrightarrow MeSi(OCH\!\!=\!\!CMe_2)_3 + 3Bu_3SnCl$$
$$\text{(ref. 12)}$$

The silyl halides react more readily than the alkyl halides and cooling is often necessary. Trans-alkoxylations can be effected by heating an alkoxide with excess of another alcohol [68].

$$Bu_3SnOMe + HO(CH_2)_nX \longrightarrow Bu_3SnO(CH_2)_nX + MeOH$$
$$(X = Cl, Br; \; n = 2, 3, 4)$$

An alkoxysilane may also be used [65].

$$Bu_3SnOMe + Me_3SiOBu \xrightarrow[\text{40 hr}]{145^\circ} Bu_3SnOBu + Me_3SiOMe$$

A similar reaction occurs under milder conditions when an alkoxide is treated with an ester [71].

$$Et_3SnOMe + C_6H_{11}OAc \xrightarrow{25^\circ} Et_3SnOC_6H_{11} + MeOAc$$

When an alkoxide reacts with an enol ester then the O-stannyl derivative of the enol form and/or the C-stannyl derivative of the keto form is obtained [64, 73].

$$RR'C\!\!=\!\!CR''OAc + Bu_3SnOMe \longrightarrow \left.\begin{array}{c} RR'C\!\!=\!\!CR''OSnBu_3 \\ + \\ RR'C(SnBu_3)CR''\!\!=\!\!O \end{array}\right\} + MeOAc$$

The reaction of β-propionolactone with trimethyltin methoxide (and other Me_3SnX compounds) has been studied in detail [44a, 45]. The dominant product is that resulting from acyl-oxygen bond cleavage (X) but the product obtained from alkyl-oxygen bond cleavage (XI) can also be detected, the ratio of the products depending upon the polarity of the solvent used.

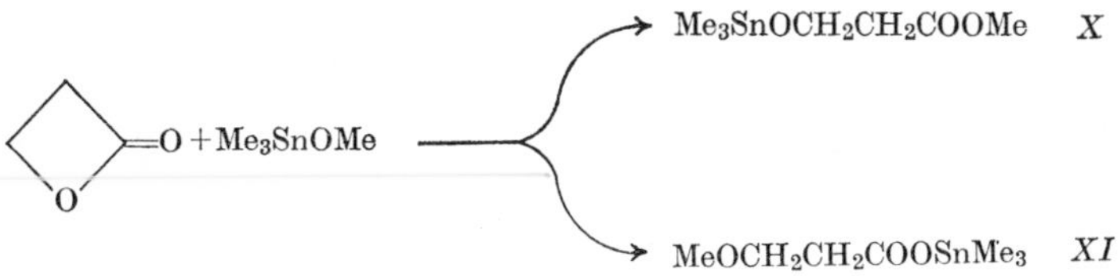

Stannylated celluloses are prepared by reaction between cellulose (or a derivative) and an organotin alkoxide [112].

The organotin phenoxides, $R_nSn(OAr)_{4-n}$, are conveniently prepared by reaction between an oxide and a phenol in the same manner as the alkoxides[101].

$$(Bu_3Sn)_2O + 2o\text{-}PhC_6H_4OH \longrightarrow 2Bu_3SnOC_6H_4Ph\text{-}o + H_2O$$

Among other methods, treatment of an organotin chloride with a phenoxide may be used[21].

$$Ph_3SnCl + p\text{-}NO_2C_6H_4ONa \longrightarrow p\text{-}NO_2C_6H_4OSnPh_3 + NaCl$$

Dibutyltin dimethoxide reacts with 2-pyridone as follows[98]:

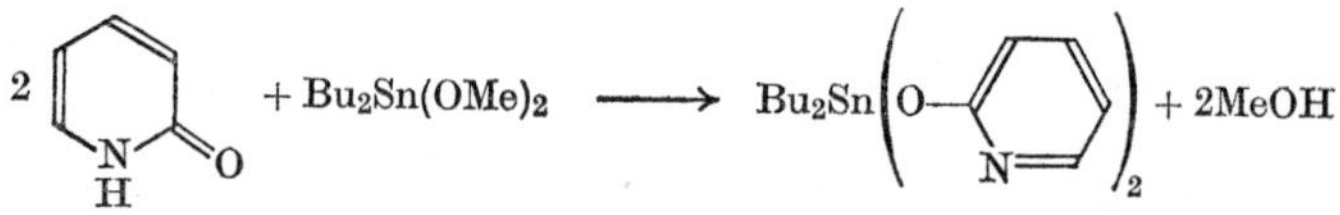

Reaction between triphenyltin hydroxide and p-nitrosophenol gave the p-nitrosophenoxide which, from the infrared and visible spectra of its solutions, was considered to be in equilibrium with the tautomeric quinoxime[53].

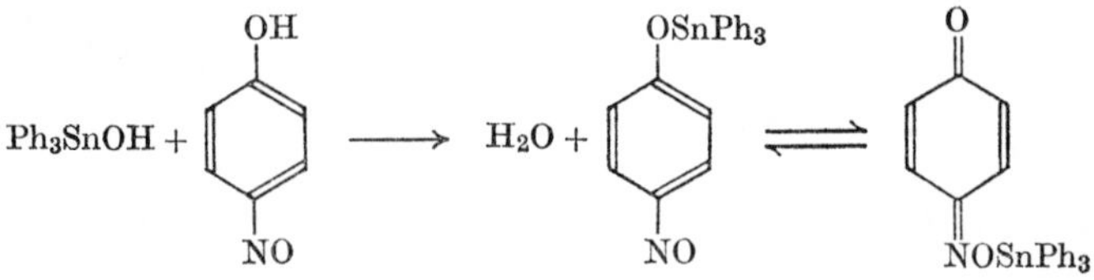

The crystalline product consisted solely of the quinoxime form.

The phenoxides, like the alkoxides, are generally sensitive to moisture but compounds derived from more acidic phenols (e.g. nitrophenols) are less readily hydrolysed[21, 93]. High thermal stability is a property of a number of phenoxides for example compounds XII[96] and XIII[97] and also of compounds derived from dihydric phenols such as XIV and XV[41].

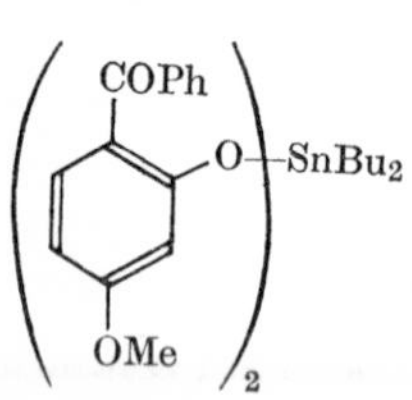

XII

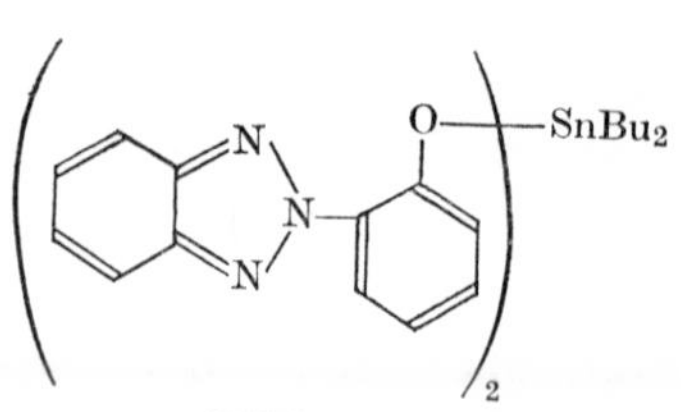

XIII

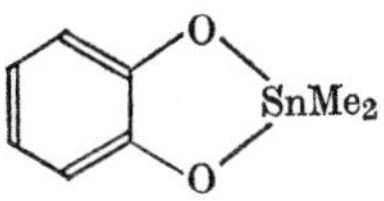

XIV

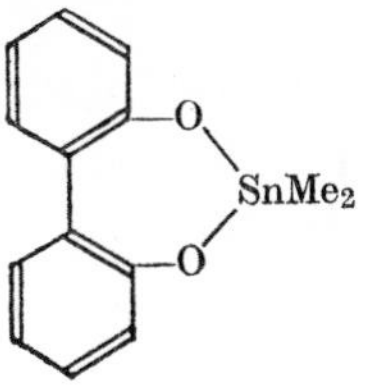

XV

A polymeric phenoxide, having the repeating unit $-OSnPh_2OC_6H_4-$, was prepared from diphenyltin dichloride and the dipotassium salt of hydroquinone[102]. A number of polymeric products have also been made by heating together, in toluene, dialkytin oxides and dihydric phenols of the type *XVI*[104].

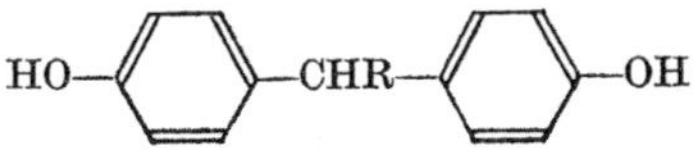

XVI

TABLE III

Physical properties of some organotin alkoxides and phenoxides

Compound	m.p.	b.p.	n_D^{20}	d^{20}
Me_3SnOBu	—	175–6°	1·4575	1·2656
$Et_3SnOCH_2CH_2OCH{=}CH_2$	—	90–90·5°/3 mm	1·4810	1·2560
Pr_3SnOMe	—	87–88°/3 mm	—	—
Pr_3SnOPh	—	145–147°/1 mm	1·5284	1·2167
Bu_3SnOMe	—	97–97·5°/0·06 mm	1·4710 (25°)	1·1294 (1·1690)
Bu_3SnOBu	—	124–128°/3 mm	1·4690	1·0189
Bu_3SnOPh	—	152°/1 mm	1·5171	1·1666
$Bu_2Sn(OMe)_2$	—	126–128°/0·05 mm	1·4852 (25°)	—
Ph_3SnOMe	65–66°	—	—	—

5.5. ORGANOTIN PEROXIDES

Most of the organotin peroxides known are mono(alkyl peroxides) and the first compound of this type was made from an organotin chloride and the sodium derivative of a hydroperoxide[76].

$$R_3SnCl + R'OONa \longrightarrow R_3SnOOR' + NaCl$$

Condensations such as this, although proceeding to give high yields, are best avoided since the sodium chloride is produced in a very finely divided condition and is difficult to remove. A more convenient route is *via* the alkoxides[5].

$$R_3SnOMe + R'OOH \longrightarrow R_3SnOOR' + MeOH$$

Alkyl stannyl peroxides have also been prepared from the oxides and hydroxides[5, 18], for example:

$$(Bu_3Sn)_2O + 2t\text{-}BuOOH \longrightarrow 2Bu_3SnOOBu\text{-}t + H_2O.$$

When t-butyl peroxide reacted with dibutyltin oxide the distannoxane was formed[18].

$$2Bu_2SnO + 2t\text{-}BuOOH \longrightarrow Bu_2Sn(OOBu\text{-}t)OSnBu_2OOBu\text{-}t + H_2O$$

Tributylstannyl t-butyl peroxide has a thermal stability comparable with that of di-t-butyl peroxide with a half life of 69·3 seconds at 199·6°; it decomposes in contact with air[7]. When $PhCMe_2OOSnEt_3$ is heated in octane an induced, first order, decomposition process occurs[6]. The thermal decomposition of triethylstannyl t-butyl peroxide in decane, in the presence of tetraethyltin or hexaethylditin, has been studied and was shown to occur *via* the initial formation of t-BuO· and Et_3SnO· radicals[4]. Triphenylstannyl t-butyl peroxide adds to the C=N bond of isocyanates[13].

$$Ph_3SnOOBu\text{-}t + RNCO \longrightarrow Ph_3SnNRCOOOBu\text{-}t$$

It appears that the acyl peroxides have no stable existence and undergo spontaneous rearrangement[7, 20]; the alkoxytin carboxylate produced

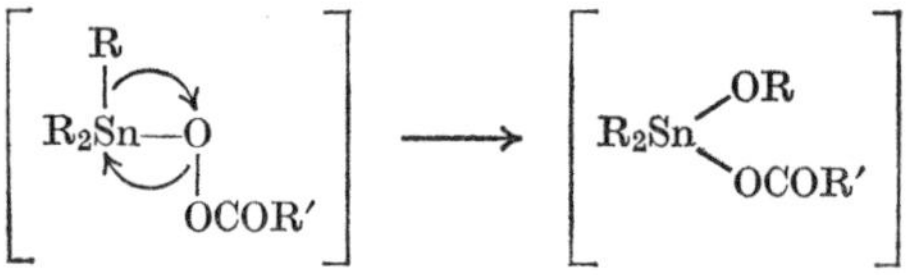

is also unstable. The kinetics of the rearrangement of a number of acyl peroxides have been studied and the Arrhenius parameters calculated[20].

Bis(triethylstannyl) peroxide was made from the corresponding oxide and hydrogen peroxide by dehydration with calcium sulphate[1].

$$(Et_3Sn)_2O + H_2O_2 \xrightarrow{\text{CaSO}_4} Et_3SnOOSnEt_3 + H_2O$$

When magnesium sulphate was used as the dehydrating agent triethyltin hydroperoxide perhydrate, $(Et_3SnOOH)_2 \cdot H_2O_2$, was isolated [3]. *Bis*(triethylstannyl) peroxide explodes on warming and, even at 0°, is decomposed within 24 hours. An attempt to make the corresponding phenyl compound by the reaction:

$$2Ph_3SnOMe + H_2O_2 \longrightarrow [Ph_3SnOOSnPh_3] + 2MeOH$$

gave only the decomposition products diphenyltin oxide and phenol [7]. Organotin hydroperoxides can be prepared from the hydroxide and hydrogen peroxide.

$$Me_3SnOH + H_2O_2 \rightleftharpoons Me_3SnOOH + H_2O$$

If the reaction is carried out in toluene the water is removed azeotropically; in ether the solubilities favour the formation of the hydroperoxide [26]. These compounds are unstable, thus Me_3SnOOH, m.p. 97–98° decomp., can only be kept for a few hours at room temperature and Ph_3SnOOH which explodes at 75° is less stable. The kinetics of decomposition of trimethylstannyl hydroperoxide have been studied, two competing reactions occur [26]:

$$2Me_3SnOOH \longrightarrow 2Me_3SnOH + O_2$$
$$Me_3SnOOH \longrightarrow Me_2SnO + MeOH.$$

TABLE IV

Physical properties of some organotin peroxides[a]

Compound	m.p.	b.p.	n_D^{20}
$Me_3SnOOEt$	95–97°	—	—
$Me_3SnOOBu-t$	—	56°/12 mm	1·4529
$Et_3SnOOBu-t$	—	56–57°/2 mm	1·4669
$Bu_3SnOOBu-t$	—	71–72°/0·001 mm	1·4608 (25°)
$Ph_3SnOOBu-t$	63–65°	—	—
$Ph_3SnOOCMe_2Ph$	111–114°	—	—
$Ph_3SnOOCPh_3$	123–125°	—	—
$Me_2Sn(OOBu-t)_2$	160–161° (decomp.)	—	—

[a] All of these compounds are more or less explosive and should be handled with caution.

5.6. STANNOXIDES OF OTHER ELEMENTS

The chemistry of the metallostannoxanes has been reviewed[43a]; the most important compounds in this class are the stannosiloxanes containing the grouping $>$Sn—O—Si$<$. Although a number of compounds, for example $Me_3SiOSnR_3$ (R = Pr, Bu), have been prepared by cohydrolysis of the mixed halides [59, 61] the alkali metal oxides are more generally used for laboratory syntheses [79].

$$Me_3SiOLi + Me_3SnCl \longrightarrow LiCl + Me_3SiOSnMe_3 \longleftarrow Me_3SnOLi + Me_3SiCl$$

By similar methods the compounds $(Me_3SiO)_2SnMe_2$, $(Me_3SnO)_2SiMe_2$ and $(Me_3SiO)_3SnMe$ were obtained as toxic, unpleasant smelling, liquids or low-melting solids [79, 89]. The reaction of stannylamines with protic species is the basis of many preparative procedures (Chapter 6) and can be applied to the preparation of stannosiloxanes [46].

$$R_3''SiOH + R_3SnNR_2' \longrightarrow R_3SnOSiR_3'' + R_2'NH$$

Benzene, toluene or ligroin solutions of organotin oxides and hydroxides react with hydroxy-silanes and -germanes to give water (which is collected in a Dean and Stark separator) and the stannosiloxanes or stannogermoxanes [33], for example:

$$(Ph_3Sn)_2O + 2Me_3SiOH \longrightarrow 2Ph_3SnOSiMe_3 + H_2O$$
$$(Bu_3Sn)_2O + Ph_2Si(OH)_2 \longrightarrow Ph_2Si(OSnBu_3)_2 + H_2O$$
$$BuSnOOH + 3Ph_3SiOH \longrightarrow BuSn(OSiPh_3)_3 + 2H_2O.$$

On treatment with a Grignard reagent the stannosiloxanes are alkylated at the tin atom, for example $Et_3SnOSiEt_3$ on treatment with EtMgBr gives, eventually, $Et_4Sn + Et_3SiOH$ [82]. Much of the work in this area has been directed at obtaining modified organopolysiloxanes in which there has been partial replacement of silicon by tin. Although such polymeric compounds have been made [10] their stabilities and physical properties are inferior to those of the organopolysiloxanes [37].

Reaction between metallic or organometallic halides and organotin oxides is a general method of preparing organohalogenostannometalloxanes [31].

$$M—X + R_2SnO \longrightarrow MOSnR_2X$$

Examples of compounds prepared by this method are $PhHgOSnBu_2Cl$, $(C_6F_5)_2TlOSnBu_2Cl$, $ClMe_2SiOSnBu_2Cl$, $Cl_2MeSiOSnMe_2Cl$, $ClMe_2GeOSnBu_2Cl$ and $ClBu_2PbOSnBu_2Cl$. Reaction between diethylcadmium and triethyltin hydroxide gave impure $Et_3SnOCdEt$ [63]; compounds containing Sn—O—P linkages have been made [44].

$$Bu_3SnCl + Bu_2POK \longrightarrow Bu_2POSnBu_3 + KCl$$

Some stannoxyzirconium alkoxides have also been isolated [107].

$$Zr(OBu)_4 + Bu_3SnOAc \longrightarrow Bu_3SnOZr(OBu)_3 + AcOBu$$

TABLE V

Physical properties of some stannoxides

Compound	m.p.	b.p.	n_D^{20}	d^{20}
$Me_3SnOSiMe_3$	—	144°	—	—
$Me_3SnOSiEt_3$	—	49°/1 mm	—	—
$Et_3SnOSiMe_3$	—	99°/20 mm	—	—
$Et_3SnOSiEt_3$	—	114–115°/4 mm	1·4635	1·1149
$Et_3SnOGePh_3$	—	167–169°/0·04 mm	—	—
$Bu_3SnOPBu_2$	—	165–167°/1 mm	—	—
$Bu_3SnOZr(OBu)_3$	—	—	1·4892	1·2020 (24°)
$Et_2Si(OSnBu\text{-}i_3)_2$	—	203°/3 mm	1·4800	1·1443
$Ph_3SnOSiMe_3$	—	140°/0·1 mm	—	—
$Ph_3SnOSiPh_3$	139–140°	—	—	—
$Ph_2Si(OSnPh_3)_2$	94·5–96·5°	—	—	—

References

1. (a) Aleksandrov, Yu. A., Brilkina, T. G. and Shushunov, V. A., *Tr. po Khim. i Khim. Tekhnol.*, 1959, **2**, 623 (*Chem. Abs.*, 1962, **56**, 14314).
   (b) Aleksandrov, Yu. A., *Tr. po Khim. i Khim. Tekhnol.*, 1960, **3**, 342 (*Chem. Abs.*, 1961, **55**, 25570).
   (c) Aleksandrov, Yu. A. and Shushunov, V. A., *Dokl. Akad. Nauk S.S.S.R.*, 1961, **140**, 595 (*Chem. Abs.*, 1962, **56**, 10176).
   (d) Aleksandrov, Yu. A. and Shushunov, V. A., *Tr. po Khim. i Khim. Tekhnol.*, 1962, **4**, 644 (*Chem. Abs.*, 1964, **60**, 543; 3452).
2. Aleksandrov, Yu. A. and Radbil, B. A., *Zh. Obshch. Khim.*, 1968, **38**, 499 (*Chem. Abs.*, 1968, **69**, 66686).
3. Aleksandrov, Yu. A. and Shushunov, V. A., *Zh. Obshch. Khim.*, 1965, **35**, 115.
4. Aleksandrov, Yu. A. and Sul'din, B. V., *Zh. Obshch. Khim.*, 1967, **37**, 2350.
5. Aleksandrov, Yu. A., Sul'din, B. V. and Kokurina, S. N., *Tr. po Khim. i Khim. Tekhnol.*, 1965, **3**, 228 (*Chem. Abs.*, 1967, **67**, 3128s).
6. Aleksandrov, Yu. A., Sul'din, B. V. and Kokurina, S. N., *Zh. Obshch. Khim.*, 1966, **36**, 2198.
7. Alleston, D. L. and Davies, A. G., *J. Chem. Soc.*, 1962, 2465.
8. Alleston, D. L., Davies, A. G. and Hancock, M., *J. Chem. Soc.*, 1964, 5744.
9. Alleston, D. L., Davies, A. G., Hancock, M. and White, R. F. M., *J. Chem. Soc.*, 1963, 5469.

10. Andrianov, K. A., Ganina, T. N. and Khrustaleva, E. T., *Izv. Akad. Nauk S.S.S.R.*, *Otd. Khim. Nauk*, 1956, 798 (*Chem. Abs.*, 1957, **51**, 3487).

11. Atavin, A. S., Dubova, R. I. and Vasil'ev, N. P., *Zh. Obshch. Khim.*, 1966, **36**, 1506.

12. Bellegarde, B., Pereye, M. and Valade, J., *Compt. Rend. Ser. C*, 1967, **264**, 340.

13. Bloodworth, A. J., *J. Chem. Soc. (C)*, 1968, 2380.

14. Bloodworth, A. J. and Davies, A. G., *J. Chem. Soc.*, 1965, 5238.

15. Bloodworth, A. J. and Davies, A. G., *J. Chem. Soc.*, 1965, 6245.

16. Bloodworth, A. J. and Davies, A. G., *J. Chem. Soc.*, 1965, 6858.

17. Bloodworth, A. J. and Davies, A. G., *J. Chem. Soc. (C)*, 1966, 299.

18. Bloodworth, A. J., Davies, A. G. and Graham, I. F., *J. Organometallic Chem.*, 1968, **13**, 351.

19. Bloodworth, A. J., Davies, A. G. and Vasishtha, S. C., *J. Chem. Soc. (C)*, 1967, 1309.

20. Brilkina, T. G., Safonova, M. K. and Shushunov, V. A., *Tr. Khim. Tekhnol.*, 1965, **67**, 74 (*Chem. Abs.*, 1967, **66**, 54769p; 54770g).

21. Brilkina, T. G., Safonova, M. K. and Sokolov, N. A., *Zh. Obshch. Khim.*, 1966, **36**, 2202.

22. Brown, M. P., Okawara, R. and Rochow, E. G., *Spectrochim. Acta*, 1960, **16**, 595.

23. Considine, W. J., *J. Organometallic Chem.*, 1966, **5**, 263.

24. Considine, W. J., Ventura, J. J., Gibbons, A. J. and Ross, A., *Canad. J. Chem.*, 1963, **41**, 1239.

25. Considine, W. J., Ventura, J. J., Kushlefsky, B. G. and Ross, A., *J. Organometallic Chem.*, 1964, **1**, 299.

26. Dannley, R. L. and Aue, W. A., *J. Org. Chem.*, 1965, **30**, 3845.

27. Davies, A. G. and Graham, I. F., *Chem. and Ind.*, 1963, 1622.

28. Davies, A. G. and Harrison, P. G., *J. Chem. Soc. (C)*, 1967, 298.

29. Davies, A. G. and Harrison, P. G., *J. Chem. Soc. (C)*, 1967, 1313.

30. Davies, A. G. and Harrison, P. G., *J. Organometallic Chem.*, 1967, **7**, P13.

31. Davies, A. G. and Harrison, P. G., *J. Organometallic Chem.*, 1967, **10**, P31.

32. Davies, A. G., Harrison, P. G. and Palan, P. R., *J. Organometallic Chem.*, 1967, **10**, P33.

33. Davies, A. G., Harrison, P. G. and Silk, T. A. G., *Chem. and Ind.*, 1968, 949.

34. Davies, A. G., Palan, P. R. and Vasishtha, S. C., *Chem. and Ind.*, 1967, 229.

35. Davies, A. G. and Symes, W. R., *J. Chem. Soc. (C)*, 1967, 1009.

36. Davies, A. G. and Symes, W. R., *J. Organometallic Chem.*, 1966, **5**, 394.

37. Delman, A. D., Stein, A. A., Simms, B. B. and Katzenstein, R. J., *J. Polymer Sci. Pt. A-1*, 1966, **4**, 2307.

38. Devaud, M., *Compt. Rend. Ser. C.*, 1966, **262**, 702.

39. Devaud, M., *Rev. Chim. Miner.*, 1967, **4**, 921 (*Chem. Abs.*, 1968, **69**, 87122).

40. Dunn, P. and Norris, T., *Report No. 269 Australian Defence Scientific Service*, Defence Standards Laboratories, 1964, 12.

41. Emeléus, H. J. and Zuckerman, J. J., *J. Organometallic Chem.*, 1964, **1**, 328.

42. Foldesi, I., *Acta Chim. Acad. Sci. Hung.*, 1965, **45**, 237 (*Chem. Abs.*, 1965, **63**, 16376e).

42a. Geissler, H., Radegia, R. and Kriegsmann, H., *J. Organometallic Chem.*, 1968, **15**, 349.

43. Gverdtsiteli, I. and Adamiya, S. V., *Soobshch. Akad. Nauk Gruz. S.S.R.*, 1967, **47**, 55 (*Chem. Abs.*, 1968, **68**, 29817).

43a. Harrison, P. G., *Organometallic Chem. Revs., Sect. A*, 1969, **4**, 379.

44. Issleib, K. and Walther, B., *Angew. Chem. Intern. Ed. Engl.*, 1967, **6**, 88.

44a. Itoh, K., Kato, Y. and Ishii, Y., *J. Org. Chem.*, 1969, **34**, 459.

45. Itoh, K., Kobayashi, S., Sakai, S. and Ishii, Y., *J. Organometallic Chem.*, 1967, **10**, 451.

46. Jones, K. and Lappert, M. F., *Organomet. Chem. Rev.*, 1966, **1**, 67.

47. Khrapov, V. V., Gol'danskii, V. I., Prokof'ev, A. K., Rochev, V. Ya. and Kostyanovskii, R. G., *Izv. Akad. Nauk S.S.S.R. Ser. Khim.*, 1968, 1261 (*Chem. Abs.*, 1968, **69**, 96850).

48. Kocheshkov, K. A., Panov, E. M. and Zemlyanskii, N. N., *Izv. Akad. Nauk S.S.S.R. Otd. Khim. Nauk*, 1961, 2255.

49. Luijten, J. G. A., *Rev. Trav. Chim.*, 1966, **85**, 873.

50. Mehrotra, R. C. and Gupta, V. D., *J. Organometallic Chem.*, 1965, **4**, 145.

51. Mirskov, R. G. and Vlasov, V. M., *Zh. Obshch. Khim.*, 1966, **36**, 166.

52. Mufti, A. S. and Poller, R. C., *J. Chem. Soc.*, 1965, 5055.

53. Nesmeyanov, A. N., Al'Fonso, L., Kravtsov, D. N. and Loshkin, B. V., *Izv. Akad. Nauk S.S.S.R. Ser. Khim.*, 1968, 296 (*Chem. Abs.*, 1968, **69**, 52238).

54. Neumann, W. P. and Kleiner, F. G., *Tetrahedron Letters*, 1964, 3779.

55. Noltes, J. G., *Rec. Trav. Chim.*, 1965, **84**, 799.

56. Noltes, J. G., Creemer, H. M. J. C. and van der Kerk, G. J. M., *J. Organometallic Chem.*, 1968, **11**, P21.

57. Ohara, M., Okawara, R. and Nakamura, Y., *Bull. Chem. Soc. Japan*, 1965, **38**, 1379.

58. Okawara, R., *Proc. Chem. Soc.*, 1961, 383.

59. Okawara, R. and Sugita, K., *J. Amer. Chem. Soc.*, 1961, **83**, 4480.

60. Okawara, R. and Wada, M., *J. Organometallic Chem.*, 1963, **1**, 81.

61. Okawara, R., White, D. G., Fujitani, K. and Sato, H., *J. Amer. Chem. Soc.*, 1961, **83**, 1342.

62. Okawara, R. and Yasuda, K., *J. Organometallic Chem.*, 1964, **1**, 356.

63. Pankratova, V. N. and Stepovik, L. P., *Zh. Obshch. Khim.*, 1968, **38**, 844 (*Chem. Abs.*, 1968, **69**, 77395).

64. Pereyre, M., Bellegarde, B., Mendelsohn, J. and Valade, J., *J. Organometallic Chem.*, 1968, **11**, 97.

65. Pereyre, M., Bellegarde, B. and Valade, J., *Compt. Rend. Ser. C.*, 1967, **265**, 939.

66. Poller, R. C., *J. Chem. Soc.*, 1963, 706.

67. Poller, R. C., Unpublished observation.

68. Pommier, J. C., Delmont, B. and Valade, J., *Tetrahedron Letters*, 1967, 5289.

69. Pommier, J. C., Pereyre, M. and Valade, J., *Compt. Rend.*, 1965, **260**, 6397.

70. Pommier, J. C. and Valade, J., *Compt. Rend.*, 1965, **260**, 4549.

71. Ponomarev, S. V., Baukov, Yu. I., Dudikina, O. V., Petrosyan, I. V. and Petrovskaya, L. I., *Zh. Obshch. Khim.*, 1967, **37**, 2204.

7

72. Ponomarev, S. V., Machigin, E. V. and Lutsenko, I. F., *Zh. Obshch. Khim.*, 1966, **36**, 548.
73. Ponomarev, S. V., Rogachev, B. G. and Lutsenko, I. F., *Zh. Obshch. Khim.*, 1966, **36**, 1348.
74. Rees, R. G. and Webb, A. F., *J. Organometallic Chem.*, 1968, **12**, 239.
75. Reifenberg, G. H. and Considine, W. J., *J. Organometallic Chem.*, 1967, **10**, 279.
76. Rieche, A. and Bertz, T., (a) *Angew. Chem.*, 1958, **70**, 507; (b) German Patent 1,081,891, May 19, 1960 (*Chem. Abs.*, 1961, **55**, 19214).
77. Rubinchik, G. F. and Manulkin, Z. M., *Zh. Obshch. Khim.*, 1968, **38**, 841 (*Chem. Abs.*, 1968, **69**, 77394).
78. Schmidbauer, H. and Hussek, H., *Angew. Chem.*, 1963, **75**, 575.
79. Schmidbauer, H. and Hussek, H., *J. Organometallic Chem.*, 1964, **1**, 244.
80. Schweitzer, G. K. and McCarty, S. W., *J. Inorg. Nucl. Chem.*, 1965, **27**, 191.
81. Shostakovskii, M. F., Atavin, A. S., Vyalykh, E. P. and Trofimov, B. A., *Zh. Obshch. Khim.*, 1965, **35**, 751.
82. Shostakovskii, M. F., Komarov, N. V., Sklyanova, A. M. and Suvorova, A. V., *Dokl. Akad. Nauk S.S.S.R.*, 1967, **176**, 356.
83. Shostakovskii, M. F., Mirskov, R. G., Vlasov, V. M. and Tarpischev, S. H. I., *Zh. Obshch. Khim.*, 1967, **37**, 1738.
84. Shostakovskii, M. F., Vlasov, V. M. and Mirskov, R. G., *Zh. Obshch. Khim.*, 1965, **35**, 1121.
85. Shostakovskii, M. F., Vlasov, V. M., Mirskov, R. G. and Petrova, V. N., *Zh. Obshch. Khim.*, 1965, **35**, 47.
86. Smith, T. A. and Kipping, F. S., *J. Chem. Soc.*, 1913, 2034.
87. Sukhani, D., Gupta, V. D. and Mehrotra, R. C., *Aust. J. Chem.*, 1968, **21**, 1175.
88. Tanaka, T., Ueeda, R., Wada, M. and Okawara, R., *Bull. Chem. Soc. Japan*, 1964, **37**, 1554.
89. Thies, C. and Kinsinger, J. B., *Inorg. Chem.*, 1964, **3**, 551.
90. Voronkov, M. G. and Romadane, I., *Khim. Geterotsikl. Soedin.*, 1966, 892 (*Chem. Abs.*, 1967, **67**, 32751v).
91. Wada, M., Nishino, M. and Okawara, R., *J. Organometallic Chem.*, 1965, **3**, 70.
92. Zemlyanskii, N. N., Panov, E. M., Shamagina, O. P. and Kocheshkov, K. A., *Zh. Obshch. Khim.*, 1965, **35**, 1029.
93. French Patent 1,405,428, July 9, 1965 (*Chem. Abs.*, 1965, **63**, 14904d).
94. French Patent 1,427,563, Feb. 4, 1966 (*Chem. Abs.*, 1966, **65**, 8962f).
95. French Patent 1,438,111, March 28, 1966 (*Chem. Abs.*, 1967, **66**, 38050s).
96. French Patent 1,454,605, Oct. 7, 1967 (*Chem. Abs.*, 1967, **67**, 32782f).
97. French Patent 1,457,966, Nov. 4, 1966 (*Chem. Abs.*, 1967, **67**, 54265s).
98. French Patent 1,475,896, April 7, 1967 (*Chem. Abs.*, 1968, **68**, 13178).
99. Japanese Patent 6172, April 1, 1966 (*Chem. Abs.*, 1966, **65**, 5490d).
100. Japanese Patent 10,888, June 20, 1966 (*Chem. Abs.*, 1966, **65**, 12240f).
101. Japanese Patent 12,415, July 14, 1967 (*Chem. Abs.*, 1968, **68**, 49777).
102. U.S. Patent 3,184,430, May 18, 1965 (*Chem. Abs.*, 1965, **63**, 5774d).
103. U.S. Patent 3,213,119, Oct. 19, 1965 (*Chem. Abs.*, 1966, **64**, 3602g).
104. U.S. Patent 3,221,036, Nov. 30, 1965 (*Chem. Abs.*, 1966, **64**, 5138h).

105. U.S. Patent 3,275,659, Sept. 27, 1966 (*Chem. Abs.*, 1966, **65**, 20164e).
106. U.S. Patent 3,347,890, Oct. 17, 1967 (*Chem. Abs.*, 1968, **68**, 49775).
107. U.S. Patent 3,361,775, Jan. 2, 1968 (*Chem. Abs.*, 1968, **68**, 50618).
108. U.S. Patent 3,376,328 ,April 2, 1968 (*Chem. Abs.*, 1968, **68**, 105364).
109. U.S.S.R. Patent 173,760, Aug. 6, 1965 (*Chem. Abs.*, 1966, **64**, 3602e).
110. U.S.S.R. Patent 175,505, Oct. 9, 1965 (*Chem. Abs.*, 1966, **64**, 5138d).
111. U.S.S.R. Patent 180,593, March 26, 1966 (*Chem. Abs.*, 1966, **65**, 12240a).
112. U.S.S.R. Patent 199,883, July 13, 1967 (*Chem. Abs.*, 1968, **68**, 41235).

6

COMPOUNDS CONTAINING TIN–NITROGEN BONDS

The literature to early 1965 on organotin nitrogen compounds has been summarised in two comprehensive review articles [18, 26]. Essentially two types of compound will be considered in this chapter, the organostannylamines $R_{4-n}Sn(NR'R'')_n$ and the organostannylamides $R_{4-n}Sn(NR'COR'')_n$, the latter includes N-stannyl derivatives of urea and urethans.

6.1. ORGANOSTANNYLAMINES

The first compounds in this class were described in 1962 and an intense exploitation of the synthetic applications of these reactive substances has ensued. A widely used method for the preparation of stannylamines is by a trans-metallation reaction between the lithium derivative of the amine and an organotin halide [16, 51].

$$R_{4-n}SnCl_n + nLiNR'R'' \longrightarrow R_{4-n}Sn(NR'R'')_n + nLiCl$$

The method works well when the nitrogen atom carries two organic substituents and also where $R' = H$ and $R'' = aryl$. However, in those cases where $R' = H$ and $R'' = alkyl$ the product may disproportionate thus [18]:

$$Me_3SnCl + LiNHMe \longrightarrow [Me_3SnNHMe] + LiCl$$
$$2[Me_3SnNHMe] \longrightarrow (Me_3Sn)_2NMe + MeNH_2.$$

The *tris*(trialkylstannyl)amines have also been prepared by this procedure.

$$3R_3SnCl + Li_3N \longrightarrow (R_3Sn)_3N + 3LiCl$$

Amino derivatives of other metals have been used in similar transmetallations.

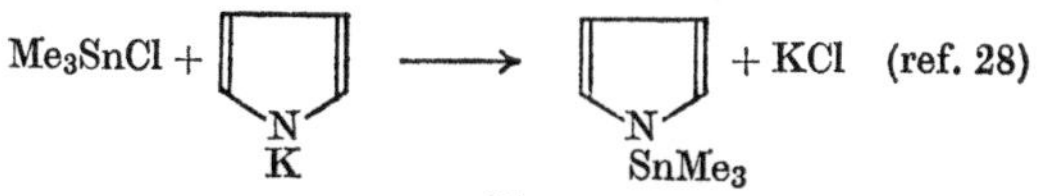

$$Me_3SnCl + \quad \longrightarrow \quad + KCl \quad (ref. 28)$$

$$R_3SnCl + Et_2NMgBr \longrightarrow R_3SnNEt_2 + MgBrCl \qquad \text{(ref. 45)}$$

The same method is used to make organotin dicyanamides.

$$Ph_3SnBr + AgN(CN)_2 \longrightarrow Ph_3SnN(CN)_2 + AgBr \qquad \text{(ref. 20)}$$

$$Me_2SnCl_2 + 2KN(CN)_2 \longrightarrow Me_2Sn[N(CN)_2]_2 + 2KCl \qquad \text{(ref. 21)}$$

Reaction between lithium benzamidine and trimethyltin chloride gives the stannylbenzamidine [39].

$$PhC(NMe_2)=NLi + Me_3SnCl \longrightarrow PhC(NMe_2)=NSnMe_3 + LiCl$$

By using lithium derivatives of silylamines, compounds in which a nitrogen atom is bound to both tin and silicon have been prepared [40, 41, 42].

$$Me_{4-n}SnCl_n + nMe_3Si(Me)NLi \longrightarrow Me_{4-n}Sn[N(Me)SiMe_3]_n + nLiCl$$
$$(n = 1, 2, 3)$$

Extensions of these reactions have made possible the preparation of related compounds containing the SiNSiNSn system [38].

$$\begin{array}{c} Me \\ | \\ Me_3SiN-Si-N-Li + Me_3SnCl \\ |\ \ |\ \ | \\ Me\ Me\ Me \end{array} \longrightarrow \begin{array}{c} Me \\ | \\ Me_3Si-N-Si-N-SnMe_3 + LiCl \\ |\ \ |\ \ | \\ Me\ Me\ Me \end{array}$$

Preparation of structures in which a single nitrogen atom has tin, germanium and silicon atoms attached to it has also been achieved.

$$\begin{array}{c} Et_3SiNLi + Me_3SnCl \\ | \\ GeMe_3 \end{array} \longrightarrow \begin{array}{c} Et_3Si-N-SnMe_3 + LiCl \\ | \\ GeMe_3 \end{array} \qquad \text{(ref. 37)}$$

$$\begin{array}{c} Me \\ | \\ Me_3SiN-Si-N-Li + Me_3SnCl \\ |\ \ |\ \ | \\ Me\ Me\ GeMe_3 \end{array} \longrightarrow \begin{array}{c} Me \\ | \\ Me_3Si-N-Si-N-SnMe_3 + LiCl \\ |\ \ |\ \ | \\ Me\ Me\ GeMe_3 \end{array}$$

$$\text{(ref. 36)}$$

A number of organotin nitramines have been described [50] and are best prepared by one of the two following procedures.

$$R_3SnX + R'N(NO_2)Ag \longrightarrow R'N(NO_2)SnR_3 + AgX \qquad (X = Cl, I)$$

$$(R_3Sn)_2O + 2R'HNNO_2 \longrightarrow 2R'N(NO_2)SnR_3 + H_2O$$

Although the low yields do not make it an attractive synthetic method it is of interest that the nitramines are sufficiently acidic to effect Sn—C bond cleavage.

$$R_4Sn + R'HNNO_2 \longrightarrow R'N(NO_2)SnR_3 + RH$$

Another general method of preparation is the trans-amination reaction whereby one organostannylamine is converted to another by treatment with an amine or ammonia. In cases where one secondary amine group ($-NR_2'$) is displaced by another ($-NR_2''$) the equilibrium reaction:

$$R_3SnNR_2' + R_2''NH \rightleftharpoons R_3SnNR_2'' + R_2'NH$$

proceeds in a direction such that the most volatile amine is displaced[16, 18]. When both primary and secondary amines are involved in the trans-amination the reaction appears to be under steric rather than volatility control and the course of the reaction depends upon the particular amine used[16, 18]. For example, aniline will displace dimethylamine in trimethylstannyldimethylamine.

$$Me_3SnNMe_2 + PhNH_2 \longrightarrow Me_3SnNHPh + Me_2NH$$

However, when the same reaction is attempted using ethylamine only the distannylamine is obtained.

$$2Me_3SnNMe_2 + EtNH_2 \longrightarrow (Me_3Sn)_2NEt + 2Me_2NH$$

This procedure is used to prepare N-trimethylstannylethyleneimine[22].

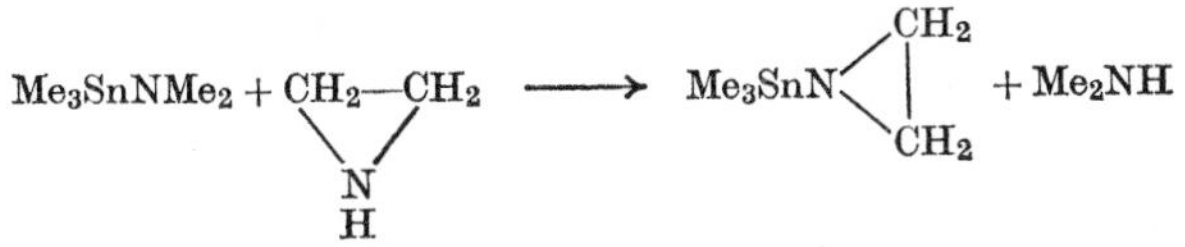

Cyclic stannylamines containing alternating tin and nitrogen atoms are also readily obtained[16, 53], for example:

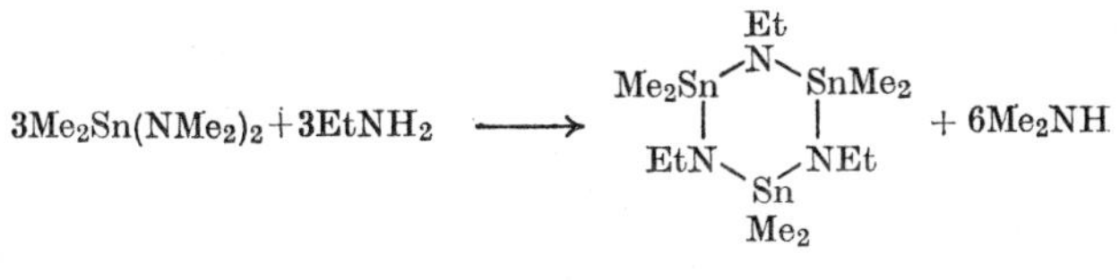

By the use of diamines as trans-amination reagents, ring systems containing Sn, N and C atoms have been made[43].

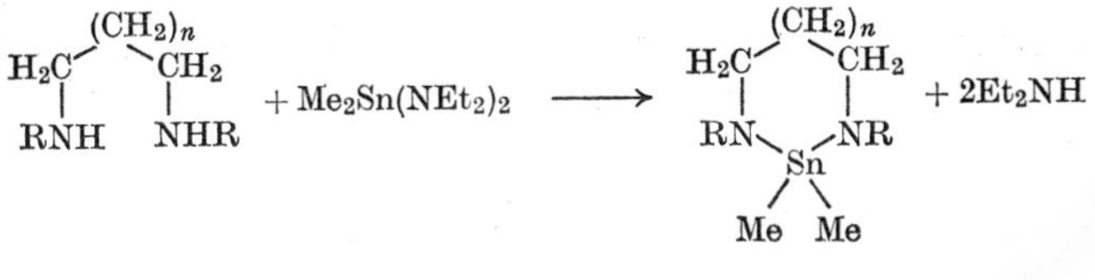

The relative ease of trans-amination for compounds $R_2M(NR_2')_2$ is in the order $M = Sn > Ge > Si$ whereas the ease of replacement of the chlorine atoms, by amines, in compounds of the type R_2MCl_2 is in the reverse order, no replacement occurring with the organotin dichlorides. Possible causes for these differences and the mechanisms of the reactions involved have been discussed[52].

Other methods of preparing organostannylamines include the addition of hydrides to azomethines[30].

$$p\text{-}MeC_6H_4N{=}CHPh + Et_3SnH \longrightarrow p\text{-}MeC_6H_4N(CH_2Ph)SnEt_3$$

N-Triphenylstannyl-N,N'-diphenylhydrazine has been prepared by the addition of triphenyltin hydride to azobenzene[31].

$$Ph_3SnH + PhN{=}NPh \longrightarrow Ph_3SnN(Ph)N(Ph)H$$

A number of derivatives of heterocyclic amines have been prepared by condensation between an organotin oxide or hydroxide and the amine[23, 25, 28], for example:

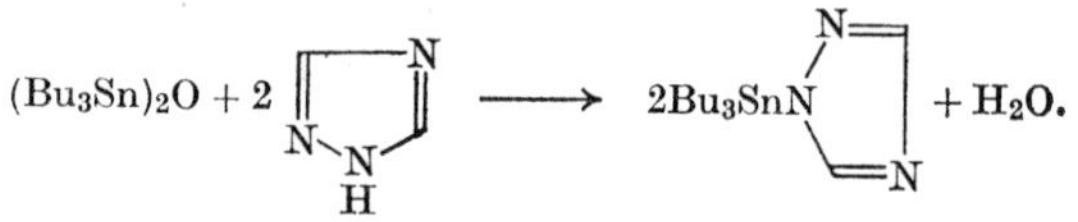

6.1.1. Reactions of organostannylamines

In general the stannylamines are very sensitive to moisture and carbon dioxide and must be protected from the atmosphere. Water is only one of a wide range of protic species which attack the stannylamines as indicated by the general equation:

$$R_3SnNR_2' + HA \longrightarrow R_3SnA + R_2'NH.$$

This is the basis of a large number of synthetically useful reactions[17, 18], Table I gives an indication of the scope of the method. It is possible that the formation of bis(trimethylstannyl)diazomethane also occurs by a protolysis reaction[24,24a].

$$2Me_3SnNMe_2 + CH_2N_2 \longrightarrow (Me_3Sn)_2CN_2 + 2Me_2NH$$

Addition reactions are becoming increasingly important in organotin chemistry and this point is well illustrated by the diversity of compounds which can be obtained by additions of stannylamines to unsaturated systems.

$$Me_3SnNMe_2 + A{=}B \longrightarrow Me_3SnABNMe_2$$

TABLE I

Reactions of organostannylamines, $R_3SnNR'_2$, with protic reagents

Reagent	Organotin product	Reference
H_2O	R_3SnOH	—
$R''OH$	R_3SnOR''	26, 44
R''_3SiOH	$R_3SnOSiR''_3$	18
HN_3	R_3SnN_3	26
HCl	R_3SnCl	18, 44
HCN	R_3SnCN	18
R''_2PH	$R_3SnPR''_2$	19
PH_3	$(R_3Sn)_3P$	18
R''_2AsH	$R_3SnAsR''_2$	19
$R''C{\equiv}CH$	$R_3SnC{\equiv}CR''$	19
$HC{\equiv}CH$	$R_3SnC{\equiv}CSnR_3$	18
 ⬠	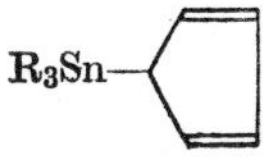 $R_3Sn{-}\diagdown$	19
$HC(N_2)COOEt$	$R_3SnC(N_2)COCEt$	24b
$R''R'''CHCN$	$R_3SnC(CN)R''R'''$	18
R''_3SnH	$R_3SnSnR''_3$	47
R''_3GeH	$R_3SnGeR''_3$	18
$R''SH$	R_3SnSR''	1, 44
$CHCl_3$	R_3SnCCl_3	6, 8
$Cl_2C{=}CHCl$	$R_3SnCCl{=}CCl_2$	6
$CHBr_3$	R_3SnCBr_3	8

The readiness with which this aminostannation reaction occurs has been attributed, in part, to the low thermochemical energy of the Sn—N bond and a concerted mechanism (*I*) has been suggested [10].

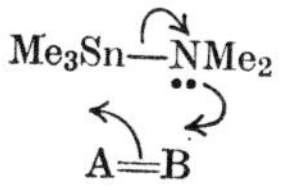

$$Me_3Sn{-}NMe_2$$
$$A{=}B$$

I

Lappert and his coworkers [4] have shown that these additions are not confined to substrates in which the double bond is between unlike elements but also occurs with reactive olefins and acetylenes as indicated in Table II. Not all of the olefins gave addition products however, for example [4]:

$Me_3SnNMe_2 + MeCOOCH{=}CH_2 \longrightarrow Me_3SnOCOMe + Me_2NCH{=}CH_2$

$Me_3SnNMe_2 + CH_2{=}C(Me)COOMe \longrightarrow Me_3SnOMe + CH_2{=}C(Me)CONMe_2.$

TABLE II

Addition reactions of organostannylamines

Stannylamine	Reagent	Product	Reference
Me_3SnNMe_2	CO_2	$Me_3SnOCONMe_2$	10
Me_3SnNMe_2	CS_2	$Me_3SnSCSNMe_2$	10
Me_3SnNMe_2	$PhNCO$	$Me_3SnN(Ph)CONMe_2$	10
Me_3SnNMe_2	$PhNCS$	$Me_3SnN(Ph)CSNMe_2$	10
Me_3SnNMe_2	$PhCN$	$Me_3SnN{=}C(Ph)NMe_2$	10
Me_3SnNMe_2	SO_2	$Me_3SnOSONMe_2$	10
Me_3SnNMe_2	$PhNSO$	$Me_3SnN(Ph)SONMe_2$	10
Me_3SnNMe_2	$EtOCOC{\equiv}CCOOEt$	$Me_3SnC(COOEt){=}C(COOEt)NMe_2$	4
Me_3SnNMe_2	$PhC{\equiv}CCl$	$Me_3SnC(Cl){=}C(Ph)NMe_2$	4
Me_3SnNMe_2	$CH_2{=}C(Me)CN$	$Me_3SnC(Me)(CN)CH_2NMe_2$	4
Me_3SnNMe_2	$CH_2{=}CO$	$Me_3SnCH_2CONMe_2$	34
Me_3SnNEt_2	$p\text{-}MeC_6H_4N{=}C{=}NC_6H_4Me\text{-}p$	$p\text{-}MeC_6H_4N(SnMe_3)C(NEt_2){=}NC_6H_4Me\text{-}p$	10
Et_3SnNMe_2	$RCH{=}CHCOR'$	$Et_3SnCH(COR')CH(R)NMe_2$	4

(R = H, R' = OMe; R = H, R' = H; R = Me, R' = H; R = Ph, R' = H)

The adduct from *bis*(trimethylstannyl)methylamine and carbon disulphide was unstable and only its decomposition products were isolated[15].

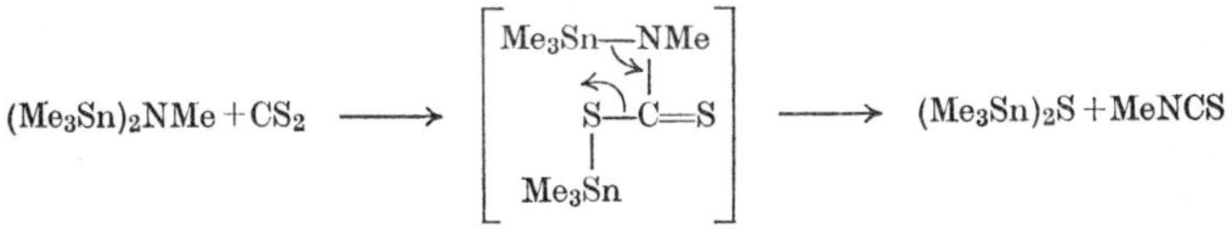

$$(Me_3Sn)_2NMe + CS_2 \longrightarrow \left[\begin{array}{c} Me_3Sn{-}NMe \\ \swarrow \searrow \\ S{-}C{=}S \\ | \\ Me_3Sn \end{array} \right] \longrightarrow (Me_3Sn)_2S + MeNCS$$

It can be seen from Table II that addition of Me_3SnNMe_2 to phenyl-isothiocyanate occurs at the $C{=}N$ bond, the distannylamine, however, adds to the $C{=}S$ bond[15].

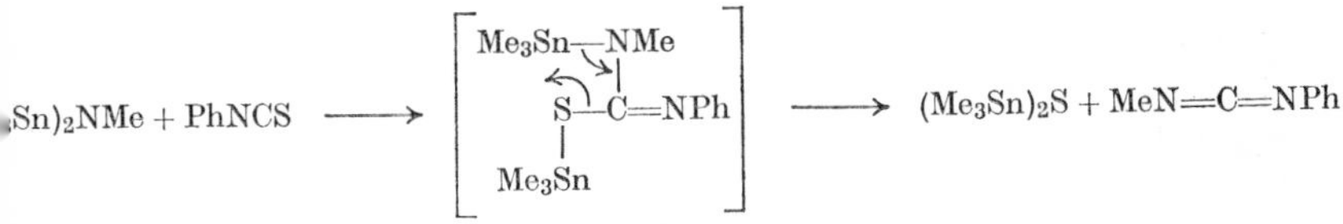

$$(Me_3Sn)_2NMe + PhNCS \longrightarrow \left[\begin{array}{c} Me_3Sn{-}NMe \\ \swarrow \searrow \\ S{-}C{=}NPh \\ | \\ Me_3Sn \end{array} \right] \longrightarrow (Me_3Sn)_2S + MeN{=}C{=}NPh$$

Trimethylstannyldimethylamine gives 1,4 addition products with benzoyl isocyanate and benzoyl-t-butylcarbodiimide[15a].

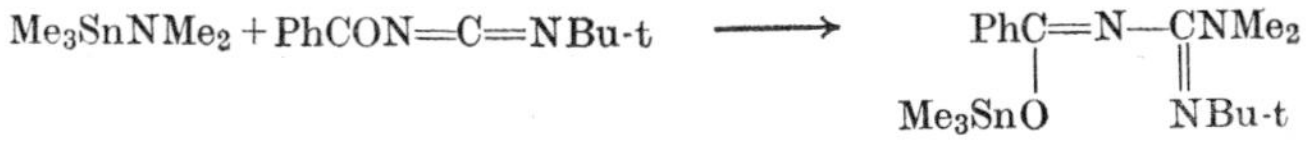

$$Me_3SnNMe_2 + PhCON{=}C{=}O \longrightarrow \begin{array}{c} PhC{=}N{-\!-\!-}CNMe_2 \\ | \quad\quad \| \\ OSnMe_3 \quad O \end{array}$$

$$Me_3SnNMe_2 + PhCON{=}C{=}NBu\text{-}t \longrightarrow \begin{array}{c} PhC{=}N{-}CNMe_2 \\ | \quad\quad \| \\ Me_3SnO \quad NBu\text{-}t \end{array}$$

Similar studies have been reported of reactions between stannyl-amines and benzoyl isothiocyanate[15a] and also thiobenzoyl isocyanate[15b]. Unstable adducts are also formed by addition to the $C{=}S$ bonds of carbon oxysulphide, thioketones and thioamides so that the overall equations for these reactions are[14]:

$$(Me_3Sn)_2NMe + OCS \longrightarrow (Me_3Sn)_2S + MeNCO$$
$$(Me_3Sn)_2NMe + PhCSR \longrightarrow (Me_3Sn)_2S + PhC(R){=}NMe.$$
$$(R = Ph, NMe_2)$$

The adduct from trichloroacetaldehyde and a stannylamine breaks down to give the trichloromethyltin compound[13].

$$Cl_3CCHO + Me_3SnNR_2 \longrightarrow$$
$$[Cl_3CCH(NR_2)OSnMe_3] \longrightarrow Cl_3CSnMe_3 + HCONR_2$$
$$(R = Me, Et)$$

A similar reaction occurs on the addition of alkoxides to hexa
chloroacetone (p. 77). The addition product can, however, b
isolated from the reaction between hexafluoroacetone and trimethyl
stannyldimethylamine [2].

$$Me_3SnNMe_2 + (CF_3)_2CO \longrightarrow Me_3SnOC(CF_3)_2NMe_2$$

The stannylamines react with epoxides, in the presence of a catalyst
as follows:

$$Bu_3SnNEt_2 + RCH\!\!-\!\!CHR' \xrightarrow{LiNEt_2} Bu_3SnOCHRCHR'NEt_2.$$
$$\diagdown\!\diagup$$
$$O$$

This reaction has been demonstrated with but-1-ene oxide, styrene
oxide and cyclohexene oxide [49]. The products from the reaction
between trimethylstannyldiethylamine and β-propionolactone indicate that acyl-oxygen bond cleavage has occurred [12].

$$Me_3SnNEt_2 + \underset{\lfloor\text{---}O\text{---}\rfloor}{CH_2CH_2CO} \longrightarrow Me_3SnOCH_2CH_2CONEt_2$$

The stannylamines can also function as aminating reagents whereby
an amino group is transferred from tin to another element in a
metathetical reaction [11, 35].

$$R_3SnNMe_2 + A\!\!-\!\!B \longrightarrow R_3SnA + BNMe_2$$

The element A must have nucleophilic character and it is commonly
halogen as in the following examples [11].

$$3R_3SnNMe_2 + BF_3 \cdot OEt_2 \longrightarrow B(NMe_2)_3 + Et_2O + 3R_3SnF$$
$$R_3SnNMe_2 + CFCl\!\!=\!\!CF_2 \longrightarrow CFCl\!\!=\!\!CFNMe_2 + R_3SnF$$
$$R_3SnNMe_2 + C_6F_6 \longrightarrow C_6F_5NMe_2 + R_3SnF$$

In contrast to the protolysis reactions discussed earlier, hydride
compounds react with stannylamines to give organotin hydrides [11,23a].

$$R_3SnNEt_2 + Bu_2AlH \longrightarrow R_3SnH + Bu_2AlNEt_2$$
$$2R_3SnNEt_2 + B_2H_6 \longrightarrow 2R_3SnH + 2H_2BNEt_2$$

Trimethylstannyldimethylamine reacted with decaborane, however,
to form the adduct $Me_3SnNMe_2 \cdot B_{10}H_{14}$ [5].

The N-stannylketenimines, $RCH_2C(CN)\!\!=\!\!C\!\!=\!\!NSnR_3'$ react with
compounds containing labile halogen to introduce the $RCH_2C(CN)_2\!\!-$
group [46], for example:

$$RCH_2C(CN)\!\!=\!\!C\!\!=\!\!NSnEt_3 + CH_2\!\!=\!\!CHCH_2Br \longrightarrow$$
$$RCH_2C(CN)_2CH_2CH\!\!=\!\!CH_2 + Et_3SnBr.$$

TABLE III

Physical properties of some organostannylamines

Compound	m.p.	b.p.	n_D^{20}
Me$_3$SnNMe$_2$	—	126°	—
Me$_3$SnNEt$_2$	—	162°	1·4651
Me$_3$SnNHEt	—	153°	1·4689
(Me$_3$Sn)$_2$NEt	—	93°/15 mm	—
Me$_2$Sn(NMe$_2$)$_2$	—	138°	—
(Et$_3$Sn)$_3$N	21–22°	192–194°/2 mm	—
Bu$_3$SnNMe$_2$	—	86°/0·1 mm	—
Ph$_3$SnNMe$_2$	—	128°/0·2 mm	—
Me$_3$SnN(Me)SiMe$_3$	—	79–81°/30 mm	—
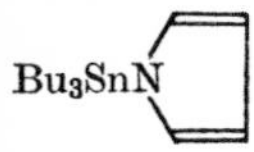 Bu$_3$SnN (pyrrolyl)	—	101·5–102°	1·5302
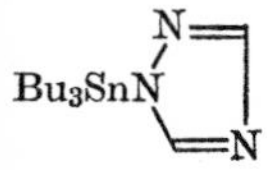 Bu$_3$SnN (triazolyl)	66–71°	—	—
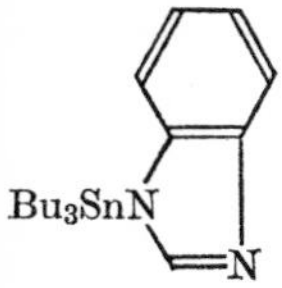 Bu$_3$SnN (benzotriazolyl)	137·5–139°	—	—
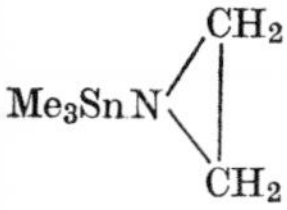 Me$_3$SnN(CH$_2$CH$_2$) (aziridinyl)	—	53–55°/16 mm	1·4950
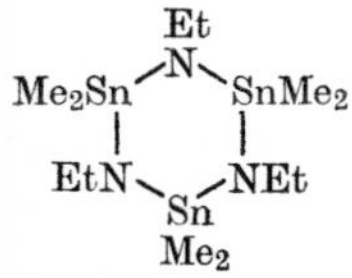 cyclic Me$_2$Sn–NEt–SnMe$_2$–NEt–SnMe$_2$	—	104°/0·05 mm	—

3.2. ORGANOSTANNYLAMIDES

Luijten and van der Kerk[27] prepared *N*-triethylstannyl-saccharin and -phthalimide by condensation reactions in which water was removed by azeotropic distillation with benzene.

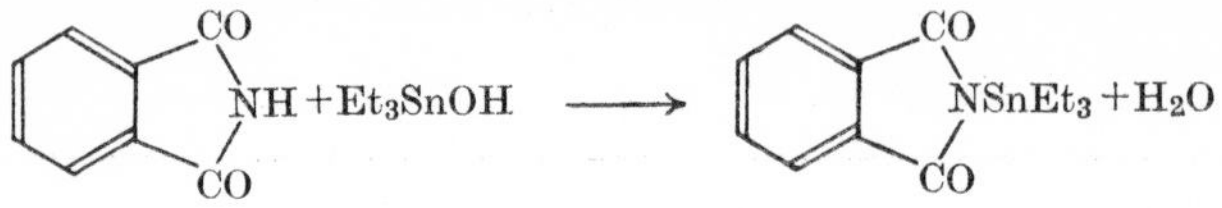

Isocyanuric acid can be stannylated in a similar manner [48, 54].

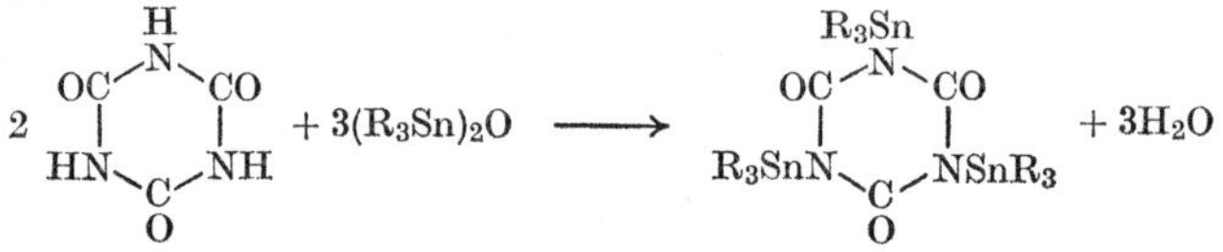

These compounds have exceptional thermal stabilities, *tris*(triphenylstannyl)isocyanurate is unmelted at 360°. However, when reaction occurred between dibutyltin oxide and a number of cyclic and acyclic imides no water was eliminated [55], for example:

$$Bu_2SnO + HN(COCH_2Ph)_2 \longrightarrow Bu_2Sn(OCOCH_2Ph)NHCOCH_2Ph.$$

The condensation reaction has also been used to prepare a number of stannylamides [9].

$$(Bu_3Sn)_2O + 2RNHCOX \longrightarrow 2Bu_3SnNRCOX + H_2O$$
$$(R = Me, X = H; R = C_6H_{13}, X = H; R = Ph, X = H; R = Ph, X = Me; R = H, X = Me;$$
$$R = Et, X = OMe; R = Ph, X = OMe)$$

In some cases the *N*-stannylamide reacted further and the following scheme was thought to be the most likely route to the products isolated [9].

$$Bu_3SnNRCOX + (Bu_3Sn)_2O \longrightarrow Bu_3SnX + Bu_3SnNRCOOSnBu_3$$
$$Bu_3SnNRCOOSnBu_3 + Bu_3SnNRCOX \longrightarrow$$
$$Bu_3SnNRCONRSnBu_3 + Bu_3SnX + CO_2$$

Examples of compounds which reacted in this manner are *N*,*N'*-diphenylurea (R = Ph, X = NHPh) which gave $Bu_3SnN(Ph)CON(Ph)SnBu_3$ and *N*-1-naphthyl-*N'*-phenylurea (R = 1-$C_{10}H_7$, X = NHPh) which gave $Bu_3SnN(1-C_{10}H_7)CON(1-C_{10}H_7)SnBu_3$. Some simple primary amides also failed to give the *N*-stannyl derivatives, for example:

$$2NH_2CONH_2 + (Bu_3Sn)_2O \longrightarrow 2Bu_3SnNCO + 2NH_3 + H_2O$$

Organotin alkoxides are effective reagents for the stannylation of amides. Triethyltin methoxide reacts with amides at 150–160° in an equilibrium reaction which gives the required products by distilling off the methanol formed[32].

$$Et_3SnOMe + RNHCOX \longrightarrow Et_3SnN(R)COX + MeOH$$

(R = Ph, X = H; R = C_6H_{13}, X = H; R = Ph, X = Me; R = Ph, X = OMe; R = C_6H_{13}, X = OMe)

This procedure was found to be unsuitable for the preparation of the corresponding triphenyltin compounds and these were obtained by reaction between a triphenylstannylamine and the appropriate amide[7].

$$Ph_3SnNEt_2 + PhNHCOH \longrightarrow Ph_3SnN(Ph)COH + Et_2NH$$

This reaction occurs at temperatures below 100° and is a good general method for the preparation of organostannylamides.

Another widely used method for the preparation of N-stannyl-amides, -ureas and -urethans is by the use of addition reactions. A number of these reactions have already been discussed in Chapter 5 and at the beginning of the present chapter, they are summarised in the following equations.

$$(Bu_3Sn)_2O + RNCO \xrightarrow[\text{temp.}]{\text{room}} Bu_3SnN(R)COOSnBu_3 \qquad \text{(p. 70)}$$

$$(Bu_3Sn)_2O + 2RNCO \xrightarrow{\text{heat}} Bu_3SnN(R)CON(R)SnBu_3 \qquad \text{(p. 70)}$$

$$Et_3SnOEt + ArNCO \longrightarrow Et_3SnN(Ar)COOEt \qquad \text{(p. 77)}$$

$$Me_3SnNMe_2 + ArNCO \longrightarrow Me_3SnN(Ar)CONMe_2 \qquad \text{(p. 96)}$$

Addition of hydrides to isocyanates is a convenient method of preparing N-stannyl-N-arylformamides[33].

$$R_3SnH + ArNCO \longrightarrow HCON(Ar)SnR_3$$

A number of N-stannylsulphonamides were made for testing as sternutatory agents[29]; they can be prepared by reaction between the organotin halide and the N-sodiosulphonamide.

$$R_3SnX + R'SO_2NHNa \longrightarrow R_3SnNHOSOR' + NaCl$$

N-(Triethylstannyl)toluene-p-sulphonamide has powerful sternutatory properties and should be handled with caution[27].

6.2.1. *Reactions of organostannylamides*

Like the stannylamines these compounds are generally sensitive to water and other protic reagents although the derivatives of the more acidic imides and sulphonamides show enhanced hydrolytic stability.

Acetic acid, hydrogen sulphide and ethanol cause Sn—N bond cleavage in the stannylcarbamates within a few seconds[3].

$$Bu_3SnN(Ph)COOMe + HA \longrightarrow PhNHCOOMe + Bu_3SnA$$
$$(A = AcO, SH, EtO)$$

Amines react more slowly and, after 3 days, the products from treatment with ethylamine indicated that fission of both Sn—N and C—O bonds had occurred.

$$Bu_3SnN(Ph)COOMe + EtNH_2 \longrightarrow PhNHCONHEt + Bu_3SnOMe$$

Methyl N-tributylstannyl-N-phenylcarbamate underwent a metathetical reaction with acetic anhydride to give tributyltin acetate and the acetylcarbamate[3].

$$Bu_3SnN(Ph)COOMe + Ac_2O \longrightarrow Bu_3SnOAc + AcN(Ph)COOMe$$

An important method of forming tin—tin bonds is by the hydrostannolysis of stannylamides.

$$R_3SnH + R_3'SnN(Ph)CHO \longrightarrow R_3SnSnR_3' + PhNHCHO$$

Various modifications of this reaction allow the synthesis of polystannanes containing up to 6 catenated tin atoms and the method is discussed in detail in Chapter 9.

TABLE IV

Physical properties of some organostannylamides and related compounds

Compound	m.p.	b.p.
$Bu_3SnN(Me)CHO$	37–37·5°	136–140°/0·4 mm
$Bu_3SnN(Ph)CHO$	64–67°	170°/0·1 mm
$Bu_3SnN(Ph)COMe$	—	104–106°/0·03 mm
p-OHCN(SnEt_3)C_6H_4N(SnEt_3)CHO	191–195°	—
$Et_3SnN(Bu)COOEt$	—	57·5°/0·01 mm
$Bu_3SnN(Et)COOMe$	—	89–90°/0·05 mm
$Bu_3SnN(Ph)COOMe$	—	98–100°/0·01 mm
$Bu_3SnN(Ph)CON(Ph)SnBu_3$	—	130°/0·05 mm
$\begin{matrix} CH_2CO \\ | \quad\quad NSnEt_3 \\ CH_2CO \end{matrix}$	58–60°	—
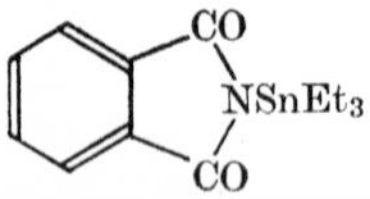	71–73°	—
$Et_3SnNHO_2SC_6H_4Me$-p	69·5–71°	—

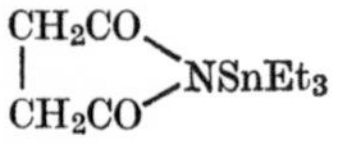

References

1. Abel, E. W., Armitage, D. A. and Brady, D. B., *J. Organometallic Chem.*, 1966, **5**, 130.
2. Abel, E. W. and Crow, J. P., *J. Chem. Soc. (A)*, 1968, 1361.
3. Bloodworth, A. J. and Davies, A. G., *J. Chem. Soc.*, 1965, 5238.
4. Chandra, G., George, T. A. and Lappert, M. F., *Chem. Comm.*, 1967, 116.
5. Chandra, G. and Lappert, M. F., *J. Chem. Soc. (A)*, 1968, 1940.
6. Chivers, T. and David, B., *J. Organometallic Chem.*, 1968, **13**, 177.
7. Creemers, H. M. J. C., Ph.D. Thesis, University of Utrecht, 1967.
8. Davies, A. G. and Mitchell, T. N., *J. Organometallic Chem.*, 1966, **6**, 568.
9. Davies, A. G., Mitchell, T. N. and Symes, W. R., *J. Chem. Soc. (C)*, 1966, 1311.
10. George, T. A., Jones, K. and Lappert, M. F., *J. Chem. Soc.*, 1965, 2157.
11. George, T. A. and Lappert, M. F., *Chem. Comm.*, 1966, 463.
12. Ishii, Y., *Asahi Garasu Kogyo Gijutsu Shorei-kai Kenkyu Hokoku*, 1967, **13**, 479 (*Chem. Abs.*, 1968, **69**, 96842).
13. Itoh, K., Fukui, M. and Ishii, Y., *Tetrahedron Letters*, 1968, 3867.
14. Itoh, K., Fukumoto, Y. and Ishii, Y., *Tetrahedron Letters*, 1968, 3199.
15. Itoh, K., Lee, I. K., Matsuda, I., Sakai, S. and Ishii, Y., *Tetrahedron Letters*, 1967, 2667.
15a. Itoh, K., Matsuda, I. and Ishii, Y., *Tetrahedron Letters*, 1969, 2675.
15b. Itoh, K., Katsuura, T. and Ishii, Y., *J. Organometallic Chem.*, 1969, **19**, 347.
16. Jones, K. and Lappert, M. F., *J. Chem. Soc.*, 1965, 1944.
17. Jones, K. and Lappert, M. F., *J. Organometallic Chem.*, 1965, **3**, 295.
18. Jones, K. and Lappert, M. F., *Organometallic Chem. Rev.*, 1966, **1**, 67.
19. Jones, K. and Lappert, M. F., *Proc. Chem. Soc.*, 1964, 22.
20. Koehler, H. and Beck, W., *Z. Anorg. Allg. Chem.*, 1968, **359**, 241.
21. Koehler, H. and Seifert, B., *J. Organometallic Chem.*, 1968, **12**, 253.
22. Kostyanovskii, R. G. and Prokof'ev, A. K., *Izv. Akad. Nauk S.S.S.R. Ser. Khim.*, 1967, 473.
22a. Kula, M. R., Lorberth, J. and Amberger, E., *Chem. Ber.*, 1964, **97**, 2087.
23. Kupchik, E. J. and McInerney, E. F., *J. Organometallic Chem.*, 1968, **11**, 291.
24. Lappert, M. F. and Lorberth, J., *Chem. Comm.*, 1967, 836.
24a. Lappert, M. F. and Poland, J. S., *Chem. Comm.*, 1969, 156.
24b. Lorberth, J., *J. Organometallic Chem.*, 1968, **15**, 251.
25. Luijten, J. G. A., Janssen, M. J. and van der Kerk, G. J. M., *Rec. Trav. Chim.*, 1962, **81**, 202.
26. Luijten, J. G. A., Rijkens, F. and van der Kerk, G. J. M., *Advances in Organometallic Chemistry*, Academic Press, New York and London, 1965, **3**, 397.
27. Luijten, J. G. A. and van der Kerk, G. J. M., *Investigations in the Field of Organotin Chemistry*, Tin Research Institute, Greenford, 1955, 97.
28. Luijten, J. G. A. and van der Kerk, G. J. M., *Rec. Trav. Chim.*, 1963, **82**, 1181.
29. McCombie, H. and Saunders, B. C., *Nature*, 1947, **159**, 491.
30. Neumann, W. P., *Angew. Chem. Intern. Ed. Engl.*, 1963, **2**, 100.
31. Noltes, J. G., *Rec. Trav. Chim.*, 1964, **83**, 515.
32. Noltes, J. G., *Rec. Trav. Chim.*, 1965, 84, 799.
33. Noltes, J. G. and Janssen, M. J., *J. Organometallic Chem.*, 1964, **1**, 346.
34. Ponomarev, S. V., Lisina, Z. M. and Lutsenko, I. F., *Zh. Obshch. Khim.*, 1966, **36**, 1818.

35. Scherer, O. J., *Organometallic Chem. Rev. Section A*, 1968, **3**, 281.
36. Scherer, O. J. and Biller, D., *Angew. Chem. Intern. Ed. Engl.*, 1967, **6**, 446.
37. Scherer, O. J. and Biller, D., *Z. Naturforsch.*, 1967, **B20**, 1079.
38. Scherer, O. J., Biller, D. and Schmidt, M., *Inorg. Nucl. Chem. Letters*, 1966, **2**, 103.
39. Scherer, O. J. and Hornig, P., *Chem. Ber.*, 1968, **101**, 2533.
40. Scherer, O. J. and Hornig, P., *J. Organometallic Chem.*, 1967, **8**, 465.
41. Scherer, O. J. and Schmidt, M., *Intern. Symp. Organosilicon Chem., Sci. Commun.*, Prague, 1965, 315 (*Chem. Abs.*, 1966, **65**, 10606h).
42. Scherer, O. J. and Schmidt, M., *J. Organometallic Chem.*, 1965, **3**, 156.
43. Scherer, O. J., Schmidt, J., Wokulat, J. and Schmidt, M., *Z. Naturforsch.*, 1965, **B20**, 183.
44. Schmid, G. D., Dissertation, Munich, 1963.
45. Sisido, K. and Kozima, S., *J. Org. Chem.*, 1962, **27**, 4051.
46. Sommer, R. and Neumann, W. P., *Angew. Chem. Intern. Ed. Engl.*, 1966, **5**, 515.
47. Sommer, R., Neumann, W. P. and Schneider, B., *Tetrahedron Letters*, 1964, 3875.
48. Stamm, W., *J. Org. Chem.*, 1965, **30**, 693.
49. Tzschach, A. and Reiss, E., *J. Organometallic Chem.*, 1967, **8**, 255.
50. Winters, L. J. and Hill, D. T., *Inorg. Chem.*, 1965, **4**, 1433.
51. Wright, C. M. and Muetterties, E. L., *Inorg. Syn.*, 1967, **10**, 137.
52. Yoder, C. H. and Zuckerman, J. J., *J. Amer. Chem. Soc.*, 1966, **88**, 4831.
53. British Patent 1,026,405, April 20, 1966 (*Chem. Abs.*, 1966, **65**, 2299b).
54. Netherlands Patent 6,411,318, April 2, 1965 (*Chem. Abs.*, 1965, **63**, 13316c).
55. U.S. Patent 3,355,469, Nov. 28, 1967 (*Chem. Abs.*, 1968, **68**, 49782).

7

ORGANOTIN HYDRIDES

Tin hydride or stannane, SnH_4, is unstable, decomposing at room temperature into tin and hydrogen. Partial replacement of hydrogen by organic groups confers increased stability but the organotin hydrides remain highly reactive and are used to synthesise an extensive range of organotin products of great diversity. In addition the hydrides are being increasingly used in organic synthesis as specialised reducing agents. The subject was reviewed in 1964[38] and more recent articles[14, 47, 99] dealing with specific types of reactions also include summaries of the general properties of organotin hydrides.

7.1. PREPARATION OF ORGANOTIN HYDRIDES

The most widely used method is the reduction of an organotin halide by lithium aluminium hydride. The reaction is conveniently

$$4R_3SnCl + LiAlH_4 \longrightarrow 4R_3SnH + LiAlCl_4$$

carried out by adding the organotin halide to a solution of lithium aluminium hydride in ether at room temperature. The reaction is allowed to proceed, without heating, for about 2 hours, excess $LiAlH_4$ is removed by cautious treatment with water and the ether layer is dried and evaporated to give the required hydride. This method is suitable for all the hydrides R_nSnH_{4-n} where $n = 0$ to 3 but, as indicated above, the stability depends upon the number of organic groups attached to tin giving the stability sequence $R_3SnH > R_2SnH_2 > RSnH_3 > SnH_4$, and the alkyl compounds are more stable than their aryl analogues. On storage the mode of decomposition depends upon conditions but, in general, a redistribution reaction into tetralkyltin and tin hydride occurs with the latter then decomposing further into tin and hydrogen.

$$4R_nSnH_{4-n} \longrightarrow nR_4Sn + (4-n)Sn + (8-2n)H_2$$

Silicone grease, metals, acids and aluminium halides all catalyse the decomposition[38] so that rigorous purification is necessary and it is claimed that the hydrides are stable at room temperature for an indefinite period in the absence of catalysts[62, 115]. Decomposition in the presence of amines follows a rather different course and reactions of the following type have been exploited in the synthesis of catenated tin compounds and are discussed in Chapter 9.

$$2R_3SnH \xrightarrow{\text{amine}} R_3SnSnR_3 + H_2$$

Other aluminium compounds which have been used for the preparation of organotin hydrides are the much cheaper sodium aluminium hydride[14] and the dialkylaluminium hydrides[71].

$$R_nSnX_{4-n} + (4-n)R_2'AlH \longrightarrow R_nSnH_{4-n} + (4-n)R_2'AlX$$

Although the usual starting materials are the organotin halides, *bis*(tributyltin) oxide was reduced by lithium aluminium hydride to give Bu_3SnH in high yield[13].

Hydrogen transfer from boron to tin is an alternative, though less widely used, preparative method and low temperature reduction of organotin alkoxides with diborane gives very good yields of the corresponding hydrides. More recently, reductions of organotin chlorides by excess sodium borohydride in either ethylene glycol dimethyl ether (monoglyme) or diethylene glycol dimethyl ether (diglyme) have been used[9].

There is considerable current interest in a procedure involving hydrogen transfer from a monomeric or polymeric silicon hydride to the tin atom of an organotin oxide or alkoxide[6, 32, 116, 117, 121], the essential reaction is as follows.

$$\text{>Sn-O-} + \text{>Si-H} \longrightarrow \text{>Sn-H} + \text{>Si-O-}$$

The relative reactivities of the organotin compounds are in the sequence $Bu_2Sn(OEt)_2 > Bu_3SnOEt > (Bu_3Sn)_2O > (Pr_3Sn)_2O > Bu_2SnO > (Ph_3Sn)_2O > Bu_3SnOSiBu_3 > Bu_3SnOSiPh_3$[28]. The reactivity sequence for the silicon hydrides is $[MeSi(H)O]_x > (Ph_2SiH)_2O > Ph_3SiH > (HSiO_{3/2} + Me_2SiO) \gg Bu_3SiH$[28]. By use of a secondary phosphine oxide, hydrogen transfer from phosphorus to the tin atom of an organostannoxane may be effected[31].

$$R_2P(O)H + R_3'SnOSnR_3' \longrightarrow R_2P(O)OSnR_3' + R_3'SnH$$
$$(R = Ph, R' = Et; R = Ph, R' = Bu; R = Ph, R' = Ph; R = Bu, R' = Bu)$$

An unusual method of making trialkyltin hydrides is the decarboxylation of trialkyltin formates by heating at $100\text{–}170°$ under reduced pressure[86, 119].

$$R_3SnOCOH \longrightarrow R_3SnH + CO_2$$

Colorimetric methods have been devised for the determination of organotin hydrides at concentrations down to 10^{-4}M [23]. The labile nature of the hydrides has already been referred to and all manipulations of these compounds must be carried out under nitrogen or, preferably, argon.

7.2. THE HYDROSTANNATION OF OLEFINS AND ACETYLENES

The term hydrostannation is now widely used for reactions of the general type:

$$\text{>Sn—H} + \text{A}\text{=}\text{B} \longrightarrow \text{>Sn—A—B—H} .$$

The additions which have received most study are those to olefinic and acetylenic compounds though as will be seen there is increasing interest in additions to other unsaturated systems.

Addition of an organotin hydride to a carbon–carbon double bond was first demonstrated by van der Kerk and his coworkers in 1956 [105] and reactions of the type:

$$R_3SnH + CH_2\text{=}CHX \longrightarrow R_3SnCH_2CH_2X$$

soon became established as a major new method of forming tin–carbon bonds. By the use of appropriately substituted olefins this method can be used for attaching to tin organic radicals which contain functional groups. As noted in Chapter 2 the traditional methods of synthesis of organotin compounds involving transfer of organic groups from metals such as magnesium and lithium to tin cannot be used with functionally substituted organic groups. Some examples are shown in Table I. (The progress of hydrostannation reactions can be monitored by observing the disappearance of the characteristic ν(Sn—H) band at ~1800 cm^{-1} in the infrared spectrum of the reaction mixture.)

The method fails, however, when certain functional groups are present and with allyl bromide and crotyl chloride the halogens are replaced by hydrogen on treatment with triphenyltin hydride [40, 107]. Reaction between dibutyltin dihydride and allyl alcohol gave none of the product $Bu_2Sn(CH_2CH_2CH_2OH)_2$ instead the alkoxide, $Bu_2Sn(OCH_2CH\text{=}CH_2)_2$, was isolated together with the cyclic compound I. It was suggested that either $Bu_2Sn(H)OCH_2CH\text{=}CH_2$ or

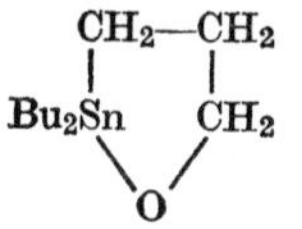

$$
\begin{array}{c}
\text{CH}_2\text{—CH}_2 \\
| \qquad\quad | \\
\text{Bu}_2\text{Sn} \qquad \text{CH}_2 \\
\diagdown \qquad \diagup \\
\text{O}
\end{array}
$$

I

TABLE I

Addition reactions between organotin hydrides and functionally substituted olefins

Hydride	Olefin	Product	Reference
Et_3SnH	$CH_2{=}CHCOOMe$	$Et_3SnCH_2CH_2COOMe$	71
Et_2SnH_2	$CH_2{=}CHCOOMe$	$Et_2Sn(CH_2CH_2COOMe)_2$	71
$BuSnH_3$	$CH_2{=}CHCOOMe$	$BuSn(CH_2CH_2COOMe)_3$	106
Ph_3SnH	$CH_2{=}CH(CH_2)_8COOMe$	$Ph_3Sn(CH_2)_{10}COOMe$	123
Ph_3SnH	$(CH_2{=}CHCO)_2O$	$(Ph_3SnCH_2CH_2CO)_2O$	106
Ph_3SnH	$CH_2{=}CHOAc$	$Ph_3SnCH_2CH_2OAc$	107
Et_3SnH	$CH_2{=}CHCONH_2$	$Et_3SnCH_2CH_2CONH_2$	71
Pr_3SnH	$CH_2{=}CHCN$	$Pr_3SnCH_2CH_2CN$	107
Et_2SnH_2	$CH_2{=}CHCN$	$Et_2Sn(CH_2CH_2CN)_2$	71
Ph_2SnH_2	$CH_2{=}CHCN$	$Ph_2Sn(CH_2CH_2CN)_2$	92
Bu_3SnH	$CH_2{=}CHCH_2OH$	$Bu_3SnCH_2CH_2CH_2OH$	71
Ph_3SnH	$CH_2{=}CHCH_2OH$	$Ph_3SnCH_2CH_2CH_2OH$	107
Ph_3SnH	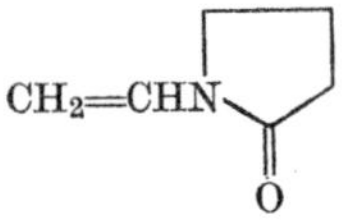 $CH_2{=}CHN{\displaystyle\diagdown}$ (pyrrolidinone) O	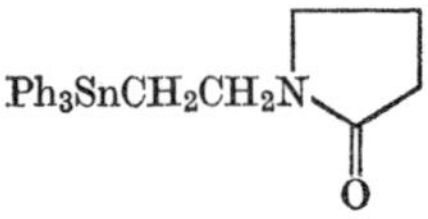 $Ph_3SnCH_2CH_2N{\displaystyle\diagdown}$ (pyrrolidinone) O	107
Ph_3SnH	$p\text{-}CH_2{=}CHC_6H_4NHAc$	$p\text{-}Ph_3SnCH_2CH_2C_6H_4NHAc$	106
Me_3SnH	$\begin{array}{l}CF_2{-}CF\\ \|\quad\quad\|\|\\ CF_2{-}CF\end{array}$	$\begin{array}{l}CF_2{-}CFH\\ \|\quad\quad\|\\ CF_2{-}CFSnMe_3\end{array}$	20
Ph_3SnH	$CH_2{=}CHMPh_3$	$Ph_3SnCH_2CH_2MPh_3$ $(M = Si, Ge, Sn)$	29
Ph_2SnH_2	$p\text{-}CH_2{=}CHC_6H_4MPh_3$	$Ph_2Sn(CH_2CH_2C_6H_4MPh_3\text{-}p)_2$ $(M = Ge, Sn, Pb)$	84

$Bu_2Sn(H)CH_2CH_2CH_2OH$ were intermediates in the formation of I[46]. Similarly no addition occurs between dibutyltin dihydride and allyl mercaptan, the sole product being the dimercaptide[46].

$$Bu_2SnH_2 + 2CH_2{=}CHCH_2SH \longrightarrow Bu_2Sn(SCH_2CH{=}CH_2)_2 + 2H_2$$

In general addition occurs to a monosubstituted olefin, $CH_2 = CHX$, so that the tin atom becomes attached to the methylene carbon atom irrespective of the nature of X, i.e. the direction of addition is governed by steric factors. (Some exceptions are discussed below.) In the majority of cases the hydrostannation of olefins occurs by a radical mechanism[47, 77]. The addition reactions are accelerated in the presence of certain radical sources such as azo*bis*isobutyronitrile (but not by diacyl peroxides, see Section 7.6) and retarded by radical scavengers such as phenoxyl[77]. In some instances, such as in the addition of trialkyltin hydrides to unsubstituted alkenes, e.g oct-1-ene, reaction will only occur in the presence of a radical generator. The mechanism of the radical addition reaction is as follows.

$$R_3SnH + S\cdot \longrightarrow R_3Sn\cdot + SH \qquad (S\cdot = \text{a radical initiator})$$
$$R_3Sn\cdot + CH_2{=}CHR' \rightleftharpoons R_3SnCH_2CHR'\cdot$$
$$R_3SnCH_2CHR'\cdot + R_3SnH \longrightarrow R_3SnCH_2CH_2R' + R_3Sn\cdot$$

Recent experiments with 1-deuteriohex-1-ene[42] and penta-1,3-diene[66] have shown that, as indicated above, the second step in the sequence is reversible. In both experiments the method used was to treat an olefin of known configuration with a deficiency of hydride, examination of the unreacted olefin showed that isomerisation had occurred by rotation about the CH_2—CHR' bond in $R_3SnCH_2CHR'\cdot$. This was followed by loss of $R_3Sn\cdot$ to regenerate the olefin $CH_2{=}CHR'$. In the case of 1-deuteriohex-1-ene this led to the recovery of isomerized olefin and reversible addition at C_1 of penta-1,3-diene gave a mesomeric radical in which free rotation caused a change in configuration at the 3,4 double bond in the recovered olefin.

Hydrostannation of acrylonitrile is exceptional in that, as well as the normal β-adduct formed by the radical addition, lesser amounts of the α-adduct are produced by an accompanying ionic reaction[47, 57, 88].

$$R_3SnH + CH_2{=}CHCN \left\{ \begin{array}{l} \longrightarrow R_3SnCH_2CH_2CN \quad \beta\text{-adduct} \\ \longrightarrow CH_3CH(CN)SnR_3 \quad \alpha\text{-adduct} \end{array} \right.$$

The β-adducts react further with the hydride to give the distannane and saturated nitrile[88].

$$CH_3CH(CN)SnR_3 + R_3SnH \longrightarrow R_3SnSnR_3 + CH_3CH_2CN$$

Hence the overall effect is that the β-adduct is obtained together with a small amount of the reduced olefin e.g. when Bu_3SnH and $CH_2{=}CHCN$ were allowed to react for 5 hours at 150°, 7% of ethyl cyanide was obtained.

Organotin hydrides will add to non-terminal olefins when there is an electron-withdrawing group attached to an unsaturated carbon atom as in the following examples.

$$Me_3SnH + MeCH{=}CHCOOMe \longrightarrow Me_3SnCH(Me)CH_2COOMe \quad \text{(ref. 47)}$$

$$Bu_3SnH + MeOCOCH{=}CHCOOMe \longrightarrow Bu_3SnCH(COOMe)CH_2COOMe \quad \text{(ref. 87)}$$

$$Bu_3SnH + MeCH{=}C(COOEt)_2 \longrightarrow Bu_3SnCH(Me)CH(COOEt)_2 \quad \text{(ref. 87)}$$

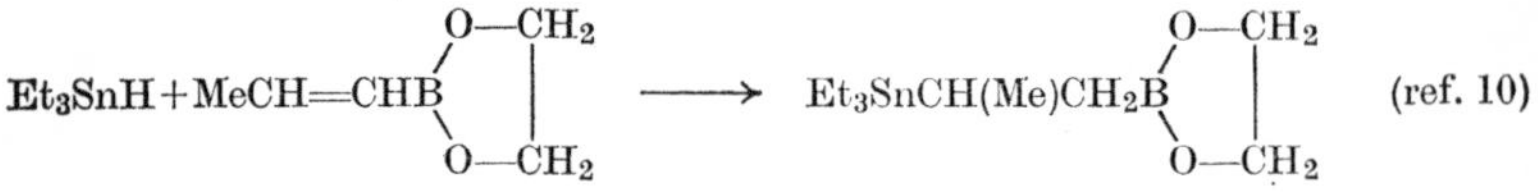

$$Et_3SnH + MeCH{=}CHB\begin{matrix} O-CH_2 \\ | \\ O-CH_2 \end{matrix} \longrightarrow Et_3SnCH(Me)CH_2B\begin{matrix} O-CH_2 \\ | \\ O-CH_2 \end{matrix} \quad \text{(ref. 10)}$$

Moreover a slow reaction occurs between trimethyltin hydride and simple internal olefins such as but-2-ene, pent-2-ene, cyclopentene, etc., and becomes rapid when the reactants are irradiated with ultraviolet light [42, 104].

Hydrostannation of allene gives two products.

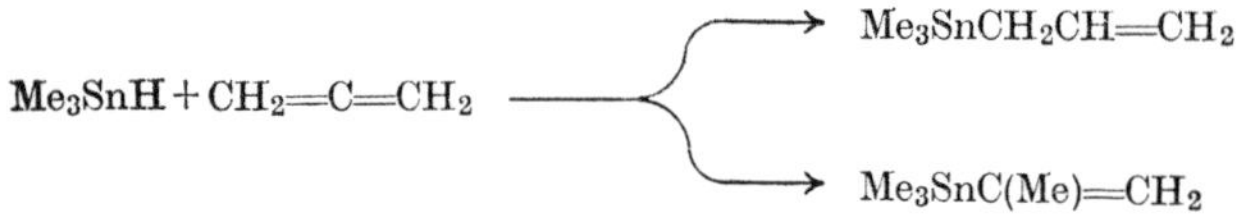

$$Me_3SnH + CH_2{=}C{=}CH_2 \longrightarrow \begin{cases} Me_3SnCH_2CH{=}CH_2 \\ \\ Me_3SnC(Me){=}CH_2 \end{cases}$$

However, with unsymmetrically substituted allenes the number of products increases and from the corresponding reaction with $MeCH{=}C{=}CH_2$ five products were isolated and characterised [41]. The factors which influence the ratio of products resulting from 1,2- and 1,4-additions in conjugated dienes have been investigated [78].

In addition to the free radical sources already mentioned other compounds such as aluminium chloride [120] and organoaluminium compounds [70] are sometimes used as catalysts in the hydrostannation of olefins. The catalytic function of the organoaluminium compounds is explained in terms of the following reactions in which an aluminium hydride adds to the olefin.

$$\geqslant SnH + \geqslant AlR \longrightarrow \geqslant SnR + \geqslant AlH$$

$$\geqslant AlH + CH_2{=}CHR' \longrightarrow \geqslant AlCH_2CH_2R'$$

$$\geqslant AlCH_2CH_2R' + \geqslant SnH \longrightarrow \geqslant AlH + \geqslant SnCH_2CH_2R'$$

When trimethyltin hydride reacts with trimethylperfluorovinyltin, reduction occurs and the main products are $cis\text{-}Me_3SnCF{=}CFH$ and trimethyltin fluoride [8]. However, addition reactions do occur to the perfluorovinyltin compounds and, also, to their germanium and silicon analogues [1, 2]. The stability of these adducts depends upon the

$$\text{Me}_3\text{SnH} + \text{Me}_3\text{MCF}{=}\text{CF}_2 \quad \begin{cases} \rightarrow \text{Me}_3\text{MCFHCF}_2\text{SnMe}_3 \quad \textit{II} \\ \rightarrow \text{Me}_3\text{MCF(SnMe}_3)\text{CF}_2\text{H} \quad \textit{III} \end{cases}$$

$$(\text{M} = \text{Si, Ge, Sn})$$

element M in the sequence M = Si > Ge > Sn and the symmetrical adducts *II* are more stable than the unsymmetrical compounds *III*. Decomposition of the adducts occurs by elimination of trimethyltin fluoride giving the reduced olefin, for example:

$$\text{Me}_3\text{SnCFHCF}_2\text{SnMe}_3 \longrightarrow \text{Me}_3\text{SnF} + cis\text{-Me}_3\text{SnCF}{=}\text{CFH}.$$

Organotin hydrides add more readily to acetylenes than olefins and, about 10 years ago, van der Kerk and Noltes [106] reported a number of addition reactions of the following type.

$$\text{R}_3\text{SnH} + \text{R}'\text{C}{\equiv}\text{CH} \longrightarrow \text{R}_3\text{SnCH}{=}\text{CHR}'$$

Acetylene itself reacts with 2 molar proportions of triphenyltin hydride to give the symmetrical distannylethane.

$$2\text{Ph}_3\text{SnH} + \text{HC}{\equiv}\text{CH} \longrightarrow \text{Ph}_3\text{SnCH}_2\text{CH}_2\text{SnPh}_3$$

Some earlier structural assignments for 2:1 adducts have had to be revised with the demonstration that, when 2 molecules of a hydride

TABLE II

Addition reactions between organotin hydrides and acetylenes

Hydride	Acetylene	Product	Reference
Et_3SnH	$\text{HC}{\equiv}\text{CC}{\equiv}\text{CH}$	$\text{Et}_3\text{SnCH}{=}\text{CHC}{\equiv}\text{CH}$	114
Et_3SnH	$\text{HC}{\equiv}\text{CCH}{=}\text{CHMe}$	$\text{Et}_3\text{SnCH}{=}\text{CHCH}{=}\text{CHMe}$	61
Et_3SnH	$\text{HC}{\equiv}\text{COEt}$	$\text{Et}_3\text{SnCH}{=}\text{CHOEt}$	35
Et_3SnH	$\text{HC}{\equiv}\text{CCH}_2\text{OCHMeOBu}$	$\text{Et}_3\text{SnCH}{=}\text{CHCH}_2\text{OCHMeOBu}$	108
R_3SnH	$\text{HC}{\equiv}\text{CSiR}_3'$	$\text{R}_3\text{SnCH}{=}\text{CHSiR}_3'$	36, 65
Bu_3SnH	$\text{HC}{\equiv}\text{CGeBu}_3$	$\text{Bu}_3\text{SnCH}{=}\text{CHGeBu}_3$	63
Bu_3SnH	$\text{HC}{\equiv}\text{CSnBu}_3$	$\text{Bu}_3\text{SnCH}{=}\text{CHSnBu}_3$	63
Bu_2SnH_2	$\text{HC}{\equiv}\text{CSnBu}_3$	$\text{Bu}_2\text{Sn(CH}{=}\text{CHSnBu}_3)_2$	64
R_3SnH	$\text{HC}{\equiv}\text{CCH}_2\text{OSnR}_3'$	$\text{R}_3\text{SnCH}{=}\text{CHCH}_2\text{OSnR}_3'$	125

add to a monosubstituted acetylene, the unsymmetrical β,β-adduct is usually obtained [21a, 55a].

$$2R_3SnH + R'C{\equiv}CH \longrightarrow R'CH_2CH(SnR_3)_2$$

Some recent examples are listed in Table II to indicate the range of olefins which have been obtained by addition reactions.

Detailed studies of the structure and stereochemistry of the products of hydrostannation of acetylenes were initiated by Fulton [26] and extended by Leusink and coworkers [47, 49, 52, 55] who have also investigated the mechanism of the reaction. When addition occurs to a monosubstituted acetylene there are three main products which are labelled α, cis-β and trans-β (N.B. this cis-β product arises from a trans addition and the trans-β product from a cis-addition).

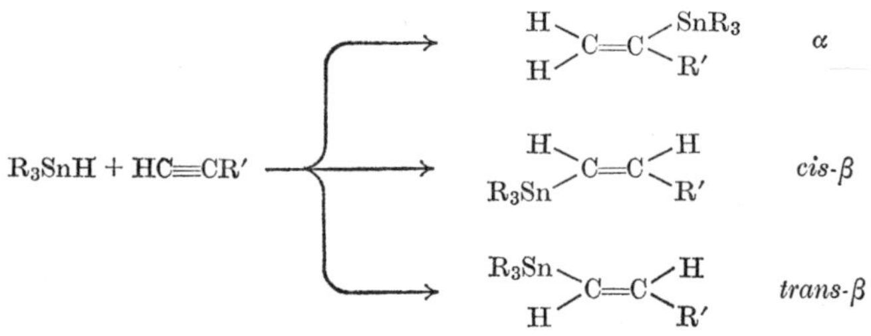

If R′ is strongly electron-withdrawing (e.g. R′ = COOEt, CN) then a mainly ionic reaction occurs and relatively large amounts of the α-adduct are formed. For this ionic reaction the order of reactivity is $Ph_3SnH \ll Me_3SnH < Et_3SnH < Bu_3SnH$ and the first and rate-determining step is attack by hydride hydrogen on carbon [54]:

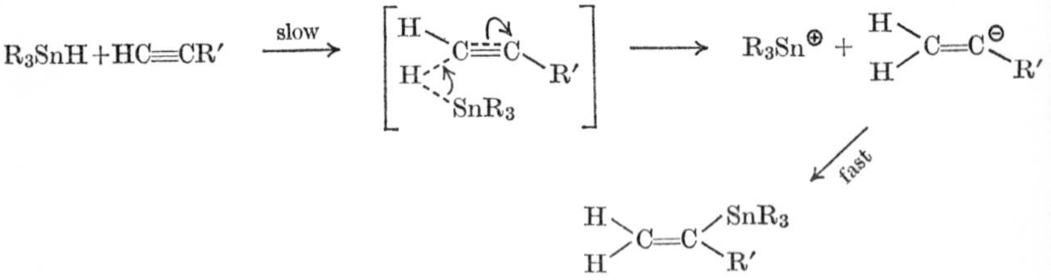

For the alkyl propiolates, e.g. HC≡CCOOEt, the α-adducts are favoured by decreasing size of the alkyl groups on tin but substantial amounts of the β-adducts are also formed together with some unidentified products of high boiling point and small amounts of

$R_3SnCH_2CH_2COOEt$ and $R_3SnC\equiv CCOOEt$ [47, 60]. The nature of the products isolated from the reactions between organotin hydrides and hexafluorobut-2-yne was analysed using nmr spectroscopy, the

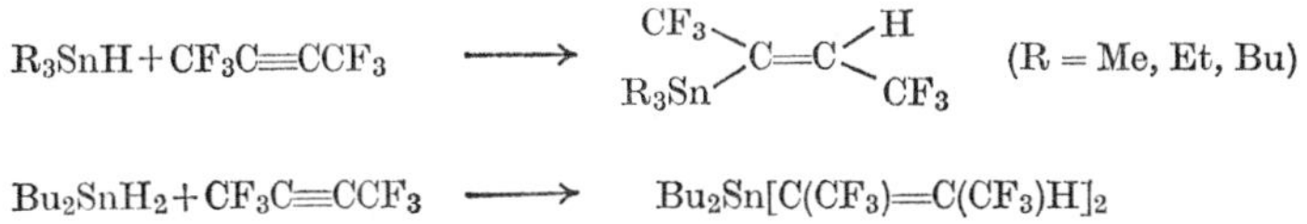

$$R_3SnH + CF_3C\equiv CCF_3 \longrightarrow \underset{R_3Sn}{\overset{CF_3}{\diagdown}}C=C\underset{CF_3}{\overset{H}{\diagup}} \quad (R = Me, Et, Bu)$$

$$Bu_2SnH_2 + CF_3C\equiv CCF_3 \longrightarrow Bu_2Sn[C(CF_3)=C(CF_3)H]_2$$

trans–trans configuration being most likely for the adduct from the dihydride[19]. The relative proportions of the three 1:1 adducts obtained from trimethyltin hydride and trifluoromethylacetylene were determined.

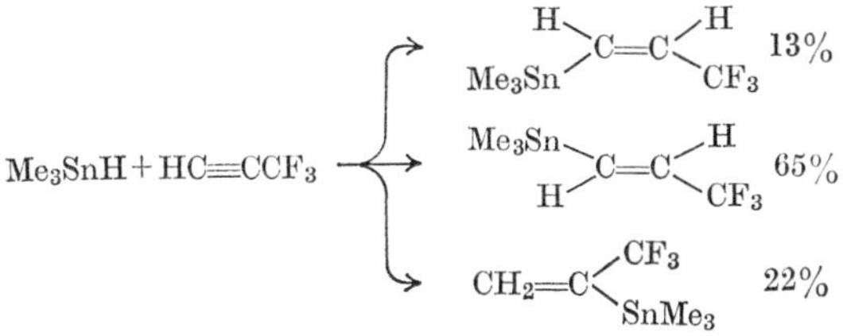

$$Me_3SnH + HC\equiv CCF_3 \longrightarrow
\begin{cases}
\underset{Me_3Sn}{\overset{H}{\diagdown}}C=C\underset{CF_3}{\overset{H}{\diagup}} & 13\% \\[2ex]
\underset{H}{\overset{Me_3Sn}{\diagdown}}C=C\underset{CF_3}{\overset{H}{\diagup}} & 65\% \\[2ex]
CH_2=C\underset{SnMe_3}{\overset{CF_3}{\diagdown}} & 22\%
\end{cases}$$

The amount of α-adduct formed depends upon how well the group R' can stabilise the negative charge on the intermediate $CH_2=\overset{\ominus}{C}R'$ and the order found is $CN > COOR > CH_2OH \sim CF_3 > CH_3 \sim Ph$.

When the acetylenic substituent is electron-releasing (e.g. $R' = Bu$, OEt) (or when both electron-releasing and electron-withdrawing substituents are present) the addition proceeds by a radical process and the primary product is the *cis-β* adduct[47, 48].

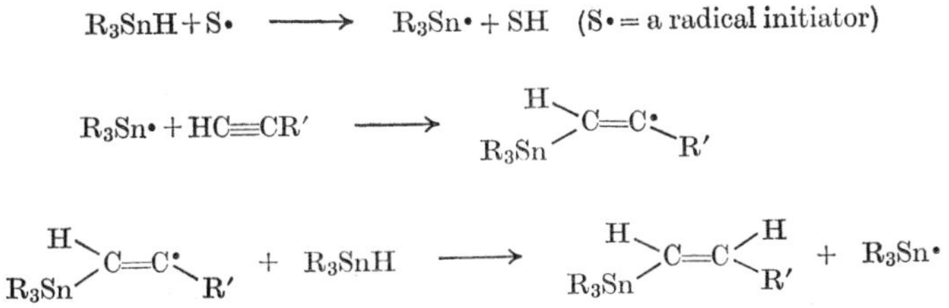

$$R_3SnH + S\cdot \longrightarrow R_3Sn\cdot + SH \quad (S\cdot = \text{a radical initiator})$$

$$R_3Sn\cdot + HC\equiv CR' \longrightarrow \underset{R_3Sn}{\overset{H}{\diagdown}}C=C\underset{R'}{\overset{\cdot}{\diagup}}$$

$$\underset{R_3Sn}{\overset{H}{\diagdown}}C=C\underset{R'}{\overset{\cdot}{\diagup}} + R_3SnH \longrightarrow \underset{R_3Sn}{\overset{H}{\diagdown}}C=C\underset{R'}{\overset{H}{\diagup}} + R_3Sn\cdot$$

In practice a mixture of *cis-β*- and *trans-β*-adducts is obtained, the latter arising from an isomerisation of the *cis-β* product which is catalysed by $R_3Sn\cdot$ radicals[26, 47, 50, 102].

Although emphasis has been placed on the addition reactions of the monohydrides an important method of preparing polymeric organotin compounds is by reaction between dihydrides and diolefins or diacetylenes [30, 82, 83, 85].

$$xR_2SnH_2 + xCH_2=CHXCH=CH_2 \longrightarrow \text{\textlbrackdbl}R_2SnCH_2CH_2XCH_2CH_2\text{\textrbrackdbl}_x$$
$$xR_2SnH_2 + xCH\equiv CXC\equiv CH \longrightarrow \text{\textlbrackdbl}R_2SnCH=CHXCH=CH\text{\textrbrackdbl}_x$$

To consider a specific example, diphenyltin dihydride reacted with o-divinylbenzene in benzene at reflux temperature in the following manner [59]:

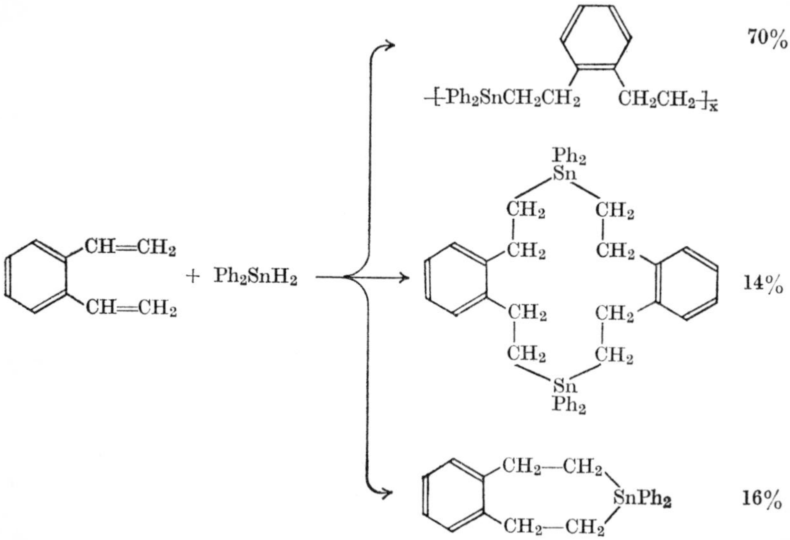

With the corresponding o-diethynylbenzene and dimethyltin dihydride a 75% yield of polymeric product was obtained. Organo-aluminium compounds are useful catalysts for the preparation of polymers from α,ω-diolefins and organotin dihydrides although the autocatalytic decomposition of the latter is a troublesome side-reaction [74]. To overcome this difficulty addition of the bis(mono-hydride) $HSn(Bu)_2(CH_2)_6Sn(Bu)_2H$ has been advocated [74]. Another approach to the synthesis of organotin polymers is by the homopolymerisation of an unsaturated organotin monohydride $R_2Sn(H)(CH_2)_nCH=CH_2$ [8, 51a, 74]; compounds of this type can be prepared by either of the following reactions.

$$R_2SnH_2 + CH\equiv CR' \longrightarrow R_2Sn(H)CH=CHR'$$
$$R_2SnH_2 + CH_2=CH(CH_2)_nCH=CH_2 \longrightarrow R_2Sn(H)(CH_2)_{n+2}CH=CH_2$$

Polymers containing organotin groups have also been prepared by addition reactions between organotin hydrides and unsaturated polyesters[124] and by copolymerising butadiene and styrene in the presence of dibutyltin dihydride[122].

7.3. HYDROSTANNATION OF C=O, C=N, C=S AND N=N BONDS

Reaction of organotin hydrides with aldehydes or ketones leads to the products of hydrostannation or reduction or both.

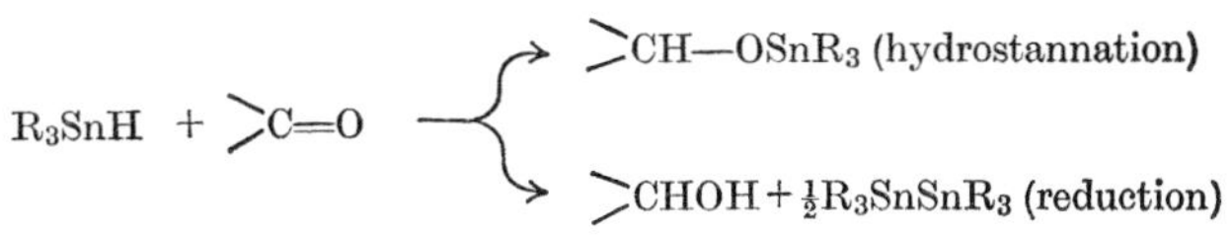

$$R_3SnH \ + \ {>}C{=}O \ \begin{cases} {>}CH{-}OSnR_3 \text{ (hydrostannation)} \\ \\ {>}CHOH + \tfrac{1}{2}R_3SnSnR_3 \text{ (reduction)} \end{cases}$$

With trialkyltin hydrides hydrostannation predominates when the reaction is carried out at 60–80° in the presence of a radical source (e.g. azo*bis*isobutyronitrile) or under irradiation with ultraviolet light or in the presence of zinc chloride or methanol as catalysts[47, 91]. With other hydrides reduction is the normal reaction, the relative ease of reduction showing the following dependence on the nature of the hydride, $Bu_2SnH_2 > BuSnH_3 > Ph_3SnH > Bu_3SnH$[38]. The selectivity of reduction by organotin hydrides has been examined with reference to ketosteroids containing carbonyl groups in different environments[7] and the stereochemistry of reduction of substituted cyclohexanones has been discussed[38]. The selectivity of reduction is comparable with that of sodium borohydride and it seems likely that the organotin hydrides will come into more general use as carbonyl reducing agents.

When α,β-unsaturated ketones are treated with triphenyltin hydride the saturated ketones are formed[58, 89, 90], for example:

$$2Ph_3SnH + Me_2C{=}CHCOMe \longrightarrow Me_2CHCH_2COMe + Ph_3SnSnPh_3.$$

When tributyltin hydride was used the O-stannyl derivative of the enol was the dominant product.

$$Bu_3SnH + Me_2C{=}CHCOMe \longrightarrow Me_2CHCH{=}C(Me)OSnBu_3$$

It is considered that both types of product arise from a 1,4-addition.

$$R_3SnH + {>}C{=}C{-}C{=}O \longrightarrow {>}CH{-}C{=}C{-}OSnR_3$$

The adduct is either isolated as such, or when $R = Ph$, it undergoes further reaction with the hydride[58].

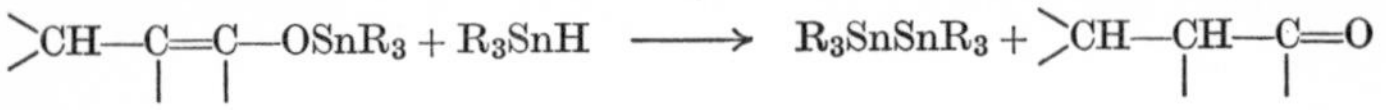

$$>CH\text{---}C\text{=}C\text{---}OSnR_3 + R_3SnH \longrightarrow R_3SnSnR_3 + >CH\text{---}CH\text{---}C\text{=}O$$

The scope of the carbonyl-hydrostannation reaction is wide, as the examples in Table III indicate. These reactions may be compli-

TABLE III

Hydrostannation reactions of aldehydes and ketones

Hydride	Carbonyl compound	Product	% yield	Reference
Et_3SnH	$PhCHO$	$PhCH_2OSnEt_3$	87	67
Et_3SnH	$p\text{-}MeOC_6H_4CHO$	$p\text{-}MeOC_6H_4CH_2OSnEt_3$	95	67
Et_3SnH	C_6F_5CHO	$C_6F_5CH_2OSnEt_3{}^a$	90	47
Et_3SnH	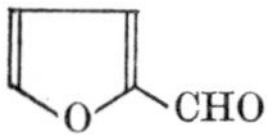 (furfural)	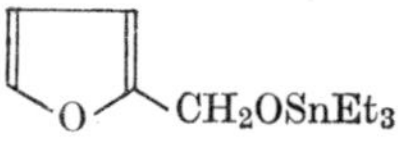 (furfuryl CH_2OSnEt_3)	63	67
Bu_3SnH	Me_2CHCHO	$Me_2CHCH_2OSnBu_3$	89	67
Me_3SnH	Cl_3CCHO	$Cl_3CCH_2OSnMe_3{}^a$	30	47
Et_3SnH	$PhCOCF_3$	$PhCH(CF_3)OSnEt_3$	73	47
Et_3SnH	$MeCOMe$	$Me_2CHOSnEt_3$	60	67
Bu_3SnH	$MeCOEt$	$MeCH(Et)OSnBu_3$	85	67
Me_3SnH	CF_3COCF_3	$(CF_3)_2CHOSnMe_3{}^a$	100	18
Me_2SnH_2	CF_3COCF_3	$Me_2Sn[OCH(CF_3)_2]_2{}^a$	—	18

^{a} These products may undergo further reaction.

cated by addition of the initial product to a second molecule of the carbonyl compound[53].

$$R_3SnOCH_2CCl_3 + Cl_3CCHO \longrightarrow R_3SnOCH(CCl_3)OCH_2CCl_3$$

Or by reduction with a second molecule of hydride.

$$R_3SnOCH_2CCl_3 + R_3SnH \longrightarrow R_3SnOCH_2CHCl_2 + R_3SnCl$$

A few examples are known where the hydrostannation product undergoes spontaneous rearrangement, for example when salicylaldehyde reacted with triethyltin hydride it was the phenoxide rather than the alkoxide which was isolated[47, 67].

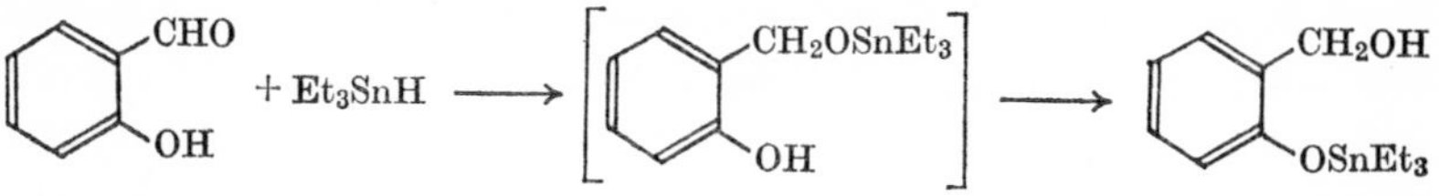

Hydrostannation of the C=O bond may proceed by a radical or by an ionic mechanism depending upon the carbonyl compound and the nature of any catalyst which is present [47, 67]. When a powerful electron-withdrawing group, such as trifluoromethyl, is attached to the carbonyl group hydrostannation occurs by an ionic reaction in which the rate-determining step is a nucleophilic attack by hydride hydrogen on carbon [47, 51].

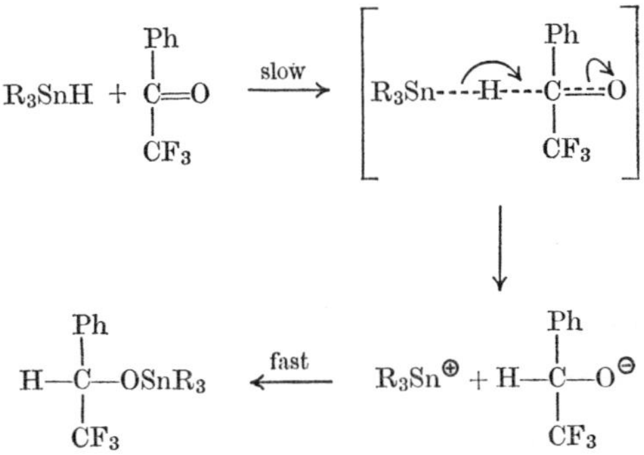

Examples of additions to other unsaturated systems are given in Table IV from which it can be seen that organotin hydrides add to

TABLE IV

Miscellaneous hydrostannation reactions

Hydride	Unsaturated compound	Product	Reference
Et_3SnH	$p\text{-}XC_6H_4N{=}C{=}O$ ($X = H, Cl, NO_2$)	$p\text{-}XC_6H_4N(SnEt_3)CHO$	81
Et_3SnH	$C_6H_{13}N{=}C{=}O$	$C_6H_{13}N(SnEt_3)CHO$	47, 81
Et_2SnH_2	$PhN{=}C{=}O$	$Et_2Sn[N(Ph)CHO]_2$	81
Et_3SnH	$PhN{=}C{=}S$	$Et_3SnSCH{=}NPh$	81
Bu_3SnH	$PhCH{=}NPh$	$PhN(SnBu_3)CH_2Ph$	68
Et_3SnH	$PhNO$	$Et_3SnONHPh^a$	68
Et_3SnH	$C_6H_{11}N{=}C{=}NC_6H_{11}$	$C_6H_{11}N{=}CHN(SnEt_3)C_6H_{11}$	68
Ph_3SnH	$PhN{=}NPh$	$PhN(SnPh_3)NHPh$	80

^{a} Obtained by evaporation, at 0°, of a solution of the reactants in acetone.

the C=N bond of isocyanates, anils and carbodiimides to give Sn—N links whereas addition to isothiocyanates occurs across the C=S bond. Although an adduct was isolated from triethyltin hydride and nitrosobenzene by evaporation of an acetone solution at 0°, when the same compounds were allowed to react spontaneously heat was evolved and azoxybenzene was formed [68]. The hydrostannation of isocyanates and isothiocyanates appears to occur by similar, ionic, mechanisms in which the rate-determining step is attack by hydride hydrogen on carbon [47, 56].

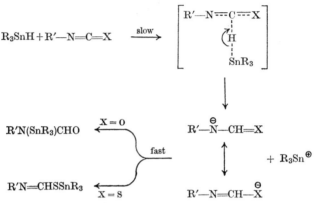

7.4. HYDROSTANNOLYSIS REACTIONS

Organotin hydrides undergo numerous metathetical reactions, often of great synthetic utility, which are summarised by the following equation.

$$R_3SnH + A—B \longrightarrow R_3SnA + BH$$

These are referred to as hydrostannolysis reactions.

The reduction of alkyl and aryl halides is the most widely studied hydrostannolysis reaction; some representative examples are given

$$R_3SnH + R'X \longrightarrow R_3SnX + R'H$$
$$(R' = alkyl, aryl; X = Cl, Br, I)$$

in Table V. It is clear that there are many circumstances in which an organotin hydride would be the most favourable reagent for the reduction of an organic halide. These reductions occur in high yields, often under mild conditions and possess the following additional advantages.

(a) Reduction of geminal di- and tri-halides occurs in a step-wise manner.

(b) Selective replacement of bromine in bromochloro compounds can be effected.

(c) Reduction of unsaturated halides occurs selectively [107].

$$CH_2{=}CHCH_2Br + Ph_3SnH \longrightarrow CH_2{=}CHMe + Ph_3SnBr$$

(d) Haloketones can also be reduced selectively [111].

$$PhCOCH_2Br + Bu_3SnH \longrightarrow PhCOMe + Bu_3SnBr$$

TABLE V

Reduction of organic halides by organotin hydrides

Hydride	Halide	Product	Reaction temp.[a]	Yield %	Reference
Bu₃SnH	PhCH₂Br	PhMe	Ex.	68	40
Bu₃SnH	PhCH₂Cl	PhMe	140°	78	111
Ph₃SnH	PhCHClMe	PhCH₂Me	Ex.	89	40
Bu₃SnH	PhCHCl₂	PhCH₂Cl	140°	70	111
Bu₃SnH	PhCCl₃	PhCHCl₂	Ex.	95	111
Bu₃SnH	CHBr₃	CH₂Br₂	Ex.	62	103
Et₃SnH	ClC=CCl(CF₂)ₙ	ClC=CH(CF₂)ₙ	100°	100	20
Bu₃SnH	C₇H₁₅I	C₇H₁₆	Ex.	96	111
Bu₃SnH	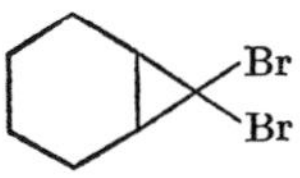 (bicyclic CHBr₂)	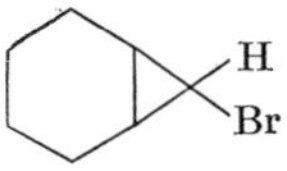 (bicyclic CHBrH)	Ex.	82	103
Bu₃SnH	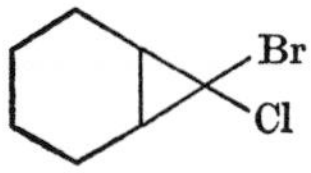 (bicyclic CHBrCl)	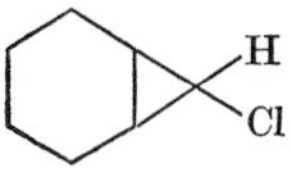 (bicyclic CHHCl)	Ex.	97	103
Ph₃SnH	PhBr	PhH	154°	90	38
Ph₃SnH	p-ClC₆H₄Br	PhCl	154°	97	38

[a] Ex. = exothermic reaction, the temperature was kept below 50°.

Among the few reactions where complications occur it is noted that attempted reduction of vicinal dibromides leads to elimination [38]

$$RCHBrCHBrR + R'_3SnH \longrightarrow RCH{=}CHR + H_2 + 2R'_3SnBr$$

and tributyltin hydride reduces the nitro group in *p*-nitrobenzyl bromide [38]. Reduction of a number of substituted propargyl chlorides with Bu₃SnH gave, in all cases, a mixture of the corresponding acetylene and the isomeric allene [22].

9

In general, bromides are reduced more readily than chlorides and aryl halides require much more vigorous conditions than their alkyl counterparts. The reactions are accelerated in the presence of free-radical sources and this and other evidence indicates that reduction occurs by a radical mechanism[38].

$$R_3SnH + S\cdot \longrightarrow R_3Sn\cdot + SH \qquad (S\cdot = \text{a radical initiator})$$
$$R_3Sn\cdot + R'X \longrightarrow R_3SnX + R'\cdot$$
$$R'\cdot + R_3SnH \longrightarrow R'H + R_3Sn\cdot$$

In this connection it is of interest that a mixture of the norbornenyl radical (*IV*) and nortricyclyl radical (*V*) could be generated by the action of triphenyltin hydride on either the norbornenyl halides (*VII*, X = Br, Cl) or the nortricyclyl halides (*VIII*, X = Br, Cl)[113]. (The mixture of radicals *IV* and *V* can, alternatively, be regarded as a single entity, i.e. the non-classical radical *VI*, unfortunately it was not possible to choose between these two possibilities.)

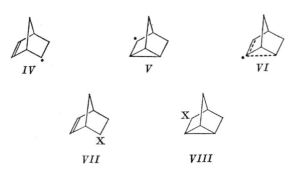

It would be anticipated that hydride reduction would not be stereospecific and, for example, the conversion of 1,1-dibromo-2,2,3-trimethylcyclopropane to the corresponding monobromo compound by tributyltin hydride gives a mixture of isomers[103]. However, some recent reductions of halofluorocyclopropanes, which are subject to catalysis by radical initiators, are stereospecific[4]:

$$(\overset{\cdot}{C}H_2)_4 \triangleright \overset{\cdot\cdot Cl}{\underset{F}{}} + Bu_3SnH \longrightarrow (\overset{\cdot}{C}H_2)_4 \triangleright \overset{\cdot\cdot H}{\underset{F}{}} + Bu_3SnCl$$

and it was suggested that the cyclopropyl radical may be constrained in a pyramidal configuration by coordination from fluorine to tin. Alternative explanations must be sought for the fact that the radical catalysed reduction of the mixed epimers of 7-chloro-7-phenyl-norcarane gave 80% of the *endo-* and only 1% of the *exo-*isomer[33].

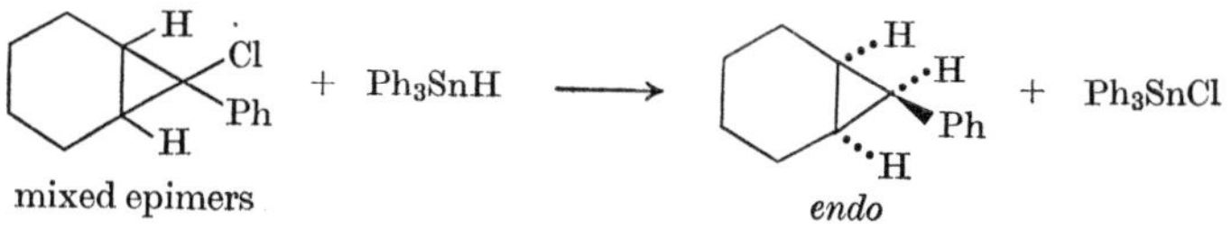

mixed epimers   *endo*

When either *cis-* or *trans-*9-chlorodecalin is reduced by one of a number of organotin hydrides the same mixture of *cis-* and *trans-*decalin (in which the *trans-*form predominates) results. It was concluded that, in the step:

$$R'\cdot + R_3SnH \longrightarrow R'H + R_3Sn\cdot$$

the activation energy for the abstraction of hydrogen from the hydride exceeds the energy barrier for isomerisation of the *cis-*lecalin radical to the *trans-*radical[27].

Reaction between organotin deuterides and aliphatic halides is an excellent specific method for the introduction of deuterium[27, 110]

$$Bu_3SnD + RX \longrightarrow Bu_3SnX + RD$$

and the method has been extended by using organotin deuterides or the reduction of carbonyl compounds[37].

Since lithium aluminium hydride rapidly converts organotin halides to the corresponding hydrides but reduces organic halides only slowly (by a nucleophilic process) it is often advantageous, in the reduction of an organic halide by $LiAlH_4$, to add a few mole percent of an organotin halide. The effective reducing agent is then the organotin hydride and, in many cases, there is a dramatic increase in the rate of reaction[39].

$$LiAlH_4 + 4R_3SnCl \longrightarrow LiAlCl_4 + 4R_3SnH$$
$$R_3SnH + R'X \longrightarrow R_3SnX + R'H \qquad (X = Cl, Br, I)$$

Organotin hydrides are far more effective in the reduction of halogeno compounds than are the corresponding organogermanium- and organosilicon-hydrides. This is clearly demonstrated when the insertion of the dichloromethylene group into an M—H bond (M = Si, Ge, Sn) using phenylbromodichloromethylmercury is attempted. The insertion reaction occurs readily with the silicon and germanium compounds:

$$Ph_3MH + PhHgCCl_2Br \longrightarrow Ph_3MCHCl_2 + PhHgBr \qquad (M = Si, Ge)$$

but fails with the tin hydride which preferentially reduces th
reagent [100, 101].

$$Bu_3SnH + PhHgCCl_2Br \longrightarrow Bu_3SnBr + PhHgCHCl_2$$

(Insertion can be achieved by the use of sodium trichloroacetate [12].

$$Bu_3SnH + Cl_3CCOONa \longrightarrow Bu_3SnCHCl_2 + CO_2 + NaCl)$$

The reduction of acyl halides by organotin hydrides is a much les
attractive synthetic procedure since a mixture of aldehyde and este
is formed.

$$R_3SnH + R'COCl \quad \Big\langle \begin{array}{l} \longrightarrow R'CHO \\ \\ \longrightarrow R'CH_2OCOR' \end{array}$$

There is some disagreement as to the factors which influence th
relative proportions of the two products. Thus it has been stated tha
when reductions are carried out in a solvent, the aldehyde pre
dominates [38]. Other workers report that reduction of a range of acy
halides by triphenyltin hydride with or without solvent gives th
ester as principal product [43, 44, 45]. It is generally agreed that free
radical reactions are involved and the formation of the aldehyde
thought to occur by a similar mechanism to that proposed for th
reduction of alkyl halides.

$$R_3SnH + S\cdot \longrightarrow R_3Sn\cdot + SH \text{ (}S\cdot = \text{a radical initiator)}$$
$$R_3Sn\cdot + R'COCl \longrightarrow R_3SnCl + R'\dot{C}O$$
$$R'\dot{C}O + R_3SnH \longrightarrow R'CHO + R_3Sn\cdot$$

The ester is considered to arise by attack of the acyl radical on th
aldehyde [112].

$$R'\dot{C}O + R'CHO \longrightarrow R'COO\dot{C}HR'$$
$$R'COO\dot{C}HR' + R_3SnH \longrightarrow R'COOCH_2R' + R_3Sn\cdot$$

By reactions which are thought to occur by a similar mechanis
it is possible to convert a ketone to an ester in quantitative yield b
treatment with an acyl halide and triphenyltin hydride [34], fo
example:

$$PhCOMe + MeCOCl + Ph_3SnH \longrightarrow PhMeCHOCOMe + Ph_3SnCl.$$

Another widely exploited group of hydrostannolysis reactions a
those which are used to form tin–tin bonds.

$$R_3SnH + R'_3SnA \longrightarrow R_3SnSnR' + HA$$

This type of reaction is fully discussed in Chapter 9 though we may note here that A in the above equation is usually nitrogen but can also be oxygen or certain other elements. Some illustrative examples are as follows [14].

$$Me_3SnN(Ph)CHO + Ph_3SnH \longrightarrow Me_3SnSnPh_3 + HN(Ph)CHO$$

$$2Me_3SnN(Ph)CHO + Ph_2SnH_2 \longrightarrow Me_3SnSnPh_2SnMe_3 + 2HN(Ph)CHO$$

$$Et_3SnN(Ph)CHO + Et_2SnH_2 \longrightarrow Et_3SnSnEt_2H + HN(Ph)CHO$$

There is good evidence that, in marked contrast to the nucleophilic hydrostannation reactions discussed earlier in this chapter, hydrostannolysis of organotin nitrogen compounds involves electrophilic attack by the hydride at the nitrogen atom [17].

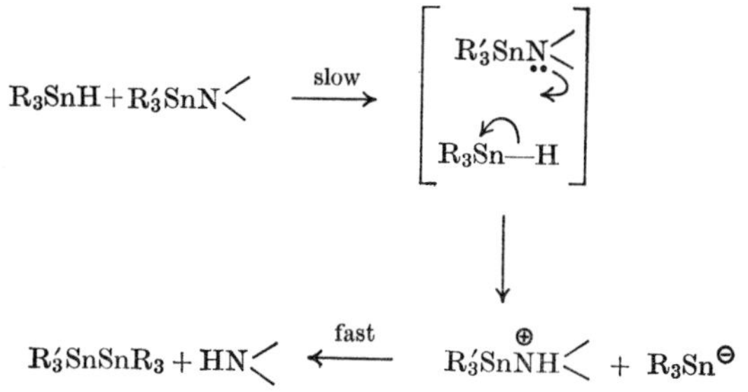

A final example of a hydrostannolysis reaction is the alkylation of organotin hydrides by organoaluminium compounds [98].

$$R_3SnH + R_3'Al \longrightarrow R_3SnR' + R_2'AlH$$

$$R_2SnH_2 + 2R_3'Al \longrightarrow R_2SnR_2' + 2R_2'AlH$$

$$RSnH_3 + 3R_3'Al \longrightarrow RSnR_3' + 3R_2'AlH$$

7.5. EXCHANGE REACTIONS

When an organotin hydride is treated with some other organotin species then an exchange reaction may occur instead of, or in addition to, the hydrostannolysis reaction referred to above.

$$R_3SnH + R_3'SnA \rightleftharpoons R_3SnA + R_3'SnH$$

In this equation A may be halogen [72], $R''COO$ [95], CN [75], OR'' [16], SR'' [16], $OSnR_3''$ [75], $SSnR_3''$ [75], PR_2'' [17, 75], AsR_2'' [17] and deuterium [76]. The relative rates of the hydrostannolysis and exchange reactions for a number of $R_3'SnA$ species have been examined [17]. When A = nitrogen, only hydrostannolysis is observed and when A = SR'' exchange occurs exclusively but where A = PPh$_2$ or AsPh$_2$ the two reactions occur at comparable rates and mixtures of products are obtained.

Perhaps the most intensively studied exchange reaction is that occurring between dihydrides and dihalides.

$$R_2SnH_2 + R_2SnX_2 \rightleftharpoons 2R_2SnXH$$

This reaction occurs when X = F, Cl, Br and I, and, for R = Bu, nmr studies indicate that the equilibria lie almost entirely to the right[62]. Reactions of monohydrides with dichlorides have also been studied[94].

$$Bu_3SnH + R_2SnCl_2 \rightleftharpoons Bu_3SnCl + R_2SnClH$$

When a dihydride and a dichloride which have different groups attached to tin react the following equilibria occur[96].

$$R_2SnH_2 + R_2'SnCl_2 \rightleftharpoons R_2SnClH + R_2'SnClH \rightleftharpoons R_2SnCl_2 + R_2'SnH_2$$

The Bu_2SnXH compounds are rather less stable than dibutyltin dihydride at room temperature; at 100° decomposition is complete over periods ranging from 20 minutes for X = F to 6 hours for X = Br[93]. It is thought that the chloro compound decomposes to give a distannane and hydrogen[93].

$$2Bu_2SnClH \longrightarrow Bu_2Sn(Cl)SnBu_2Cl + H_2$$

The halohydrides can also be prepared by reaction between the hydrides and a hydrogen halide.

$$Bu_2SnH_2 + HCl \longrightarrow Bu_2SnClH + H_2 \qquad \text{(ref. 93)}$$
$$EtSnH_3 + HBr \longrightarrow EtSnBrH_2 + H_2 \qquad \text{(ref. 25)}$$

At −78° hydrogen bromide effects tin–carbon bond cleavage in triphenyltin hydride to give diphenyltin bromohydride which decomposes on warming to room temperature[25].

$$Ph_3SnH + HBr \longrightarrow Ph_2SnBrH + PhH$$

The halohydrides, like other organotin hydrides, undergo olefin addition, thus di-isobutyltin chlorohydride adds to acrylonitrile in the presence of azo*bis*isobutyronitrile[118].

$$(i\text{-}Bu)_2SnClH + CH_2{=}CHCN \longrightarrow (i\text{-}Bu)_2Sn(Cl)CH_2CH_2CN$$

Carbon–mercury bond fission occurs when certain organomercury compounds are treated with the chlorohydrides[5].

$$Et_2SnClH + Hg(CH_2COOMe)_2 \longrightarrow Et_2Sn(Cl)CH_2COOMe + Hg + MeCOOMe$$

Although the equilibrium:

$$R_3SnH + R_3'PbX \rightleftharpoons R_3SnX + R_3'PbH$$

may not favour the formation of the lead hydride the fact that the latter undergoes addition reactions much more readily than the tin compound can be utilised in the preparation of lead hydride adducts [15].

$$Bu_3PbOAc + Ph_3SnH + HC{\equiv}CCN \nearrow\searrow \begin{array}{l} Bu_3PbCH{=}CHCN + Ph_3SnOAc \\ \\ CH_2{=}C(CN)PbBu_3 + Ph_3SnOAc \end{array}$$

7.6. MISCELLANEOUS REACTIONS

Neumann and his coworkers have investigated the ready decomposition of diacyl peroxides which occurs in the presence of organotin hydrides [73]. By determining the relative proportions of the products of this decomposition it was shown that the principal reactions occurring are as follows.

$$RCOOOCOR \longrightarrow 2RCOO\cdot$$
$$RCOO\cdot + R_3'SnH \longrightarrow RCOOH + R_3'Sn\cdot$$
$$RCOOOCOR + R_3'Sn\cdot \longrightarrow RCOOSnR_3' + RCOO\cdot$$
$$RCOOH + R_3'SnH \longrightarrow RCOOSnR_3' + H_2$$

In fact the carboxylic acid RCOOH successfully competes with the acid radical RCOO $\cdot$ for the hydride so the production of hydrogen and $RCOOSnR_3'$ is a major pathway in the decomposition [73] and this is the reason why, for example, benzoyl peroxide does not catalyse the addition of triphenyltin hydride to olefins [12]. Photolysis of R_3SnH in vacuo gave the radicals $R_3Sn \cdot$ (R = Ph, i-Bu) which were observed by electron spin resonance measurements [97].

The reaction of dibutyltin dihydride with benzoyl peroxide depends upon the molar ratio of the reactants [109].

$$Bu_2SnH_2 + PhCOOOCOPh \longrightarrow H_2 + Bu_2Sn(OCOPh)_2$$
$$2Bu_2SnH_2 + PhCOOOCOPh \longrightarrow 2H_2 + Bu_2Sn(OCOPh)Sn(OCOPh)Bu_2$$

A number of products are formed when organotin hydrides react with azo compounds; thus equimolar proportions of triethyltin hydride and 1,1'-dicyanoazocyclohexane reacted in toluene at 100° to give the following products [79].

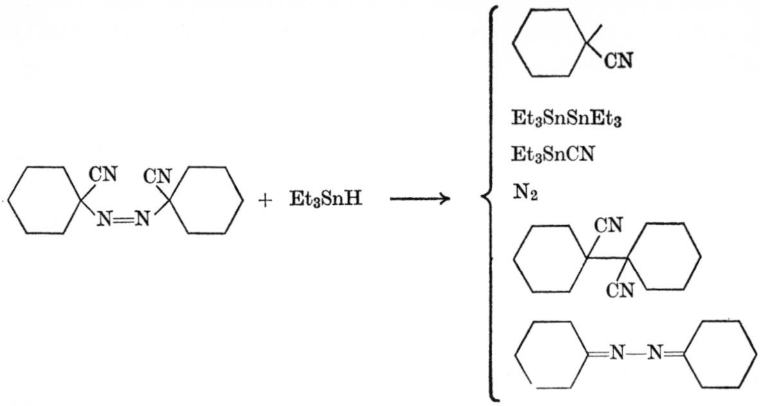

Rather more than half of the azo compound decomposed with loss of nitrogen giving cyclohexyl cyanide as principal product. A radical mechanism seems likely for this reaction and a radical mechanism has also been proposed for the decomposition of phenyl phenylazo-sulphone induced by triethyltin hydride [69].

$$PhN{=}NSO_2Ph + Et_3Sn\cdot \longrightarrow Ph\cdot + N_2 + PhSO_2SnEt_3$$
$$Et_3SnH + Ph\cdot \longrightarrow PhH + Et_3Sn\cdot$$

Decomposition is accompanied by a certain amount of the addition reaction.

$$PhN{=}NSO_2Ph + Et_3SnH \longrightarrow PhNHN(SnEt_3)SO_2Ph$$

TABLE VI

Physical properties of some organotin hydrides

Compound	m.p.	b.p.	n_D^{20}	d^{20}
Me₃SnH	−70°	59°	—	—
Et₃SnH	—	146°	1·4707	1·258
Bu₃SnH	—	76–81°/0·7–0·9 mm	1·4721 (22°)	—
Ph₃SnH	29°	168–172°/0·5 mm	1·6342 (25°)	1·3771 (25°)
Me₂SnH₂	—	33–40°	1·4480	1·4766
Et₂SnH₂	—	96–98°	—	—
Bu₂SnH₂	—	85°/19 mm	—	—
Ph₂SnH₂	—	89–93°/0·3 mm	1·5950	—
MeSnH₃	—	1·4°	—	—
EtSnH₃	—	25°	1·4491	—
BuSnH₃	—	99–101°	1·4609	—
PhSnH₃	—	35°/2·5 mm	—	—
Et₃SnSnEt₂H	—	—	1·5876	—

Nitrogen is evolved when benzoyl azide reacts with tributyltin hydride and the initial product is the N-stannylbenzamide[24].

$$PhCON_3 + Bu_3SnH \longrightarrow N_2 + PhCONHSnBu_3$$

In a rather different type of reaction organotin hydrides interact with epoxides and, from 1,2-epoxybut-3-en and triethyltin hydride two isomeric products are formed in equal amounts[11].

$$2CH_2\text{---}CHCH=CH_2 + 2Et_3SnH \longrightarrow MeCH=CHCH_2OH + CH_2=CHCH_2CH_2OH$$
$$\diagdown O \diagup$$
$$+ Et_3SnSnEt_3$$

References

1. Akhtar, M. and Clark, H. C., *Can. J. Chem.*, 1968, **46**, 633.
2. Akhtar, M. and Clark, H. C., *Can. J. Chem.*, 1968, **46**, 2165.
3. Amberger, E. and Kula, M. R., *Chem. Ber.*, 1963, **96**, 2560.
4. Ando, T., Namigata, F., Yamanaka, H. and Funasaka, W., *J. Amer. Chem. Soc.*, 1967, **89**, 5719.
5. Baukov, Yu. I., Belavin, I. Yu. and Lutsenko, I. F., *Zh. Obshch. Khim.*, 1965, **35**, 1092.
6. Bellegarde, B., Pereyre, M. and Valade, J., *Bull. Soc. Chim. France*, 1967, 3082.
7. Beumel, O. F., Ph.D. Thesis, University of New Hampshire, Durham, New Hampshire, 1960.
8. Beveridge, A. D., Clark, H. C. and Kwon, J. T., *Can. J. Chem.*, 1966, **44**, 179.
9. Birnbaum, E. R. and Javora, P. H., *J. Organometallic Chem.*, 1967, **9**, 379.
10. Braun, J., *Compt. Rend.*, 1965, **260**, 218.
11. Bryskovskaya, A. V., Al'bitskaya, V. M. and Petrov, A. A., *Zh. Organ. Khim.*, 1965, **1**, 1898 (*Chem. Abs.*, 1966, **64**, 3446d).
12. Choa-Lun Ts'eng, Jen-Hsi Cho and Shun-Chun Ma, *K'o Hsueh T'ung Pao*, 1966, **17** (2), 77 (*Chem. Abs.*, 1967, **66**, 28862u).
13. Considine, W. J. and Ventura, J. J., *Chem. and Ind.*, 1962, 1683.
14. Creemers, H. M. J. C., Ph.D. Thesis, University of Utrecht, 1967.
15. Creemers, H. M. J. C., Leusink, A. J., Noltes, J. G. and van der Kerk, G. J. M., *Tetrahedron Letters*, 1966, 3167.
16. Creemers, H. M. J. C. and Noltes, J. G., *Rec. Trav. Chim.*, 1965, **84**, 1589.
17. Creemers, H. M. J. C., Verbeek, F. and Noltes, J. G., *J. Organometallic Chem.*, 1967, **8**, 469.
18. Cullen, W. R. and Styan, G. E., *Inorg. Chem.*, 1965, **4**, 1437.
19. Cullen, W. R. and Styan, G. E., *J. Organometallic Chem.*, 1966, **6**, 117.
20. Cullen, W. R. and Styan, G. E., *J. Organometallic Chem.*, 1966, **6**, 633.
21. Del Franco, G. J., Resnick, P. and Dillard, C. R., *J. Organometallic Chem.*, 1965, **4**, 57.
21a. Delmas, M., Maire, J. C. and Pinzelli, R., *J. Organometallic Chem.*, 1969, **16**, 83.
22. Fantazier, R. M. and Poutsma, M. L., *J. Amer. Chem. Soc.*, 1968, **90**, 5490.

23. Frankel, M., Wagner, D., Gertner, D. and Zilkha, A., *Israel J. Chem.*, 1966, **4**, 183 (*Chem. Abs.*, 1967, **66**, 52043y).
24. Frankel, M., Wagner, D., Gertner, D. and Zilkha, A., *J. Organometallic Chem.*, 1967, **7**, 518.
25. Fritz, G. and Scheer, H., *Z. Anorg. Allgem. Chem.*, 1965, **338**, 1.
26. Fulton, R. F., Ph.D. Thesis, Purdue University, Lafayette, Indiana, 1960 (*Dissertation Abs.*, 1962, **22**, 3397).
27. Greene, F. D. and Lowry, N. L., *J. Org. Chem.*, 1967, **32**, 882.
28. Hayashi, K., Iyoda, J. and Shiihara, I., *J. Organometallic Chem.*, 1967, **10**, 81.
29. Henry, M. C. and Noltes, J. G., *J. Amer. Chem. Soc.*, 1960, **82**, 558.
30. Ingham, R. K. and Gilman, H., *Inorganic Polymers*. Ed. Stone, F. G. A. and Graham, W. A. G., Academic Press, New York and London, 1961, p. 376.
31. Issleib, K. and Walter, B., *J. Organometallic Chem.*, 1967, **10**, 177.
32. Itoi, K. and Kumano, S., *Kogyo Kagaku Zasshi*, 1967, **70**, 82 (*Chem. Abs.*, 1967, **67**, 11556v).
33. Jensen, F. R. and Patterson, D. B., *Tetrahedron Letters*, 1966, 3837.
34. Kaplan, L., *J. Amer. Chem. Soc.*, 1966, **88**, 1833; 4970.
35. Kazankova, M. A., Protsenko, N. P. and Lutsenko, I. F., *Zh. Obshch. Khim.*, 1968, **38**, 106 (*Chem. Abs.*, 1968, **69**, 67501).
36. Kraihanzel, C. S. and Losee, M. L., *J. Organometallic Chem.*, 1967, **10**, 427.
37. Kuehlein, K., Neumann, W. P. and Mohring, H., *Angew. Chem. Intern. Ed. Engl.*, 1968, **7**, 455.
38. Kuivila, H. G., *Advances in Organometallic Chemistry*, Academic Press, New York and London, 1964, **1**, 47.
39. Kuivila, H. G. and Menapace, L. W., *J. Org. Chem.*, 1963, **28**, 2165.
40. Kuivila, H. G., Menapace, L. W. and Warner, C. R., *J. Amer. Chem. Soc.*, 1962, **84**, 3584.
41. Kuivila, H. G., Rahman, W. and Fish, R. H., *J. Amer. Chem. Soc.*, 1965, **87**, 2835.
42. Kuivila, H. G. and Sommer, R., *J. Amer. Chem. Soc.*, 1967, **89**, 5616.
43. Kuivila, H. G. and Walsh, E. J., *J. Amer. Chem. Soc.*, 1966, **88**, 571.
44. Kupchik, E. J. and Kiesel, R. J., *J. Org. Chem.*, 1964, **29**, 3690.
45. Kupchik, E. J. and Kiesel, R. J., *J. Org. Chem.*, 1966, **31**, 456.
46. Laliberte, B. R., Davidsohn, W. and Henry, M. C., *J. Organometallic Chem.*, 1966, **5**, 526.
47. Leusink, A. J., Ph.D. Thesis, University of Utrecht, 1966.
48. Leusink, A. J. and Budding, H. A., *J. Organometallic Chem.*, 1968, **11**, 533.
49. Leusink, A. J., Budding, H. A. and Drenth, W., *J. Organometallic Chem.*, 1967, **9**, 295.
50. Leusink, A. J., Budding, H. A. and Drenth, W., *J. Organometallic Chem.*, 1968, **11**, 541.
51. Leusink, A. J., Budding, H. A. and Drenth, W., *J. Organometallic Chem.*, 1968, **13**, 163.
52. Leusink, A. J., Budding, H. A. and Marsman, J. W., *J. Organometallic Chem.*, 1967, **9**, 285.
53. Leusink, A. J., Budding, H. A. and Marsman, J. W., *J. Organometallic Chem.*, 1968, **13**, 155.

54. Leusink, A. J. and Marsman, J. W., *Rec. Trav. Chim.*, 1965, **84**, 1123.
55. Leusink, A. J., Marsman, J. W. and Budding, H. A., *Rec. Trav. Chim.*, 1965, **84**, 689.
55a. Leusink, A. J. and Noltes, J. G., *J. Organometallic Chem.*, 1969, **16**, 91.
56. Leusink, A. J. and Noltes, J. G., *Rec. Trav. Chim.*, 1965, **84**, 585.
57. Leusink, A. J. and Noltes, J. G., *Tetrahedron Letters*, 1966, 335.
58. Leusink, A. J. and Noltes, J. G., *Tetrahedron Letters*, 1966, 2221.
59. Leusink, A. J., Noltes, J. G., Budding, H. A. and van der Kerk, G. J. M., *Rec. Trav. Chim.*, 1964, **83**, 1036.
60. Leusink, A. J., Noltes, J. G., Budding, H. A. and van der Kerk, G. J. M., *Rec. Trav. Chim.*, 1965, **84**, 567.
61. Mal'tseva, E. N., Zavgorodnii, V. S., Maretina, I. A. and Petrov, A. A., *Zh. Obshch. Khim.*, 1968, **38**, 203 (*Chem. Abs.*, 1968, **69**, 52240).
62. Moedritzer, K., *Organometallic Chem. Rev.*, 1966, **1**, 179.
63. Nesmeyanov, A. N. and Borisov, A. E., *Dokl. Akad. Nauk S.S.S.R.*, 1967, **174**, 96.
64. Nesmeyanov, A. N. and Borisov, A. E., *Izv. Akad. Nauk S.S.S.R., Ser. Khim.*, 1967, 226.
65. Nesmeyanov, A. N., Borisov, A. E. and Wang, S.-H., *Izv. Akad. Nauk S.S.S.R., Ser. Khim.*, 1967, 1141.
66. Neumann, W. P., Albert, H. J. and Kaiser, W., *Tetrahedron Letters*, 1967, 2041.
67. Neumann, W. P. and Heymann, E., *Annalen*, 1965, **683**, 11.
68. Neumann, W. P. and Haymann, E., *Annalen*, 1965, **683**, 24.
69. Neumann, W. P. and Lind, H., *Angew. Chem. Intern. Ed. Engl.*, 1967, **6**, 76.
70. Neumann, W. P., Niermann, H. and Schneider, B., *Annalen*, 1967, **707**, 15.
71. Neumann, W. P., Niermann, H. and Sommer, R., *Annalen*, 1962, **659**, 27.
72. Neumann, W. P. and Pedain, J., *Tetrahedron Letters*, 1964, 2461.
73. Neumann, W. P., Ruebsamen, K. and Sommer, R., *Chem. Ber.*, 1967, **100**, 1063.
74. Neumann, W. P. and Schneider, B., *Annalen*, 1967, **707**, 20.
75. Neumann, W. P., Schneider, B. and Sommer, R., *Annalen*, 1966, **692**, 1.
76. Neumann, W. P. and Sommer, R., *Angew. Chem.*, 1963, **75**, 788.
77. Neumann, W. P. and Sommer, R., *Annalen*, 1964, **675**, 10.
78. Neumann, W. P. and Sommer, R., *Annalen*, 1967, **701**, 28.
79. Neumann, W. P., Sommer, R. and Lind, H., *Annalen*, 1965, **688**, 14.
80. Noltes, J. G., *Rec. Trav. Chim.*, 1964, **83**, 515.
81. Noltes, J. G. and Janssen, M. J., *J. Organometallic Chem.*, 1964, **1**, 346.
82. Noltes, J. G. and van der Kerk, G. J. M., *Chimia*, 1962, **16**, 122.
83. Noltes, J. G. and van der Kerk, G. J. M., *Rec. Trav. Chim.*, 1961, **80**, 623.
84. Noltes, J. G. and van der Kerk, G. J. M., *Rec. Trav. Chim.*, 1962, **80**, 623.
85. Noltes, J. G. and van der Kerk, G. J. M., *Rec. Trav. Chim.*, 1962, **81**, 41.
86. Ohara, M. and Okawara, R., *J. Organometallic Chem.*, 1965, **3**, 484.
87. Omae, I., Ohnishi, S. and Matsuda, S., *Kogyo Kagaku Zasshi*, 1967, **70**, 1755 (*Chem. Abs.*, 1968, **68**, 87371).
88. Pereyre, M., Colin, G. and Valade, J., *Tetrahedron Letters*, 1967, 4805.
89. Pereyre, M. and Valade, J., *Bull. Soc. Chim. France*, 1967, 1928.
90. Pereyre, M. and Valade, J., *Compt. Rend.*, 1965, **260**, 581.

91. Pommier, J. C. and Valade, J., *Bull. Soc. Chim. France*, 1965, 975.
92. Reifenberg, G. H. and Considine, W. J., *J. Organometallic Chem.*, 1967, **9**, 505.
93. Sawyer, A. K., Brown, J. E. and Hanson, E. L., *J. Organometallic Chem.*, 1965, **3**, 464.
94. Sawyer, A. K., Brown, J. E. and May, G. S., *J. Organometallic Chem.*, 1968, **11**, 192.
95. Sawyer, A. K. and Kuivila, H. G., *J. Org. Chem.*, 1962, **27**, 837.
96. Sawyer, A. K., May, G. S. and Scofield, R. E., *J. Organometallic Chem.*, 1968, **14**, 213.
97. Schmidt, U., Kabitzke, K., Markau, K. and Neumann, W. P., *Chem. Ber.*, 1965, **98**, 3827.
98. Schneider, B. and Neumann, W. P., *Annalen*, 1967, **707**, 7.
99. Sevestre, J., *Peintures, Pigments, Vernis*, 1965, **41**, 361 (*Chem. Abs.*, 1965, **63**, 14896e).
100. Seyferth, D., Burlitch, J. M., Dertouzos, H. and Simmons, H. D., *J. Organometallic Chem.*, 1967, **7**, 405.
101. Seyferth, D., Simmons, H. D. and Todd, L. J., *J. Organometallic Chem.*, 1964, **2**, 282.
102. Seyferth, D. and Vaughan, L. G., *J. Organometallic Chem.*, 1963, **1**, 138.
103. Seyferth, D., Yamazaki, H. and Alleston, D. L., *J. Org. Chem.*, 1963, **28**, 703.
104. Sommer, R. and Kuivila, H. G., *J. Org. Chem.*, 1968, **33**, 802.
105. van der Kerk, G. J. M., Luijten, J. G. A. and Noltes, J. G., *Chem. and Ind.*, 1956, 352.
106. van der Kerk, G. J. M. and Noltes, J. G., *J. Appl. Chem.*, 1959, **9**, 106.
107. van der Kerk, G. J. M., Noltes, J. G. and Luijten, J. G. A., *J. Appl. Chem.*, 1957, **7**, 356.
108. Vlasov, V. M., Mirskov, R. G. and Petrova, V. N., *Zh. Obshch. Khim.*, 1967, **37**, 954.
109. Vyazankin, N. S. and Bychkov, V. T., *Zh. Obshch. Khim.*, 1965, **35**, 684.
110. Wahren, M., Haedge, P., Huebner, H. and Muehlstaedt, M., *Isotopenpraxis*, 1965, **1**, 65 (*Chem. Abs.*, 1966, **65**, 743e).
111. Walsh, E. J., Unpublished (quoted in ref. 38).
112. Walsh, E. J. and Kuivila, H. G., *J. Amer. Chem. Soc.*, 1966, **88**, 576.
113. Warner, C. R., Strunk, R. J. and Kuivila, H. G., *J. Org. Chem.*, 1966, **31**, 3381.
114. Zavgorodnii, V. S. and Petrov, A. A., *Zh. Obshch. Khim.*, 1965, **35**, 1313.
115. Belgium Patent 638,642, April 14, 1964 (*Chem. Abs.*, 1965, **62**, 11455d).
116. French Patent 1,368,522, July 31, 1964 (*Chem. Abs.*, 1965, **62**, 2794g).
117. French Patent 1,411,034, Sept. 17, 1965 (*Chem. Abs.*, 1966, **64**, 5137a).
118. German Patent 1,214,237, April 14, 1966 (*Chem. Abs.*, 1966, **65**, 5490f).
119. Japanese Patent 6737, April 18, 1966 (*Chem. Abs.*, 1966, **65**, 5490c).
120. Japanese Patent 4560, Feb. 25, 1967 (*Chem. Abs.*, 1967, **67**, 32779k).
121. Japanese Patent 10,133, Feb. 26, 1968 (*Chem. Abs.*, 1968, **69**, 106879).
122. Netherlands Patent 6,514,261, May 8, 1966 (*Chem. Abs.*, 1966, **65**, 15538d).
123. U.S. Patent 3,188,331, June 8, 1965 (*Chem. Abs.*, 1965, **63**, 5676c).
124. U.S. Patent 3,208,978, Sept. 28, 1965 (*Chem. Abs.*, 1965, **63**, 18372h).
125. U.S.S.R. Patent 196,839, May 31, 1967 (*Chem. Abs.*, 1968, **68**, 49781).

8

COMPOUNDS CONTAINING TIN–SULPHUR BONDS

The organotin sulphur derivatives include several interesting groups of compounds which attract increasing attention. Work in this field has been stimulated by the discovery in 1950 of the remarkable efficacy of compounds of the type $R_2Sn(SR')_2$ in stabilising poly(vinyl chloride) against degradation. Organometallic sulphur compounds containing silicon, germanium, tin and lead were reviewed in 1967[1] and some 270 organotin sulphur compounds were tabulated. However, many of these compounds have been reported in the patent literature only, with no characterisation or description of physical properties.

The various types of compound containing Sn—S bonds are conveniently classified into two main groups depending upon whether the tin–bound sulphur atom is attached to carbon or to some other element. A relatively small number of organotin derivatives of selenium and tellurium are known and are also discussed in this chapter.

8.1. COMPOUNDS CONTAINING Sn—S—C GROUPS

The organotin mercaptides, $R_{4-n}Sn(SR')_n$, are the largest group of organotin sulphur compounds, and there has been much interest in cyclic compounds derived from dithiols.

A widely used preparative method is the reaction between an organotin chloride and sodium mercaptide in ethanol.

$$R_{4-n}SnCl_n + nR'SNa \longrightarrow R_{4-n}Sn(SR')_n + nNaCl$$

This procedure gives good yields but suffers from the disadvantage that the sodium chloride is produced in a very finely divided form and can only be removed by centrifuging the reaction mixture. It has recently been shown that organotin chlorides and mercaptans will react satisfactorily in the presence of aqueous sodium hydroxide and a range of compounds, $R_{4-n}Sn(SR')_n$ ($n = 1, 2, 3$), has been prepared by this method[3]. Alternatively the reaction can be carried out under

TABLE I

The preparation of some organotin mercaptides

Organotin Compound	Thiol	Conditions	Product	Reference
Ph_3SnCl	$PhCOSH$	non-aqueous, in presence of organic base	$Ph_3SnSCOPh$	9
Ph_3SnOH	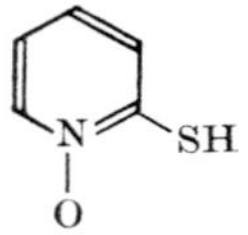 2-mercaptopyridine-SH (N-oxide)	water removed azeotropically with benzene	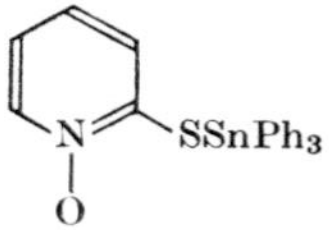 pyridine-$SSnPh_3$ (N-oxide)	51
$(Bu_3Sn)_2O$	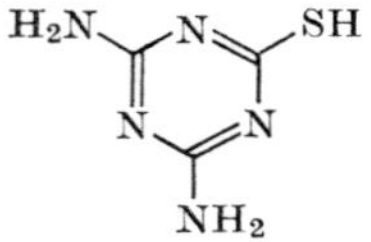 H_2N–triazine–SH, NH_2	water removed azeotropically with toluene	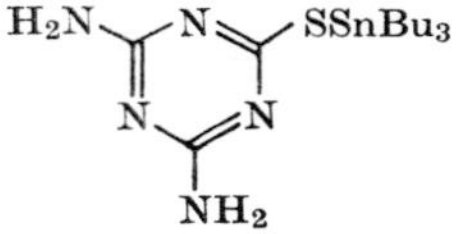 H_2N–triazine–$SSnBu_3$, NH_2	50
Ph_3SnCl	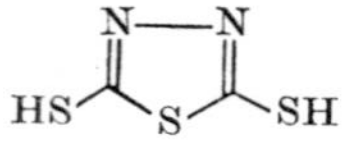 thiadiazole HS–SH	sodium mercaptide in methanol	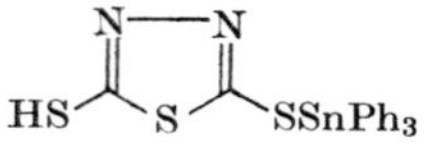 thiadiazole HS–$SSnPh_3$	56
Bu_2SnO	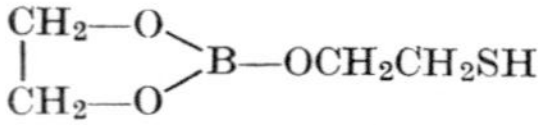 $\begin{array}{c}CH_2\text{--}O\\ \quad\quad B\text{--}OCH_2CH_2SH\\ CH_2\text{--}O\end{array}$	water removed azeotropically with toluene	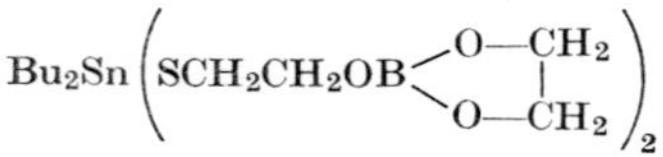 $Bu_2Sn\left(SCH_2CH_2OB\begin{array}{c}O\text{--}CH_2\\ O\text{--}CH_2\end{array}\right)_2$	54
R_2SnCl_2	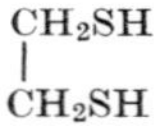 $\begin{array}{c}CH_2SH\\ CH_2SH\end{array}$	sodium mercaptide in ethanol	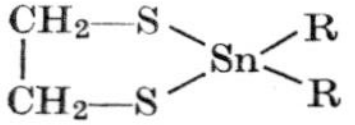 $\begin{array}{c}CH_2\text{--}S\\ \quad\quad Sn\begin{array}{c}R\\ R\end{array}\\ CH_2\text{--}S\end{array}$ ($R = Bu, Ph$)	19, 20
Ph_2SnCl_2	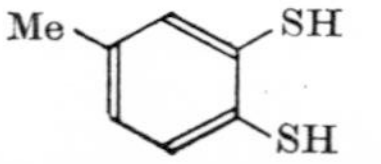 Me–benzene–SH, SH	sodium mercaptide in ethanol	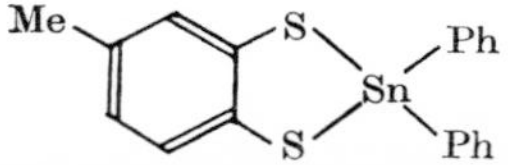 Me–benzene–S–$Sn(Ph)(Ph)$–S	19, 20

R_2SnCl_2	 $\begin{array}{c}CHSH \\ \parallel \\ CHSH\end{array}$	aqueous NaOH	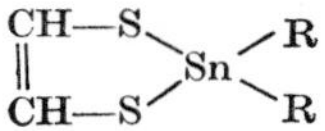 $\begin{array}{c}CH-S \\ \parallel \qquad \diagdown Sn \diagup{}^R_R \\ CH-S\end{array}$	(R = Me, Et, $CH_2=CH$)	7
Bu_2SnCl_2	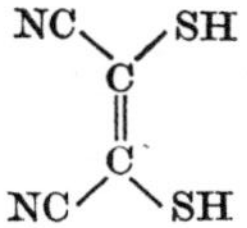 $\begin{array}{c}NC \diagdown{}_{C} \diagup {}^{SH} \\ \parallel \\ NC \diagup{}^{C} \diagdown{}_{SH}\end{array}$	sodium mercaptide in ethanol	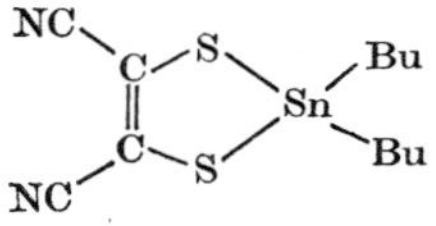 $\begin{array}{c}NC \diagdown C \diagup {}^{S} \diagdown {}^{Bu}_{Sn} \diagup {}^{Bu} \\ \parallel \qquad \diagup \\ NC \diagup C \diagdown {}_{S}\end{array}$		60
R_2SnCl_2	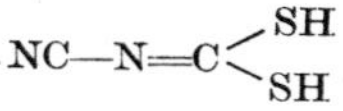 $NC-N=C \diagup{}^{SH} \diagdown{}_{SH}$	potassium salt in dimethylformamide	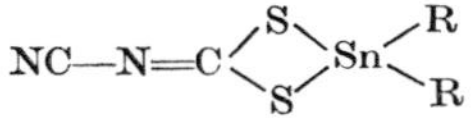 $NC-N=C \diagup{}^{S} \diagdown{}_{S} Sn \diagup{}^{R}_{R}$	(R = Bu, PhCH$_2$)	40, 61
R_2SnCl_2	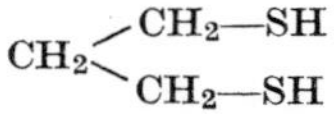 $CH_2 \diagdown{}_{CH_2-SH}^{CH_2-SH}$	sodium mercaptide in ethanol	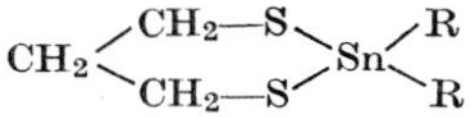 $CH_2 \diagdown{}_{CH_2-S}^{CH_2-S} Sn \diagup{}^{R}_{R}$	(R = Bu, Ph)	19, 20

non-aqueous conditions in the presence of a base such as pyridine or triethylamine, the base hydrochloride being readily removed from the reaction mixture [9, 48, 49]. A variation of this procedure, suitable for preparing dimercaptides, is reaction between a thiol and the amine complex of an organotin dichloride [55].

$$Bu_2SnCl_2 \cdot 2PhNH_2 + 2HSCH_2COOC_8H_{17}\text{-i} \longrightarrow$$
$$Bu_2Sn(SCH_2COOC_8H_{17}\text{-i})_2 + 2PhNH_2 \cdot HCl$$

Organotin oxides and hydroxides also react with mercaptans.

$$R_3SnOSnR_3 + 2R'SH \longrightarrow 2R_3SnSR' + H_2O$$

The water formed is removed azeotropically with benzene or toluene [50, 51, 54, 59] or, in the absence of solvent, by using magnesium sulphate [18]. There is a report, however, that alkyl thioglycollates react in aqueous solution with dibutyltin oxide to give the corresponding dimercaptides [53].

$$2HSCH_2COOR + Bu_2SnO \longrightarrow Bu_2Sn(SCH_2COOR)_2 + H_2O$$

Organotin mercaptides can also be made from the alkoxides [41].

$$Bu_2Sn(OEt)_2 + 2PhSH \longrightarrow Bu_2Sn(SPh)_2 + 2EtOH$$

Further examples, which indicate the range of compounds which can be prepared by these methods, are given in Table I.

Other reactions in which thiolatotin compounds are produced but which are not widely used as preparative procedures include those in which Sn—C bond cleavage is effected by a mercaptan [28].

$$Pr_4Sn + PhCH_2SH \longrightarrow Pr_3SnSCH_2Ph + PrH$$

Organotin hydrides react with thiols to give mercaptides and hydrogen [17].

$$Ph_3SnH + PhSH \longrightarrow Ph_3SnSPh + H_2$$

Trimethylstannyldiethylamine undergoes protolysis with butanethiol giving the thiolatotin compound in 92% yield [2].

$$Me_3SnNEt_2 + BuSH \longrightarrow Me_3SnSBu + Et_2NH$$

Alkylthio groups can be transferred from silicon to tin and from lead to tin [2].

$$Me_3SiSR + Me_3SnCl \longrightarrow Me_3SnSR + Me_3SiCl$$
$$Pb(SMe)_2 + 2Me_3SnCl \longrightarrow 2Me_3SnSMe + PbCl_2$$

When tetraphenyltin and sulphur interacted no phenylthiolatotin compounds were obtained but the corresponding reaction with selenium gave phenylselenatotriphenyltin in 47% yield [32].

$$Ph_4Sn + Se \longrightarrow Ph_3SnSePh$$

In addition to simple thiolatotin derivatives, other types of compounds containing Sn—S—C links include thiocarboxylates such as $R_3SnSCOR'$ (an example of which is given in Table I) and the dithiocarbamates. The latter compounds are frequently made from sodium dithiocarbamate and an organotin chloride [8, 13].

$$R_3SnCl + R_2'NCSSNa \longrightarrow R_3SnSCSNR_2' + NaCl$$
$$R_2SnCl_2 + 2R_2'NCSSNa \longrightarrow R_2Sn(SCSNR_2')_2 + 2NaCl$$

These compounds can also be obtained from an organotin oxide or hydroxide, a primary or secondary amine and carbon disulphide [8, 13].

$$(R_3Sn)_2O + 2R_2'NH + 2CS_2 \longrightarrow 2R_3SnSCSNR_2' + H_2O$$
$$R_2SnO + 2R_2'NH + 2CS_2 \longrightarrow R_2Sn(SCSNR_2')_2 + H_2O$$

It may be that these reactions occur by the transformation of the oxide to the stannylamine since it is known that stannylamines will add across a C=S bond in carbon disulphide [11], for example:

$$Me_3SnNMe_2 + CS_2 \longrightarrow Me_3SnSCSNMe_2.$$

8.1.1. *Properties of compounds containing the Sn—S—C group*

In the presence of aqueous alkali, cleavage of the Sn—S bond occurs [21].

$$Ph_3SnSPh + NaOH \longrightarrow Ph_3SnOH + PhSNa$$

The simple mercaptides are not sensitive to water, under neutral conditions, but some of the cyclic derivatives of the dithiols are attacked by moisture, particularly when dissolved in water-miscible solvents such as acetone. It is possible that the synthesis of thiolatotin compounds under aqueous conditions [3, 7] is effective because the low solubility of the products in water reduces the rate of hydrolysis.

Alkylthio and arylthio groups can be smoothly transferred from tin to other elements by treatment with a halide of the element as in the following examples.

$$3Me_3SnSBu + BCl_3 \longrightarrow B(SBu)_3 + 3Me_3SnCl \qquad (ref. 4)$$
$$3Me_3SnSPh + PCl_3 \longrightarrow P(SPh)_3 + 3Me_3SnCl \qquad (ref. 4)$$
$$Ph_3SnSPh + HgCl_2 \longrightarrow ClHgSPh + Ph_3SnCl \qquad (ref. 21)$$

10

In none of the above examples was there any evidence for the formation of an adduct between the mercaptide and the chloro compound but a chelate complex was isolated from the reaction of *bis*(methyl-thiolato)dimethyltin and chromium hexacarbonyl [5].

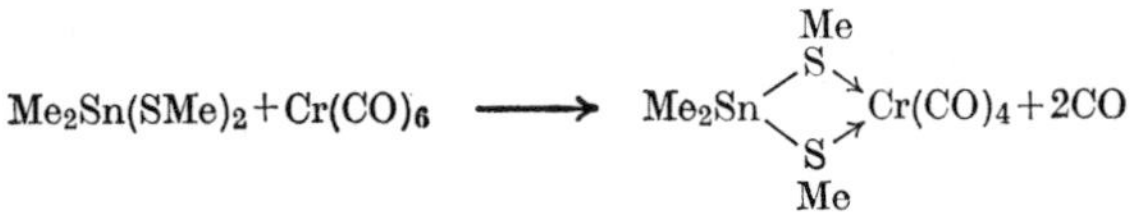

$$Me_2Sn(SMe)_2 + Cr(CO)_6 \longrightarrow Me_2Sn\begin{matrix} Me \\ S \\ \diagdown \\ S \\ Me \end{matrix}Cr(CO)_4 + 2CO$$

When a similar preparation was attempted with manganese penta-carbonyl bromide the more usual reaction of exchange between halogen and the methylthio group occurred giving a trinuclear manganese compound [6].

$$3Me_2Sn(SMe)_2 + 3Mn(CO)_5Br \longrightarrow [Mn(CO)_3SMe]_3 + 3Me_2Sn(SMe)Br + 6CO$$

Bromine and iodine effect tin–sulphur bond cleavage in thiolatotin compounds under mild conditions and the use of iodine for the quantitative estimation of Sn—S bonds has been proposed [3].

$$2R_3SnSR' + X_2 \longrightarrow 2R_3SnX + R'SSR' \quad (X = Br, I)$$

Sulphur reacts with alkylthiolatotrimethyltin compounds, Me_3SnSR, to give *bis*(trimethylstannyl) sulphide and alkyl polysulphides, RS_xR [3]. There is one report of the isolation of a stannyl sulphonium salt [3].

$$Me_3SnSMe + MeI \longrightarrow [Me_3SnSMe_2]^{\oplus}I^{\ominus}$$

Trimethylsilyl ethyl sulphide can be added across the C=O bond of chloral but, when the reaction was attempted using analogous tin compounds, substitution occurred giving trialkylthioacetaldehyde [11a].

$$3R_3SnSR' + Cl_3CCHO \longrightarrow 3R_3SnCl + (R'S)_3CCHO$$
$$(R = Me, Et; R' = Me, Et)$$

Little is known of the reactions of the stannyl dithiocarbamates. The products of thermal decomposition of these compounds depend upon conditions, thus the following reaction occurred when $Ph_2Sn(SCSNHCH_2Ph)_2$ was heated in a solvent.

$$3Ph_2Sn(SCSNHCH_2Ph)_2 \longrightarrow (Ph_2SnS)_3 + 6PhCH_2NCS + 3H_2S$$

However, diphenyltin oxide was obtained when $Ph_2Sn(SCSNPh_2)_2$ was heated in air [13].

TABLE II

Physical properties of some compounds containing Sn—S—C groups [a]

Compound	m.p.	b.p.	n_D^{20}	d^{20}
Me₃SnSMe	—	163°	1·5303	1·453
Me₃SnSBu	—	44°/0·05 mm	1·5093	1·281
Me₃SnSPh	—	69°/0·01 mm	1·5934	1·418
Pr₃SnSPh	—	157–159°/1 mm	1·5626	1·2373
Bu₃SnSMe	—	104°/0·2 mm	1·5110 (22°)	—
Bu₃SnSPh	—	147°/0·3 mm	1·5479 (22°)	—
Ph₃SnSPh	102–103°	—	—	—
Me₂Sn(SMe)₂	—	44°/0·05 mm	1·6003	1·547
Me₂Sn(SBu)₂	—	81°/0·1 mm	1·5400	1·280
Ph₂Sn(SPh)₂	66–70°		—	—
MeSn(SMe)₃	—	75°/0·01 mm	1·6352 (22°)	1·630
MeSn(SPr)₃	—	95°/0·001 mm	1·5684	1·337
Ph₃SnSCOPh	108–109°	—	—	—
Me₃SnSCSNMe₂	63°	—	—	—
Ph₃SnSCSNHCH₂Ph	124–126°	—	—	—
Me₃SnSePh	—	67–69°/0·001	1·6119	1·650
Ph₃SnSePh	87–88°	—	—	—

[a] Including some analogous selenium compounds.

3.2. COMPOUNDS CONTAINING OTHER Sn—S—X GROUPS

The most important compounds in this class are those containing one or more Sn—S—Sn sequences, i.e. the organotin sulphides of which the simplest are the $(R_3Sn)_2S$ compounds. These can be made from an organotin halide and sodium or silver sulphide.

$$2R_3SnCl + M_2S \longrightarrow (R_3Sn)_2S + 2MCl \qquad (M = Na, Ag)$$

Alternatively hydrogen sulphide may be passed into a water or acetone solution of the organotin hydroxide or oxide (which can be generated *in situ* from the chloride and sodium hydroxide).

$$(R_3Sn)_2O + H_2S \longrightarrow (R_3Sn)_2S + H_2O$$

The corresponding selenides can be obtained from the halide and sodium selenide [30].

$$2Ph_3SnCl + Na_2Se \longrightarrow (Ph_3Sn)_2Se + 2NaCl$$

Addition reactions of organotin oxides and alkoxides to compound containing $C{=}S$ bonds, such as carbon disulphide lead to the formation of organotin sulphides, for example[24]:

$$(Bu_3Sn)_2O + CS_2 \longrightarrow (Bu_3Sn)_2S + COS.$$

When triethyltin hydride was heated with sulphur in a sealed tub at 100° for 2 hours bis(triethylstannyl) sulphide was obtained; the selenide and telluride were prepared in a similar manner[42, 44]. Diethy telluride and triethyltin hydride reacted at 40° to give a high yield of bis(triethylstannyl) telluride[45].

$$Et_2Te + 2Et_3SnH \longrightarrow (Et_3Sn)_2Te + 2EtH$$

When bis(triethylgermyl) telluride was treated with triethyltin hydride at 70° the stannyl telluride was formed in 73% yield[47].

$$(Et_3Ge)_2Te + 2Et_3SnH \longrightarrow (Et_3Sn)_2Te + 2Et_3GeH$$

Moedritzer has shown that, in a solution containing a dialkyltin dihalide and a dialkyltin sulphide, there is a rapid exchange between sulphur and halogen[15, 16].

$$R_2SnCl_2 + (R_2SnS)_3 \rightleftharpoons R_2ClSn(SSnR_2)_3Cl$$

The existence of rapid exchange reactions of this nature explain some apparently conflicting reports in the literature. Thus the action of hydrogen sulphide on a series of compounds $R_2XSnOSnR_2X$ (X = halogen or pseudohalogen) gave mixtures of $(R_2SnS)_3$ and R_2SnX_2[22] whereas other workers isolated bis(chlorodibutylstanny) sulphide from the reaction between H_2S and $Bu_2ClSnOSnBu_2Cl$[14].

$$R_2XSnOSnR_2X + H_2S \longrightarrow H_2O + \begin{cases} R_2XSnSSnR_2X \\ \\ R_2SnX_2 + \dfrac{1}{3}(R_2SnS)_3 \end{cases}$$

Several examples of this type of equilibrium have been demonstrated and, for example, bis(chlorodibutyltin) sulphide has been prepared by direct reaction between dibutyltin dichloride and dibutyltin sulphide[10] and also from the dichloride and sodium sulphide[14].

$$2Bu_2SnCl_2 + Na_2S \longrightarrow Bu_2ClSnSSnBu_2Cl + 2NaCl$$

The same compound was formed when tributyltin chloride was heated with sulphur at 180–190° for 14 hours[34]; it reacts rapidly with 2,2'-bipyridyl to give Bu_2SnCl_2.bipy and $(Bu_2SnS)_3$[10].

Similar equilibria have been shown to occur in solutions of sulphides and trihalides, the resulting unsymmetrical halosulphides were isolated and shown to be air-stable[10].

$$\frac{1}{3}(R_2SnS)_3 + R'SnX_3 \rightleftharpoons XR_2SnSSnR'X_2 \qquad (X = halogen)$$

There are relatively few accounts of the reactions of the distannyl sulphides though there is an interesting recent report[33] on the preparation of a triphenylstannylcarbene from $(Ph_3Sn)_2S$ and its insertion reaction with an olefin. The relevant reactions are summarised in the following equations.

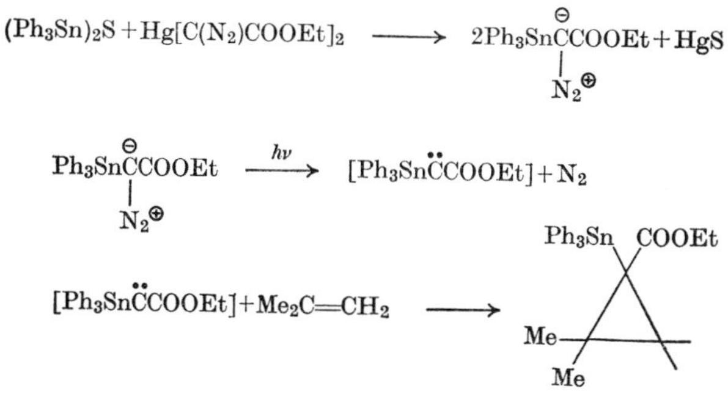

$$(Ph_3Sn)_2S + Hg[C(N_2)COOEt]_2 \longrightarrow 2Ph_3Sn\overset{\ominus}{\underset{\overset{|}{N_2^{\oplus}}}{C}}COOEt + HgS$$

$$Ph_3Sn\overset{\ominus}{\underset{\overset{|}{N_2^{\oplus}}}{C}}COOEt \xrightarrow{h\nu} [Ph_3Sn\overset{..}{C}COOEt] + N_2$$

$$[Ph_3Sn\overset{..}{C}COOEt] + Me_2C{=}CH_2 \longrightarrow$$

Bis(triethylstannyl) sulphide reacts with benzoyl peroxide which has been freed from oxygen to give triethylstannyl benzoate and sulphur (the selenide and telluride react similarly)[46].

$$(Et_3Sn)_2X + PhCOOOCOPh \longrightarrow X + 2Et_3SnOCOPh \qquad (X = S, Se, Te)$$

In the presence of freshly precipitated copper, trialkyltin sulphides interact with carbon tetrachloride to give trialkyltin chlorides in 70–90% yields, other products are cupric sulphide and tetrachloroethylene[23].

The dialkyl(aryl)tin sulphides are trimeric and are prepared using similar methods to those employed for the synthesis of bis(trialkyltin) sulphides. When an organotin dihalide is treated with excess of aqueous sulphide ion the solution must be acidified to precipitate the product, for example, $(NCCH_2CH_2)_2SnBr_2$ was treated with aqueous sodium sulphide and subsequent acidification gave $(NCCH_2CH_2)_2SnS]_3$[26]. Reichle has suggested a mechanism for the reaction which occurs when a solution containing an organotin oxide and sulphide ion is acidified; the overall equation is as follows[25].

$$\frac{1}{x}(R_2SnO)_x + 2S^{2\ominus} + 4H^{\oplus} \longrightarrow \frac{1}{3}(R_2SnS)_3 + H_2O + H_2S$$

Dialkyl(aryl)tin sulphides are also produced in good yields from organotin dichlorides and sodium thiosulphate[57]. Diphenyltin sulphide has been prepared by heating tetraphenyltin with sulphur[29, 31] and dibutyltin sulphide has been made in a similar manner[29]. The dialkyl(aryl)tin sulphides are well defined crystalline solids, generally soluble in organic solvents.

When organotin trihalides are treated with sodium sulphide the less well defined sesquisulphides are produced[12].

$$2RSnCl_3 + 3Na_2S \longrightarrow R_2Sn_2S_3 + 6NaCl$$

The structures of these compounds are discussed in Chapter 12. Very few sesquisulphides have been described, they are generally amorphous solids of varying degrees of molecular complexity including polymers.

Considering next compounds containing the group Sn—S—X where X is neither carbon nor tin, sulphur can be inserted into a tin–lithium bond to give highly reactive products containing the Sn—S—Li group[35, 36].

$$Ph_3SnLi + S \xrightarrow{\text{dry } N_2} Ph_3SnSLi$$

$$Ph_2SnLi_2 + 2S \xrightarrow{\text{dry } N_2} Ph_2Sn(SLi)_2$$

The latter product was obtained only in solution but the disodium salt of dibutyltin dithiol was obtained as a hygroscopic solid by treating dibutyltin sulphide with sodium sulphide[58].

$$\frac{1}{3}(Bu_2SnS)_3 + Na_2S \longrightarrow Bu_2Sn(SNa)_2$$

Triphenylstannyl lithium sulphide was obtained as a solid m.p. > 250°; the corresponding selenide and telluride were obtained in a similar manner but were not isolated[37]. The sulphur compounds react with benzoyl chloride to give thiobenzoates[35, 36].

$$Ph_3SnSLi + PhCOCl \longrightarrow Ph_3SnSCOPh + LiCl$$

$$Ph_2Sn(SLi)_2 + 2PhCOCl \longrightarrow Ph_2Sn(SCOPh)_2 + 2LiCl$$

Alkyl and aryl halides react with the disodium dithiolates in an analogous manner[52, 58].

$$Bu_2Sn(SNa)_2 + 2RX \longrightarrow Bu_2Sn(SR)_2 + 2NaX \qquad (X = \text{halogen})$$

These alkali metal derivatives are particularly useful for the preparation of compounds containing the group Sn—S—M (where M is another Group IVb element) together with the analogous selenium and tellurium compounds. The mixed sulphides can be prepared in two ways using either an organotin alkali metal sulphide or an organotin chloride.

$$R_3SnSLi + R_3MCl \longrightarrow R_3SnSMR_3 + LiCl \longleftarrow R_3SnCl + R_3MSLi$$

Some illustrative examples are given in the following equations.

$$Ph_3SnCl + Ph_3GeSLi \longrightarrow Ph_3SnSGePh_3 + LiCl \qquad \text{(ref. 38)}$$
$$Ph_3SnCl + Ph_3PbSLi \longrightarrow Ph_3SnSPbPh_3 + LiCl \qquad \text{(ref. 39)}$$
$$Ph_3SnSeLi + Ph_3GeBr \longrightarrow Ph_3SnSeGePh_3 + LiBr \qquad \text{(ref. 37)}$$
$$Ph_3SnTeLi + Ph_3PbCl \longrightarrow Ph_3SnTePbPh_3 + LiCl \qquad \text{(ref. 37)}$$

TABLE III

Physical properties of some compounds containing Sn—S—Sn groups[a]

Compound	m.p.	b.p.
$(Me_3Sn)_2S$	—	$118°/18$ mm
$(Et_3Sn)_2S^b$	—	$133–137°/1$ mm
$(Ph_3Sn)_2S$	$145–146°$	—
$(Me_3Sn)_2Se$	$-6°$ to $-5°$	$118°/15$ mm
$(Ph_3Sn)_2Se$	$148°$	—
$(Ph_3Sn)_2Te$	$148°$	—
$(Me_2SnS)_3$	$149–151°$	—
$(Bu_2SnS)_3$	$63–69°$	$210–214°/1$ mm
$(Ph_2SnS)_3$	$183–184°$	—
$(Me_2SnSe)_3$	$120°$	—
$(Ph_2SnSe)_3$	$176–177°$	—
$(Ph_2Sn_2S_3)_x$	decomp. $> 250°$	—

[a] Including some analogous selenium and tellurium compounds.
[b] $n_D^{20} = 1.5468$; $d^{20} = 1.429$.

However, when a similar preparation of trimethylstannyl trimethylgermyl selenide was attempted a mixture of the distannyl and digermyl selenides was obtained [27].

$$2Me_3GeSeLi + 2Me_3SnCl \longrightarrow 2LiCl + 2[Me_3GeSeSnMe_3] \longrightarrow$$
$$(Me_3Ge)_2Se + (Me_3Sn_2)Se$$

The expected product was also not obtained in a reaction between triphenylstannyl lithium selenide and diphenyltin dichloride [37].

$$2Ph_3SnSeLi + Ph_2SnCl_2 \longrightarrow 2LiCl + [Ph_3SnSeSnPh_2SeSnPh_3] \longrightarrow$$

$$\frac{1}{3}(Ph_2SnSe)_3 + (Ph_3Sn)_2Se$$

Unsymmetrical sulphides and tellurides containing tin and another Group IVb element have been successfully prepared by the use of organotin hydrides as follows [43].

$$Et_3SnH + Et_3GeSH \xrightarrow[2\ hr]{130°} Et_3SnSGeEt_3 + H_2$$

$$Et_3SnH + Et_3SiTeEt \xrightarrow[24\ hr]{20°} Et_3SnTeSiEt_3 + EtH$$

Other compounds containing Sn—S bonds, the organotin dithiophosphinates, $R_3SnSP(S)R_2'$, and dithiophosphates, $R_3SnSP(S)(OR')_2$, are discussed in Chapter 10.

TABLE IV

Physical properties of some compounds containing Sn—S—X groups (X = Li, Si, Ge, Pb)[a]

Compound	m.p.	b.p.
$Me_3SnSGeMe_3$	−8°	89–90°/12 mm
$Et_3SnTeSiEt_3[b]$	—	109–112°/1 mm
$Et_3SnSGeEt_3[c]$	—	100–103°/1 mm
$Et_3SnSeGeEt_3$	—	111°/0·5 mm
$Et_3SnTeGeEt_3$	—	126–128°/1 mm
Ph_3SnSLi	250°	—
$Ph_3SnSGePh_3$	136°	—
$Ph_3SnSeGePh_3$	144°	—
$Ph_3SnSPbPh_3$	138°	—
$Ph_3SnTePbPh_3$	136° decomp.	—

[a] Including some analogous selenium and tellurium compounds.
[b] $n_D^{20} = 1.5680$.
[c] $n_D^{20} = 1.5279$; $d^{20} = 1.299$.

References

1. Abel, E. W. and Armitage, D. A., *Advances in Organometallic Chemistry*, Academic Press, New York and London, 1967, **5**, 1.
2. Abel, E. W., Armitage, D. A. and Brady, D. B., *J. Organometallic Chem.*, 1966, **5**, 130.
3. Abel, E. W. and Brady, D. B., *J. Chem. Soc.*, 1965, 1192.

4. Abel, E. W., Brady, D. B. and Crosse, B. C., *J. Organometallic Chem.*, 1966, **5**, 260.
5. Abel, E. W. and Crosse, B. C., unpublished (see ref. 1, p. 37).
6. Abel, E. W., Crosse, B. C. and Brady, D. B., *J. Amer. Chem. Soc.*, 1965, **87**, 4397.
7. Abel, E. W. and Jenkins, C. R., *J. Chem. Soc. (A)*, 1967, 1344.
8. Bonati, F. and Ugo, R., *J. Organometallic Chem.*, 1967, **10**, 257.
9. Davidson, W. E., Hills, K. and Henry, M. C., *J. Organometallic Chem.*, 1965, **3**, 285.
10. Davies, A. G. and Harrison, P. G., *J. Organometallic Chem.*, 1967, **8**, P19.
11. George, T. A., Jones, K. and Lappert, M. F., *J. Chem. Soc.*, 1965, 2157.
11a. Itoh, K., Matsuzaki, K. and Ishii, Y., *J. Chem. Soc. (C)*, 1968, 2709.
12. Komura, M. and Okawara, R., *Inorg. Nucl. Chem. Letters*, 1966, **2**, 93.
13. Kupchik, E. J. and Calabretta, P. J., *Inorg. Chem.*, 1965, **4**, 973.
14. Migdal, S., Gertner, D. and Zilkha, A., *Can. J. Chem.*, 1967, **45**, 2987.
15. Moedritzer, K., *Organometallic Chem. Rev.*, 1966, **1**, 179.
16. Moedritzer, K. and van Wazer, J. R., *Inorg. Chem.*, 1964, **3**, 943.
17. Pang, M. and Becker, E. I., *J. Org. Chem.*, 1964, **29**, 1948.
18. Peach, M. E., *Can. J. Chem.*, 1968, **46**, 211.
19. Poller, R. C., *Proc. Chem. Soc.*, 1963, 312.
20. Poller, R. C. and Spillman, J. A., *J. Chem. Soc., (A)* 1966, 958.
21. Poller, R. C. and Spillman, J. A., *J. Organometallic Chem.*, 1966, **6**, 668.
22. Poller, R. C. and Spillman, J. A., *J. Organometallic Chem.*, 1967, **7**, 259.
23. Razuvaev, G. A., D'yachkovskaya, O. S. and Fionov, V. I., *Dokl. Akad. Nauk S.S.S.R.*, 1967, **177**, 1113.
24. Reichle, W. T., *Inorg. Chem.*, 1962, **1**, 650.
25. Reichle, W. T., *J. Org. Chem.*, 1961, **26**, 4634.
26. Reifenberg, G. H. and Considine, W. J., *J. Organometallic Chem.*, 1967, **10**, 279.
27. Ruidisch, I. and Schmidt, M., *J. Organometallic Chem.*, 1963, **1**, 160.
28. Sasin, G. S., Borror, A. L. and Sasin, R., *J. Org. Chem.*, 1958, **23**, 1366.
29. Schmidt, M., Dersin, H. J. and Schumann, H., *Chem. Ber.*, 1962, **95**, 1428.
30. Schmidt, M. and Ruf, H., *Angew Chem.*, 1961, **73**, 64.
31. Schmidt, M. and Schumann, H., *Chem. Ber.*, 1963, **96**, 462.
32. Schmidt, M. and Schumann, H., *Chem. Ber.*, 1963, **96**, 780.
33. Schoellkopf, U. and Rieber, N., *Angew. Chem. Intern. Ed. Engl.*, 1967, **6**, 884.
34. Schumann, H. and Schmidt, M., *Chem. Ber.*, 1963, **96**, 3017.
35. Schumann, H., Thom, K. F. and Schmidt, M., *J. Organometallic Chem.*, 1963, **1**, 167.
36. Schumann, H., Thom, K. F. and Schmidt, M., *J. Organometallic Chem.*, 1964, **2**, 97.
37. Schumann, H., Thom, K. F. and Schmidt, M., *J. Organometallic Chem.*, 1964, **2**, 361.
38. Schumann, H., Thom, K. F. and Schmidt, M., *J. Organometallic Chem.*, 1965, **4**, 22.
39. Schumann, H., Thom. K. F. and Schmidt, M., *J. Organometallic Chem.*, 1965, **4**, 28.

40. Seltzer, R., *J. Org. Chem.*, 1968, **33**, 3896.
41. Sukhani, D., Gupta, V. D. and Mehrotra, R. C., *J. Organometallic Chem.*, 1967, **7**, 85.
42. Vyazankin, N. S., Bochkarev, M. N. and Sanina, L. P., *Zh. Obshch. Khim.*, 1966, **36**, 166.
43. Vyazankin, N. S., Bochkarev, M. N. and Sanina, L. P., *Zh. Obshch. Khim.*, 1966, **36**, 1154.
44. Vyazankin, N. S., Bochkarev, M. N. and Sanina, L. P., *Zh. Obshch. Khim.*, 1966, **36**, 1961.
45. Vyazankin, N. S., Bochkarev, M. N. and Sanina, L. P., *Zh. Obshch. Khim.*, 1967, **37**, 1037.
46. Vyazankin, N. S., Bochkarev, M. N. and Sanina, L. P., *Zh. Obshch. Khim.*, 1967, **37**, 1545.
47. Vyazankin, N. S., Bochkarev, M. N. and Sanina, L. P., *Zh. Obshch. Khim.*, 1968, **38**, 414 (*Chem. Abs.*, 1968, **69**, 96844).
48. Wieber, M. and Schmidt, M., *J. Organometallic Chem.*, 1964, **1**, 336.
49. Wieber, M. and Schmidt, M., *J. Organometallic Chem.*, 1964, **2**, 129.
50. Belgium Patent 666,691, Jan. 10, 1966 (*Chem. Abs.*, 1966, **65**, 17004g).
51. British Patent 1,018,805, Feb. 2, 1966 (*Chem. Abs.*, 1966, **64**, 11251a).
52. British Patent 1,020,612, Feb. 23, 1966 (*Chem. Abs.*, 1966, **64**, 14219d).
53. British Patent 1,089,243, Nov. 1, 1967 (*Chem. Abs.*, 1968, **68**, 78423).
54. German Patent 1,178,853, Oct. 1 ,1964 (*Chem. Abs.*, 1965, **62**, 7796h).
55. German Patent 1,234,722, Feb. 23, 1967 (*Chem. Abs.*, 1967, **67**, 22015u).
56. Japanese Patent 22,069, Sept. 30, 1965 (*Chem. Abs.*, 1966, **64**, 3603b).
57. Japanese Patent 19,333, Nov. 9, 1966 (*Chem. Abs.*, 1967, **66**, 38049y).
58. Netherlands Patent 6,703,505, Sept. 8, 1967 (*Chem. Abs.*, 1967, **67**, 116946a).
59. U.S. Patent 3,201,408, Aug. 17, 1965 (*Chem. Abs.*, 1965, **63**, 13315e).
60. U.S. Patent, 3,317,573, May 2, 1967 (*Chem. Abs.*, 1968, **68**, 39819).
61. U.S. Patent 3,365,478, Jan. 23, 1968 (*Chem. Abs.*, 1968, **68**, 95978).

9

COMPOUNDS CONTAINING TIN–METAL BONDS

In this chapter the term 'metal' is interpreted widely so as to include compounds in which tin is bonded to elements such as silicon, arsenic and phosphorus.

Among the newer compounds in this class are those in which tin is bonded to a transition metal and these compounds are being actively studied by many research groups. Undoubtedly the oldest of these compounds are those containing tin–tin bonds but here also the advent of powerful new synthetic methods has stimulated much recent activity.

9.1. COMPOUNDS CONTAINING TIN–TIN BONDS

Two recent reviews[57,75a] of the catenated compounds of the Group IVb elements contain comprehensive sections on organotin compounds. Essentially two types of organotin compound will be discussed here, the open-chain polyorganostannanes Sn_xR_{x+2} and the cyclic compounds $(R_2Sn)_x$. It is now realised that the so-called organic derivatives of divalent tin belong to the latter class. Fully characterised compounds containing chains of up to 9 tin atoms have been made but it is clear that catenation is much more limited with tin than with silicon. The reasons for this are obscure and simple bond energy considerations, which show relatively small differences between Si—Si and Sn—Sn bonds in analogous compounds, are not illuminating[33]. In a brief discussion[33] of the subject reasons were advanced for the high reactivity of compounds containing Sn—Sn bonds but it is not easy to account for the relatively low thermodynamic stability of catenated tin compounds.

9.1.1. *Open chain polyorganostannanes*

The first general method for making these compounds was by reaction between stannylsodium derivatives and organotin halides.

$$R_3SnNa + R_3SnCl \longrightarrow R_3SnSnR_3 + NaCl$$

By combining this reaction with Wurtz-type reactions, carried out in liquid ammonia, Kraus and his associates prepared tetra- and penta-tin compounds in the 1920's, for example [72]:

$$2Me_2SnBr_2 + 6Na \longrightarrow NaMe_2SnSnMe_2Na + 4NaBr.$$
$$NaMe_2SnSnMe_2Na + 2Me_3SnBr \longrightarrow Me_3SnSn(Me)_2Sn(Me)_2SnMe_3 + 2NaBr.$$

More recently organodistannanes have been prepared by the use of lithium derivatives with tetrahydrofuran as the reaction medium [124].

$$R_3SnLi + R'_3SnCl \longrightarrow R_3SnSnR'_3 + LiCl$$

An unusual method of making hexaphenyldistannane is by treating stannous chloride with excess phenylmagnesium bromide when, presumably, the Grignard reagent effects Sn—Sn bond cleavage in the initially formed diphenyltin $(Ph_2Sn)_x$. Boeseken and Rutgers [11] studied the minor products of this reaction and isolated small quantities of dodecaphenylpentastannane to which, on the basis of its decomposition temperature, they assigned the structure $(Ph_3Sn)_4Sn$. More than forty years later the original work was repeated and dodecaphenylpentastannane obtained in 2% yield, the same product was then prepared by three other methods [58].

$$(Ph_3Sn)_3SnLi + Ph_3SnCl \longrightarrow Ph_{12}Sn_5 (70\%) + LiCl$$
$$4Ph_3SnLi + SnCl_4 \longrightarrow Ph_{12}Sn_5 (10\%) + 4LiCl$$
$$4Ph_3SnCl + 8Li + SnCl_4 \longrightarrow Ph_{12}Sn_5 (21\%) + 8LiCl$$

The products from the four reactions were shown to be identical and the original structure assignment confirmed by evidence which included iodine cleavage experiments in which triphenyltin iodide and stannic iodide were obtained.

$$(Ph_3Sn)_4Sn + 4I_2 \longrightarrow 4Ph_3SnI + SnI_4$$

Polymeric organostannanes Sn_xR_{2x+2} can be made by treatment of an alloy, of composition Mg_2Sn, with alkyl halides in the presence of a cyclic ether solvent [142].

In recent years a major new method of establishing tin–tin bonds has been developed based on the hydrostannolysis reaction whereby a bond between tin and another element, usually nitrogen or oxygen, is cleaved by an organotin hydride (see also Section 7.4).

$$R_3SnN< + R'_3SnH \longrightarrow R_3SnSnR'_3 + HN<$$

The hydrostannolysis of aminotin compounds has been developed by Neumann and his collaborators [117], some examples which indicate the scope of the method are as follows.

$$Bu_3SnNEt_2 + Et_3SnH \longrightarrow Bu_3SnSnEt_3 (78\%) + Et_2NH \qquad (\text{ref. 91})$$

$(C_6H_{11})_3SnNEt_2 + Ph_3SnH \longrightarrow (C_6H_{11})_3SnSnPh_3$ (78%) $+ Et_2NH$ (ref. 91)
$(C_6H_{11} = $ cyclohexyl$)$

$Et_2Sn(H)(CH_2)_5Sn(H)Et_2 + 2Me_3SnNEt_2 \longrightarrow$
          $Me_3SnSnEt_2(CH_2)_5SnEt_2SnMe_3 + 2Et_2NH$ (ref. 91)

$2Ph_3SnNEt_2 + (i\text{-}Bu)_2SnH_2 \longrightarrow$
          $Ph_3SnSn(i\text{-}Bu)_2SnPh_3$ (90%) $+ 2Et_2NH$ (ref. 118)

$2Me_3SnNEt_2 + H(i\text{-}Bu)_2SnSnH(i\text{-}Bu)_2 \longrightarrow$
          $Me_3SnSn(i\text{-}Bu)_2Sn(i\text{-}Bu)_2SnMe_3$ (81%) $+ 2Et_2NH$ (ref. 118)

The tristannanes can also be prepared from the *bis*(diethylamino)tin compound and the monohydride but the yields are much reduced.

$$2R_3SnH + R'_2Sn(NEt_2)_2 \longrightarrow R_3SnSnR'_2SnR_3 + 2Et_2NH$$

In all of these reactions a dialkylamine is liberated and, since organotin hydrides in the presence of amines lose hydrogen to form compounds with Sn—Sn bonds, mixtures of products sometimes arise. N-Stannylformamides may be used instead of the stannylamines[90] and Creemers[33] has made a detailed study of the hydrostannolysis of N-stannyl-N-phenylformamides, $R_3SnN(Ph)CHO$ (prepared by the addition of R_3SnH to PhNCO). Although these compounds are less reactive than the stannylamines the N-phenylformamide liberated in the hydrostannolysis does not initiate any side reactions.

$$R_3SnN(Ph)CHO + R'_3SnH \longrightarrow R_3SnSnR'_3 + PhNHCHO$$

When an N-stannyl-N-phenylformamide is treated with one equivalent of a dihydride an organodistannane monohydride results.

$$R_3SnN(Ph)CHO + R'_2SnH_2 \longrightarrow R_3SnSnR'_2H + PhNHCHO$$

This intermediate has been used to prepare a number of organo-polystannanes as indicated in the following reaction chart[33, 35].

$R_3SnSnR_2SnR_2SnR_3 + H_2$

$\uparrow$ Et_2NH

$R_3SnSnR_2H \xrightarrow{\;R_3SnN(Ph)CHO\;} R_3SnSnR_2SnR_3 + PhNHCHO$

$\downarrow$ PhNCO

$R_3SnSnR_2N(Ph)CHO \xrightarrow{\;R'_3SnH\;} R_3SnSnR_2SnR'_3 + PhNHCHO$

$\downarrow$ R_2SnH_2

$R_3SnSnR_2SnR_2H \xrightarrow{\;R_3SnSnR_2N(Ph)CHO\;} R_3SnSnR_2SnR_2SnR_2SnR_3 + PhNHCHO$

$\downarrow$ Et_2NH

$R_3SnSnR_2SnR_2SnR_2SnR_2SnR_3 + H_2$

In the above sequence, Sn—Sn bonds are formed, *inter alia*, by the base-catalysed elimination of hydrogen from monohydrides.

$$2R_3SnH \xrightarrow{\text{base}} R_3SnSnR_3 + H_2$$

Saville[100] has indicated how bases may assist reaction of organotin hydrides and, following from this, a tentative proposal for the mechanism of this reaction is made below.

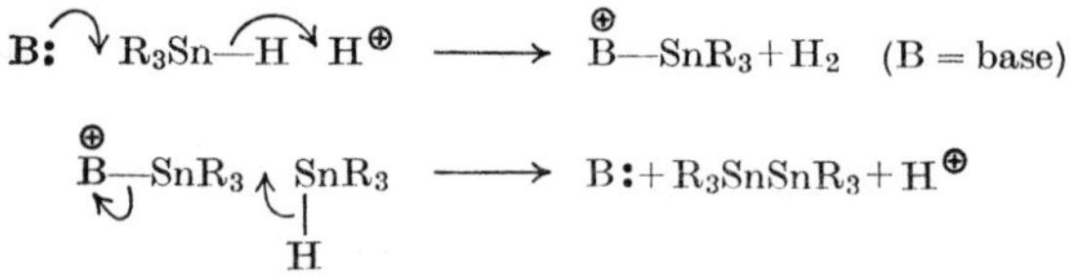

From the kinetics of the hydrostannolysis of organotin–nitrogen compounds, and by other techniques, it was shown that hydrostannolysis proceeds by electrophilic attack of the organotin hydride hydrogen on nitrogen as the first, and rate-determining, step[35, 37].

Other classes of organotin–nitrogen compounds, such as azides, react with hydrides to produce compounds with Sn—Sn bonds[75].

$$Et_3SnN_3 + Et_3SnH \longrightarrow Et_3SnSnEt_3 + HN_3$$

Similar reactions are shown by stannylhydrazines[90, 92].

$$R_3SnNR'NHR' + R''_3SnH \longrightarrow R_3SnSnR''_3 + R'NHNHR'$$

The hydrostannolysis of oxides gives organodistannanes in excellent yields.

$$(Ph_3Sn)_2O + 2Ph_3SnH \longrightarrow 2Ph_3SnSnPh_3 \ (89\%) + H_2O \qquad \text{(ref. 90)}$$
$$(Bu_3Sn)_2O + 2Bu_3SnH \longrightarrow 2Bu_3SnSnBu_3 \ (100\%) + H_2O$$
$$\text{(refs. 101, 143)}$$
$$(Ph_3Sn)_2O + 2Et_3SnH \longrightarrow 2Ph_3SnSnEt_3 \ (73\%) + H_2O \qquad \text{(ref. 90)}$$

However, attempts to use these reactions to produce compounds with more than two linked tin atoms were less successful owing to the intervention of an exchange reaction which, in the case of hexaphenyldistannoxane and diphenyltin hydride, was the sole reaction observed[118].

$$(Ph_3Sn)_2O + Ph_2SnH_2 \longrightarrow 2Ph_3SnH + Ph_2SnO$$

With other compounds, the products isolated indicated that both hydrostannolysis and exchange had occurred, for example triethyltin hydride and diethyltin oxide gave $Et_3SnSnEt_2SnEt_3$ (47%)

and $Et_3SnSnEt_3$ (25%)[118]. The latter product was obtained *via* the following reactions:

$$2Et_3SnH + Et_2SnO \longrightarrow (Et_3Sn)_2O + Et_2SnH_2$$
$$(Et_3Sn)_2O + 2Et_3SnH \longrightarrow 2Et_3SnSnEt_3 + H_2O.$$

Tristannanes with mixed organic groups, such as $Bu_3SnSnPh_2SnBu_3$, have also been prepared by this method[144]. Similar results are obtained by the hydrostannolysis of organotin hydroxides and alkoxides[91]; reactions between hydrides and alkoxides are ionic and involve electrophilic attack of the hydride hydrogen on the oxygen of the alkoxide group[36].

Apart from the classes already mentioned a number of other types of organotin compounds react with hydrides to give organodistannanes as shown in the following examples (N.B. when an acid is formed in these reactions it is neutralised by the addition of a base)[91].

$$Ph_3SnH + Bu_3SnPBu_2 \longrightarrow Ph_3SnSnBu_3 + Bu_2PH$$
$$Ph_3SnH + Me_3SnCl + Et_3N \longrightarrow Ph_3SnSnMe_3 + Et_3NHCl$$
$$Ph_3SnH + i\text{-}Bu_3SnI + Et_3N \longrightarrow Ph_3SnSn(Bu\text{-}i)_3 + Et_3NHI$$
$$Ph_3SnH + Bu_3SnOAc + Et_3N \longrightarrow Ph_3SnSnBu_3 + Et_3NHOAc$$

In some reactions of this type the yield of distannane is reduced due to an accompanying exchange reaction; exchange reactions predominate if the atom attached to tin is a second or third row element, e.g. S, P, As[37].

Polarographic reduction of triphenyltin chloride gives hexaphenyldistannane and similar reduction of diphenyltin dichloride produces 1,2-dichloro-1,1,2,2-tetraphenyldistannane which can be converted to the corresponding dibenzoate with $PhCOOAg$[41].

$$Ph_2SnCl_2 \xrightarrow[-1.6v]{e} \tfrac{1}{2}Ph_2Sn(Cl)SnPh_2Cl$$
$$Ph_2Sn(Cl)SnPh_2Cl + 2PhCOOAg \longrightarrow Ph_2Sn(OCOPh)SnPh_2(OCOPh) + 2AgCl$$

Polarographic reduction of hexaphenyldistannane gives the triphenyltin anion $Ph_3Sn^{\ominus}$ and, by the use of other electrochemical techniques the existence can be demonstrated of the corresponding cation $Ph_3Sn^{\oplus}$ and radical $Ph_3Sn\cdot$[41]. There seems little doubt that radicals such as $R_3Sn\cdot$ are involved as reactive intermediates in various areas of organotin chemistry, including oxidations of distannanes (see below). However, there is no evidence for simple dissociation of distannanes[64, 89].

$$R_3SnSnR_3 \rightleftharpoons 2R_3Sn\cdot$$

Neumann has pointed out that, if only trace amounts of radicals were formed in such equilibria, it would be expected that the action of heat on $Et_3SnSnMe_3$ would convert it to a mixture of symmetrical distannanes; no such conversion was observed[89].

The Sn—Sn bond is very rapidly attacked by halogen at room temperature.

$$R_3SnSnR_3 + X_2 \longrightarrow 2R_3SnX \qquad (X = Cl, Br, I)$$

A common method of estimation of polyorganostannanes is by titration with iodine or bromine[119]. The kinetics of iodine cleavage of organodistannanes have been studied by a number of workers and, in polar solvents, the reaction is bimolecular and is thought to proceed *via* electrophilic attack by the iodine molecule on one of the tin atoms[13, 122]. Nucleophilic assistance is rendered by attack of a solvent molecule on the second tin atom involving an intermediate of the type:

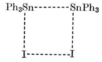

When hexaphenyldistannane is iodinated in less-polar solvents, such as cyclohexane, it is suggested[64] that the following cyclic intermediate is formed.

$$Ph_3Sn\text{-------}SnPh_3$$

Under the influence of ultraviolet light fluoroacetylenes[38] and fluoroolefins[28] can be inserted into Sn—Sn bonds, presumably by a radical reaction.

$$F_3CC{\equiv}CCF_3 + Me_3SnSnMe_3 \xrightarrow{h\nu} Me_3SnC(CF_3){=}C(CF_3)SnMe_3$$

$$CF_3CF{=}CF_2 + Me_3SnSnMe_3 \xrightarrow{h\nu} Me_3SnCF_2CF(CF_3)SnMe_3$$

Dihalocarbenes can also be inserted to give distannyldihalo-methanes[114].

$$PhHgCCl_2Br + Me_3SnSnMe_3 \longrightarrow Me_3SnCCl_2SnMe_3 + PhHgBr$$

$$PhHgCClBr_2 + Me_3SnSnMe_3 \longrightarrow Me_3SnCClBrSnMe_3 + PhHgBr$$

Since dihalocarbenes can be generated from trihalomethyltin compounds by the following reaction it was thought that it might be

$$R_3SnCCl_3 + NaI \longrightarrow R_3SnI + NaCl + :CCl_2$$

possible to produce stannylcarbenes (see p. 139) from bromochlorobis(trimethylstannyl)methane. Instead, a simple halogen exchange occurred[114].

$$Me_3SnCClBrSnMe_3 + NaI \longrightarrow Me_3SnCClISnMe_3 + NaBr$$

Trifluoroacetyl iodide and hexamethyldistannane react under mild conditions to give a 74% yield of trimethyltrifluoromethyltin, Me_3SnCF_3, other products isolated from this reaction were carbon monoxide, trimethyltin iodide and tetramethyltin[39]. When hexamethyldistannane and trifluoroacetic anhydride were heated together at 60° trimethylstannyl trifluoroacetate was formed[39].

The ultraviolet light-initiated reaction of hexaethyldistannane with oxygen gives two initial products in the primary photochemical process[4].

$$Et_3SnSnEt_3 + O_2 \xrightarrow{h\nu} \begin{cases} Et_3SnOOSnEt_3 \\ \\ Et_3SnSnEt_2OOEt \end{cases}$$

These undergo further reactions and the major products isolated are hexaethyldistannoxane and diethyltin oxide. The oxidation of hexaphenyldistannane in tetrahydrofuran by air in the presence of sodium methoxide gives triphenyltin hydroxide and, if ethyl bromide is present, Ph_3SnEt is formed; these results are accounted for in terms of a $Ph_3Sn \cdot$ radical intermediate[127]. Autoxidation of Et_6Sn_2 was shown to involve $Et_3Sn \cdot$ radicals[3].

Using a biamperometric technique a kinetic study was made of the oxidation of organodistannanes by the reaction:

$$R_3SnSnR_3 + 2Ag^{\oplus} \longrightarrow 2R_3Sn^{\oplus} + 2Ag.$$

The following order of reactivity towards the $Ag^{\oplus}$ ion was established[120, 121]: $Me_3SnSnPh_3 = Me_3SnSnMe_3 > Et_3SnSnEt_3 > Et_3SnSnPh_3 > Bu_3SnSnBu_3 > Pr_3SnSnPr_3 > Ph_3SnSnPh_3 > (i-Pr)_3SnSn(Pr-i)_3$. Amperometric titration with $Ag^{\oplus}$ ion has been proposed as a method of estimation of organodistannanes[119]. The redox couples $R_6Sn_2/R_3Sn^{\oplus}$ were found to possess electrochemical reversibility and standard potentials at a platinum electrode were

11

determined for the systems where $R = Me$, Et, Bu, Ph in methanol and methanol/benzene[50].

In the presence of Lewis acids, hexamethyldistannane disproportionates to tetramethyltin and dimethyltin[20]. The kinetics of the decomposition of hexamethyldistannane in the presence of trimethyltin chloride and triethyltin chloride have been studied, in the latter case Et_3SnMe was formed.

When R_6Sn_2 compounds reacted with sulphur dioxide products were formed having the compositions $R_3Sn \cdot SO_2$ ($R = Me$, Ph) and $R_3Sn \cdot 2SO_2$ ($R = Ph$); possible structures of these compounds have been discussed[26].

Hexaphenyldistannane forms an inclusion compound with tetraphenyltin of stoichiometry $Ph_6Sn_2 : Ph_4Sn = 1 \cdot 75 : 1$, of a number of other compounds examined only the isomorphous tetraphenylsilane and tetraphenylgermane formed similar inclusion compounds with the distannane[76].

9.1.2. Cyclic polyorganostannanes

The earlier literature contains several reports of the preparation of organic derivatives of divalent tin R_2Sn. During the last few years it has become clear that, where these preparations gave pure stable compounds, the products were cyclic oligomers $(R_2Sn)_x$. From a general knowledge of the structures of Sn (II) compounds (e.g. SnS and $SnCl_2$ both have polymeric structures in which tin shows a coordination number of 3[99]) it is to be expected that the species R_2Sn: would be unstable. In the polarographic reduction of diethyltin dichloride, the initial product is undoubtedly Et_2Sn: but it rapidly polymerises [46, 47, 78, 79]; these unstable derivatives of divalent tin have been compared with the carbenes[80].

Neumann and his coworkers have re-examined many of the earlier preparations of 'diphenyltin' and shown that, for example, the yellow product obtained from the reaction between phenylmagnesium bromide and stannous chloride in diethyl ether is a cross-linked polymer containing substantial amounts of Ph_3Sn- and $PhSn{\le}$ groups[86].

Currently used methods of preparing cyclic organostannanes usually start from dihydrides which, in the presence of catalysts, form oligomers and hydrogen.

$$xR_2SnH_2 \xrightarrow{\text{catalyst}} (R_2Sn)_x + xH_2$$

The catalysts include pyridine, dialkylamines, dimethylformamide (= DMF), alcohols and sodium alkoxides, a possible mechanism for this type of reaction was referred to on p. 148. The compounds used to promote the reactions are usually present in relatively large amounts and are often used as solvents; when amines are used small quantities of the dichloride are often added as co-catalysts. Some examples which illustrate the conditions used, and the types of ring systems formed, are as follows.

$$6(p\text{-MeC}_6H_4)_2SnH_2 \xrightarrow[70°]{\text{DMF}}$$
$$[(p\text{-MeC}_6H_4)_2Sn]_6 \ (II, R = p\text{-MeC}_6H_4) + 6H_2 \qquad \text{(ref. 86)}$$

$$4(PhCH_2)_2SnH_2 \xrightarrow[\text{trace}(PhCH_2)_2SnCl_2]{\text{DMF}}$$
$$[(PhCH_2)_2Sn]_4 \ (I, R = PhCH_2) + 4H_2 \qquad \text{(ref. 85)}$$

$$6Ph_2SnH_2 \xrightarrow[\substack{\text{absence}\\ \text{of light}}]{\text{MeOH}} 5H_2 + H(Ph_2Sn)_6H \xrightarrow{\text{pyridine}}$$
$$(Ph_2Sn)_6 \ (II, R = Ph) + H_2 \qquad \text{(ref. 86)}$$

$$9(i\text{-Bu})_2SnH_2 \xrightarrow[\text{trace}(i\text{-Bu})_2SnCl_2]{\text{pyridine}}$$
$$[(i\text{-Bu})_2Sn]_9 \ (III, R = i\text{-Bu}) + 9H_2 \qquad \text{(ref. 88)}$$

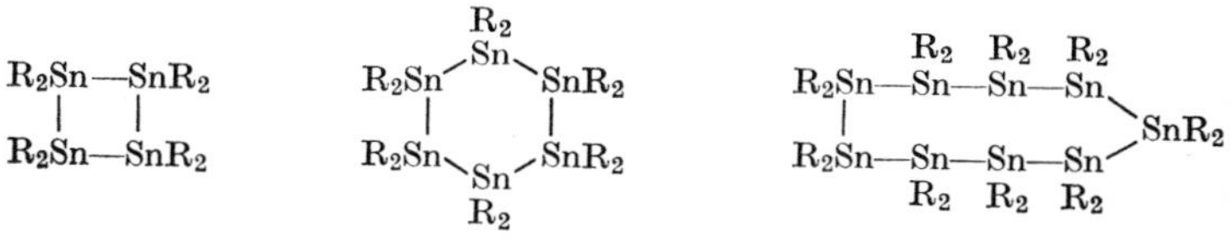

I   *II*   *III*

In addition to these catalytic decomposition reactions Neumann has shown that the cyclic stannanes can be synthesised by treating the dihydrides with *bis*(dialkylamino)tin compounds[88].

$$2(C_6H_{11})_2SnH_2 + 2(C_6H_{11})_2Sn(NEt_2)_2 \longrightarrow [(C_6H_{11})_2Sn]_4 \ (I, R = \text{cyclohexyl})$$
$$+ 4Et_2NH$$
$$3Bu_2SnH_2 + 3Bu_2Sn(NEt_2)_2 \longrightarrow (Bu_2Sn)_6 \ (II, R = Bu) + 6Et_2NH$$

In the last reaction, the same product was formed when $Bu_2Sn(OMe)_2$ was used in place of $Bu_2Sn(NEt_2)_2$.

When di-t-butyltin dichloride was treated with t-BuMgCl, tetrameric di-t-butyltin (*I*, R = t-Bu) was formed[53]. Alkylation of stannous chloride with organoaluminium compounds gave organostannanes though the degree of polymerisation was not reported[141].

Dilithioazobenzene reacts with diphenyltin dichloride to give diphenyltin, again of unknown complexity[56].

$$xPhNLiNLiPh + xPh_2SnCl_2 \longrightarrow (Ph_2Sn)_x + xPhN{=}NPh + 2xLiCl$$

Butyltin trihydride decomposed in the presence of a ketone to give a red solid of composition $(BuSn)_x$[73]. Polarographic reduction of trichlorides, $RSnCl_3$, gave $(RSn)_x$ (R = Me, Et, Ph) which were also red solids[48, 49]; the same products were obtained when the trichlorides were treated with lithium aluminium hydride and diethylamine[48]. A molecular weight determination on $(EtSn)_x$ indicated a degree of polymerisation of 8–9 and a highly crosslinked structure[48].

As noted earlier the tin–tin bond is very sensitive to iodine and this reagent has been used as an aid to structure investigations, for example[53]:

$$[(t\text{-}Bu)_2Sn]_4 + I_2 \longrightarrow I[Sn(Bu\text{-}t)_2]_4I.$$

The same bonds are readily cleaved by hydrogen halides[130].

$$(Bu_2Sn)_x + 2xHBr \longrightarrow xBu_2SnBr_2 + xH_2$$

However, with benzoic acid the distannane dibenzoate is formed.

$$2(Bu_2Sn)_x + 2xPhCOOH \longrightarrow xBu_2Sn(OCOPh)Sn(OCOPh)Bu_2 + xH_2$$

The same dibenzoate is formed when dibutyltin is treated with benzoyl peroxide[130].

TABLE I

Physical properties of some compounds containing
Sn—Sn bonds

Compound	m.p.	b.p.
$Me_3SnSnMe_3$	23°	182°
$Et_3SnSnPr_3$[a]	—	88–89°/0·07 mm
$Me_3SnSnPh_3$	107–108·5°	—
$Ph_3SnSnPh_3$	231–232°	—
$(Ph_3Sn)_4Sn$	280° decomp.	—
$[(i\text{-}Bu)_2Sn]_9$	205–208°	—
$[(i\text{-}Bu)_2Sn]_6$	200–210°	—
$[(PhCH_2)_2Sn]_4$	226–228°	—
$(PhCH_2)_8Sn_3$	163–165°	—
$(EtSn)_{8-9}$	~150° decomp.	—

[a] $n_D^{20} = 1.5219$.

Dialkyltin compounds react with alkyl halides to give trialkyltin halides[21]:

$$(R_2Sn)_x + xR'X \longrightarrow xR_2R'SnX \quad (X = Cl, Br)$$

though in some cases the major product is R_3SnX[115]. In a detailed study of the reaction of diphenyltin with alkyl bromides and iodides, RX, it was shown that, in most cases, the preponderant product is Ph_3SnX with lesser amounts of Ph_2RSnX but $PhRSnX_2$, PhR_2SnX and Ph_2SnX_2 were also formed; the relative proportions of all these products were determined[116].

9.2. COMPOUNDS CONTAINING TIN BONDED TO ANOTHER MAIN-GROUP METAL

This subject is included in a comprehensive review[135] of compounds containing metal–metal bonds.

The organotin derivatives of the alkali metals are highly reactive compounds which are used in solution soon after they are formed; they must, at all stages, be protected from the atmosphere by the use of an inert gas. In the earlier methods for synthesising this type of compound sodium in liquid ammonia was used.

$$R_3SnBr + 2Na \xrightarrow{\text{liq. NH}_3} R_3SnNa + NaBr$$
$$R_2SnBr_2 + 4Na \xrightarrow{\text{liq. NH}_3} R_2SnNa_2 + 2NaBr$$

It is now more usual to prepare lithium rather than sodium compounds and tetrahydrofuran (= THF) can be used as the reaction medium, for example[124]:

$$Ph_3SnCl + 2Li \xrightarrow{\text{THF}} Ph_3SnLi + LiCl.$$

The organotin lithium compounds can also be prepared by treating stannous chloride with 3 molar proportions of an alkyllithium compound when the following reactions occur[61].

$$2BuLi + SnCl_2 \longrightarrow [Bu_2Sn] + 2LiCl$$
$$[Bu_2Sn] + BuLi \rightleftharpoons Bu_3SnLi$$

When Me_3SnLi was made in tetrahydrofuran, by a number of methods, all attempts to remove the solvent caused decomposition giving the compound $(Me_3Sn)_3SnLi \cdot 3THF$[138].

A solution of triphenylstannyllithium in tetrahydrofuran is dark green and the solvent is slowly attacked giving $Ph_3Sn(CH_2)_4OH$[60].

The stannyllithium compounds can be estimated by a double titration technique[59] and they are usually alkylated at the tin atom on treatment with alkyl halides[24].

$$Me_3SnLi + EtBr \longrightarrow Me_3SnEt + LiBr$$

In other cases a preferential exchange reaction giving the organotin halide occurs[54].

$$Ph_3SnLi + C_6F_5Br \longrightarrow Ph_3SnBr + C_6F_5Li$$

The insertion of sulphur into the tin–lithium bond to give compounds containing the Sn—S—Li group was discussed on p. 140. The alkali metal derivatives react with organometallic chlorides, for example[87]:

$$2Ph_3SnNa + Ph_2SnCl_2 \longrightarrow Ph_3SnSnPh_2SnPh_3 + 2NaCl.$$

This method is widely used to join tin to other Group IVb elements (see below). Some interesting comparisons between the reactivities of Ph_3SnLi and $Ph_3Sn^{\ominus}Bu_4N^{\oplus}$ have been made[41]; the lithium atom can approach closely to tin and thus enhance the covalent character of the Sn—Li bond whereas the positive charge on the tetrabutylammonium group is shielded and little covalent interaction is possible in the tin–nitrogen bond.

Compounds in which tin is bonded to a Group II metal are comparatively rare though there have been several recent reports of work in this area. When a solution of triphenyltin chloride in tetrahydrofuran is treated with magnesium the hexaphenyldistannane, which is formed initially, slowly undergoes further reaction to give bis(triphenylstannyl)magnesium[125].

$$2Ph_3SnCl + Mg \longrightarrow Ph_3SnSnPh_3 + MgCl_2$$
$$Ph_3SnSnPh_3 + Mg \longrightarrow (Ph_3Sn)_2Mg$$

Reaction occurs between triphenyltin hydride and $EtMgBr \cdot NEt_3$ to give ethane and $Ph_3SnMgBr \cdot NEt_3$ as an oil. The latter compound eliminates triethylamine when heated for 5 hours at 50° and 0.01 mm to give a yellow, air sensitive solid which, apparently, arises from migration of a phenyl group from tin to magnesium[36a].

$$2Ph_3SnMgBr \cdot NEt_3 \longrightarrow \underset{Ph}{\overset{Ph_2Sn}{\diagdown}} Mg \underset{Br}{\overset{Br}{\diagdown}} Mg \underset{SnPh_2}{\overset{Ph}{\diagup}} + 2Et_3N$$

When the silylated mercury amine $Hg[N(SiMe_3)_2]_2$ reacts with two molar proportions of triphenyltin hydride at room temperature bis(triphenylstannyl)mercury is formed[51].

$$2Ph_3SnH + Hg[N(SiMe_3)_2]_2 \longrightarrow Hg(SnPh_3)_2 + 2(Me_3Si)_2NH$$

By the hydrostannolysis of chelate complexes of alkylzinc and alkylcadmium compounds products containing Sn—Zn and Sn—Cd bonds were isolated in high yield[44].

$$2Ph_3SnH + R_2Zn \cdot bipy \longrightarrow (Ph_3Sn)_2Zn \cdot bipy + 2RH$$
$$2Ph_3SnH + R_2Cd \cdot TMED \longrightarrow (Ph_3Sn)_2Cd \cdot TMED + 2RH$$

(bipy = 2,2'-bipyridyl; TMED = N,N,N',N'-tetramethylethylenediamine)

Similar treatment of uncomplexed dialkyl-zinc and -cadmium compounds with triphenyltin hydride led to deposition of metallic zinc or cadmium. The presence of the bidentate ligand is necessary to promote the required reaction and also to stabilise the product. When bis(triethylgermyl)cadmium reacted with triethyltin hydride, hexaethyldistannane and cadmium metal were isolated, probably via an unstable intermediate containing Sn—Cd bonds[134].

$$(Et_3Ge)_2Cd + 2Et_3SnH \longrightarrow 2Et_3GeH + [(Et_3Sn)_2Cd]$$
$$[(Et_3Sn)_2Cd] \longrightarrow Et_6Sn_2 + Cd$$

In the absence of chelating agents triphenyltin chloride reacted with a zinc–copper couple to form diphenyltin which disproportionated to tetraphenyltin and the metal. The diphenyltin was formed from the migration of a phenyl group from tin to zinc in an unstable organotin–zinc intermediate[45].

Rather more attention has been paid to compounds in which tin is bound to another Group IVb element and a common method of preparation utilises a stannyllithium compound and an organometallic halide[110].

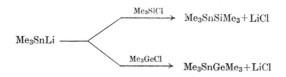

Similarly triphenylstannyllithium reacts with germanium tetrachloride to give *tetrakis*(triphenylstannyl)germane, $(Ph_3Sn)_4Ge$; the corresponding lead compound can be made by using $PbCl_2$ in place of $GeCl_4$[140]. The alternative approach to this method whereby reaction occurs between an organotin halide and a silyl- or germylalkali metal compound[19] is illustrated by the following examples.

$$Ph_3GeK + Ph_3SnCl \longrightarrow Ph_3SnGePh_3 + KCl \qquad \text{(ref. 139)}$$
$$H_3SiK + Me_3SnCl \longrightarrow Me_3SnSiH_3 + KCl \qquad \text{(ref. 5)}$$

$$\text{Li(SiPh}_2)_4\text{Li} + \text{Ph}_2\text{SnCl}_2 \longrightarrow \begin{array}{c} \text{Ph}_2\text{Si}\text{------}\text{SiPh}_2 \\ | \qquad\qquad | \\ \text{Ph}_2\text{Si}\diagdown \qquad\diagup\text{SiPh}_2 \\ \text{Sn} \\ \text{Ph}_2 \end{array} + 2\text{LiCl} \quad \textbf{(ref. 66)}$$

In this type of reaction the organotin halide can be replaced by an alkoxide[131].

$$\text{Et}_3\text{GeLi} + \text{Et}_3\text{SnOMe} \longrightarrow \text{Et}_3\text{SnGeEt}_3 + \text{MeOLi}$$

Alternatively, a hydride can be used though in this case the stannyl-germane was obtained in only 35% yield[132].

$$\text{Et}_3\text{GeLi} + \text{Et}_3\text{SnH} \longrightarrow \text{Et}_3\text{SnGeEt}_3 + \text{LiH}$$

When triethyltin hydride was heated with tetraethyllead, tetra-ethyltin and hexaethylditin were isolated together with metallic lead, these products were taken to imply the formation of unstable intermediates of the type $(\text{Et}_3\text{Sn})_n\text{PbEt}_{4-n}$[133].

Compounds containing tin–germanium bonds can be made by reaction between an amine and a hydride.

$$\diagup\!\!\!\!\!\diagdown\text{MNR}_2 + \diagup\!\!\!\!\!\diagdown\text{M'H} \rightarrow \diagup\!\!\!\!\!\diagdown\text{M---M'}\diagdown\!\!\!\!\!\diagup + \text{R}_2\text{NH}$$

(Where M = Ge and M' = Sn or M = Sn and M' = Ge.) Best results are obtained by using organostannylamines and organogermanium hydrides and the method has been ingeniously developed by Creemers and Noltes[34] to produce a wide range of compounds. Triphenylgermane and *bis*(diethylamino)diethyltin react in equi-molar proportions to give the labile intermediate *IV*. The stannyl-

$$\text{Ph}_3\text{GeH} + \text{Et}_2\text{Sn(NEt}_2)_2 \longrightarrow \underset{IV}{\text{Ph}_3\text{GeSn(NEt}_2)\text{Et}_2} + \text{Et}_2\text{NH}$$

amine, *IV*, then undergoes various standard reactions with protic species.

Ph$_3$GeSnEt$_2$OPh + Et$_2$NH Ph$_3$GeSnEt$_2$SnPh$_3$ + Et$_2$NH

PhOH Ph$_3$SnH

Et$_2$NH + Ph$_3$GeSnEt$_2$C≡CPh $\xleftarrow{\text{PhC≡CH}}$ $\underset{IV}{\text{Ph}_3\text{GeSn(NEt}_2)\text{Et}_2}$ $\xrightarrow{\text{Ph}_3\text{GeH}}$ Ph$_3$GeSnEt$_2$GePh$_3$ + Et$_2$N

$\frac{1}{2}$H$_2$O PhNHCHO

Et$_2$NH + $\frac{1}{2}$(Ph$_3$GeSnEt$_2$)$_2$O $\underset{V}{\text{Ph}_3\text{GeSnEt}_2\text{N(Ph)CHO}}$ + Et$_2$NH

Reaction between compound IV and organotin dihydrides is unsatisfactory since the diethylamine produced catalyses the decomposition of the dihydride. Instead, the transamination product V is used to prepare compounds with longer chains of tin and germanium atoms.

$$2Ph_3GeSnEt_2N(Ph)CHO + Et_2SnH_2 \longrightarrow$$
$$V$$
$$Ph_3GeSnEt_2SnEt_2SnEt_2GePh_3 + 2PhNHCHO$$

$$Ph_3GeSnEt_2N(Ph)CHO + Ph_2SnH_2 \longrightarrow Ph_3GeSnEt_2SnPh_2H + PhNHCHO$$
$$V$$
$$\downarrow Et_2NH$$
$$Ph_3GeSnEt_2SnPh_2SnPh_2SnEt_2GePh_3$$

The synthetic possibilities increase further when *tris*(diethylamino)-ethyltin is treated with one, two or three molar proportions of triphenylgermane though, in the last example, only 3% of *tris*(triphenylgermyl)ethyltin was formed. The reactions of the labile

$$EtSn(NEt_2)_3 + Ph_3GeH \longrightarrow EtSn(NEt_2)_2GePh_3 + Et_2NH$$
$$VI$$
$$EtSn(NEt_2)_3 + 2Ph_3GeH \longrightarrow Ph_3GeSnEt(NEt_2)GePh_3 + 2Et_2NH$$
$$VII$$
$$EtSn(NEt_2)_3 + 3Ph_3GeH \longrightarrow EtSn(GePh_3)_3 + 3Et_2NH$$

compounds VI and VII were exploited in a similar manner to those of compound IV. Polymeric products containing alternating tin and germanium atoms have also been prepared, for example, the product of reaction between *bis*(diethylamino)diethyltin and diphenylgermane was pale yellow with a decomposition point of 260°.

$$xPh_2Sn(NEt_2)_2 + xPh_2GeH_2 \longrightarrow [SnPh_2GePh_2]_x + 2xEt_2NH$$

A common method for making compounds having tin–phosphorus bonds is by reaction between an organotin halide and a metal derivative of a phosphine.

$$R_3SnBr + Ph_2PMgBr \longrightarrow R_3SnPPh_2 + MgBr_2 \qquad \text{(ref. 23a)}$$
$$MePLi_2 + 2Me_3SnBr \longrightarrow MeP(SnMe_3)_2 + 2LiBr \qquad \text{(refs. 7, 145)}$$

The alternative approach to this coupling reaction, using a halophosphine and a stannyllithium compound, has also been reported[146].

$$RPCl_2 + 2R_3'SnLi \longrightarrow RP(SnR_3')_2 + 2LiCl$$

In the presence of a base, organotin halides react with phosphines[107].

$$Ph_{4-n}SnCl_n + nPh_2PH + nEt_3N \longrightarrow Ph_{4-n}Sn(PPh_2)_n + nEt_3NHCl$$

In a similar reaction between triphenylphosphine and trimethyltin chloride, $(Me_3Sn)_2PPh$ was formed[111]. Various phenylstannyl-phosphines are produced when Ph_4Sn reacts directly with phosphorus[112].

Similar methods are used to prepare derivatives of other Group Vb elements as shown in the following examples.

$$Ph_2AsNa + Ph_3SnCl \longrightarrow Ph_2AsSnPh_3(42\%) + NaCl \quad \text{(ref. 108)}$$
$$Ph_2AsCl + Ph_3SnLi \longrightarrow Ph_2AsSnPh_3(81\%) + LiCl \quad \text{(ref. 108)}$$
$$SbLi_3 + 3Me_3SnCl \longrightarrow Sb(SnMe_3)_3 + 3LiCl \quad \text{(ref. 6)}$$
$$PhSbCl_2 + 2Ph_3SnLi \longrightarrow PhSb(SnPh_3)_2 + 2LiCl \quad \text{(ref. 109)}$$

Organotin hydrides react with alkylantimony compounds at 170° with evolution of alkane and formation of an Sn—Sb bond, the method is also applicable to bismuth compounds[136].

$$3Et_3SnH + Et_3M \longrightarrow 3C_2H_6 + (Et_3Sn)_3M \quad (M = Sb, Bi)$$

All of the compounds containing an Sn—M bond, where M is a Group Vb element, are sensitive to oxygen and must be protected from the atmosphere. Thus the stannyl-phosphines[104] and -arsines[108] are oxidised slowly by air, or more rapidly by hydrogen peroxide, to the corresponding stannyl-phosphinates and -arsinates.

$$Ph_3SnMPh_2 + O_2 \longrightarrow Ph_3SnOM(O)Ph_2 \quad (M = P, As)$$

Olefin and acetylene insertion into the Sn—P bond has been reported, the reactions were accelerated when azo*bis*isobutyronitrile was added[105].

$$Ph_3SnPPh_2 + CH_2=CHCH_2Cl \longrightarrow Ph_3SnCH_2CH(CH_2Cl)PPh_2$$
$$Ph_3SnPPh_2 + HC\equiv CPh \longrightarrow Ph_3SnCH=C(Ph)PPh_2$$

Similar insertion reactions occur with a wide range of unsaturated compounds (in this respect the stannylphosphines resemble the stannylamines)[103, 106].

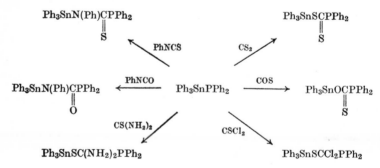

TABLE II

Physical properties of some compounds in which tin is bonded to another main group metal

Compound	m.p.	b.p.	n_D^{20}	d^{20}
$(Ph_3Sn)_4Pb$	~200° decomp.	—	—	—
$(Ph_3Sn)_4Ge$	~324° decomp.	—	—	—
$Ph_3SnGePh_3$	284–286°	—	—	—
$Me_3SnGeEt_3$	—	90–93°/14 mm	—	—
$Et_3SnGeEt_3$	—	109°/1·5 mm	—	—
$Bu_3SnSiMe_3$	—	88°/0·2 mm	1·4873	—
$Ph_3SnSiPh_3$	289–291°	—	—	—
$(Ph_3Sn)_3Bi$	138–142° decomp.	—	—	—
$(Me_3Sn)_3Sb$	39°	—	—	—
$(Ph_3Sn)_3Sb$	214–215°	—	—	—
$Ph_3SnAsPh_2$	117–119°	—	—	—
$Ph_2Sn(AsPh_2)_2$	80° decomp.	—	—	—
$PhSn(AsPh_2)_3$	85° decomp.	—	—	—
$PhAs(SnPh_3)_2$	112–115°	—	—	—
$As(SnPh_3)_3$	212–216°	—	—	—
$Et_3SnP(SEt)Me$	—	86–88°/2 mm	1·5485	1·2498
$MeP(SnMe_3)_2$	—	89–90°/3 mm	1·5778	1·5601
$(Me_3Sn)_3P$	—	136–137°/0·3 mm	1·5970	1·6769
$(Ph_3Sn)_2Zn·bipy^a$	141–144°	—	—	—
$(Ph_3Sn)_2Cd·bipy^a$	154° decomp.	—	—	—

^{a} bipy = 2,2'-bipyridyl.

Alkyl disulphides effect Sn—P bond cleavage to give alkylthiolato-tin and -phosphorus compounds[15].

$$MeP(SnEt_3)_2 + 2Et_2S_2 \longrightarrow MeP(SEt)_2 + 2Et_3SnSEt$$
$$MeP(SnEt_3)_2 + Et_2S_2 \longrightarrow MeP(SEt)SnEt_3 + 2Et_3SnSEt$$

A phosphine which has all three hydrogens replaced by tin atoms retains its donor properties and will displace carbon monoxide from transition metal complexes[113].

$$(Me_3Sn)_3P + (CO)_3Co(NO) \longrightarrow (NO)(CO)_2Co[P(SnMe_3)_3] + CO$$

9.3. COMPOUNDS IN WHICH TIN IS BONDED TO A TRANSITION METAL

This is a field which has seen intense activity during the last ten years and has reached a stage where current work is more concerned with details of structure and bonding and less with synthetic methods.

As the structures of these compounds represent a special case in that they reflect the properties of the transition metal with the tin substituent being relatively unimportant a brief discussion of some representative structures is included in this chapter. Compounds containing transition metal–tin bonds are more or less readily oxidised by atmospheric oxygen and the preparations described below generally require an inert atmosphere.

The following equation for the preparation of trimethylstannylcobalt tetracarbonyl and the corresponding triphenyl compound illustrates one of the most common synthetic procedures[14].

$$R_3SnCl + NaCo(CO)_4 \longrightarrow R_3SnCo(CO)_4 + NaCl \qquad (R = Me, Ph)$$

(Studies of the kinetics of carbonyl exchange in these products have been reported[14].) A similar reaction is used to prepare compounds containing 2 and 3 cobalt atoms joined to tin.

$$Ph_2SnCl_2 + 2NaCo(CO)_4 \longrightarrow Ph_2Sn[Co(CO)_4]_2 + 2NaCl \qquad \text{(ref. 65)}$$
$$RSnCl_3 + 3NaCo(CO)_4 \longrightarrow RSn[Co(CO)_4]_3 + 3NaCl \qquad \text{(ref. 96)}$$
$$(R = Me, Bu, CH_2{=}CH, Ph)$$

The monobutyltin compound, $BuSn[Co(CO)_4]_3$ can also be made from $BuSnCl_3$ and $Co_2(CO)_8$ and has been shown to have structure *VIII*[96]. From the same two reactants $BuSn[Co(CO)_3]_3$ was made, this compound has structure *IX* with Co—Co bonds[67].

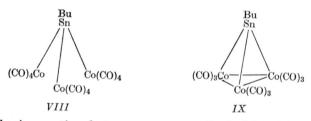

VIII   *IX*

The basic reaction between an organotin halide and a sodium derivative has been used to prepare the stannyl derivatives of a large number of transition metals, as indicated by the following examples.

$$R_2SnCl_2 + 2NaRe(CO)_5 \longrightarrow 2NaCl + R_2Sn[Re(CO)_5]_2 \qquad \text{(ref. 126)}$$
$$(R = Me, Ph)$$

$$RSnCl_3 + 3NaMn(CO)_5 \longrightarrow 3NaCl + RSn[Mn(CO)_5]_3 \qquad \text{(ref. 126)}$$
$$(R = Me, Bu, CH_2{=}CH, Ph)$$

$$PhSnCl_3 + 3NaM(CO)_5 \longrightarrow 3NaCl + PhSn[M(CO)_5]_3 \qquad \text{(ref. 81)}$$
$$(M = Mn, Re)$$

$$Me_3SnCl + C_5H_5(CO)_3MoNa \longrightarrow NaCl + Me_3SnMo(CO)_3C_5H_5 \quad (ref. 94)$$

$$Ph_3SnCl + C_5H_5(CO)_3MNa \longrightarrow NaCl + Ph_3SnM(CO)_3C_5H_5 \quad (ref. 94)$$
$$(M = W, Cr)$$

$$Me_3SnBr + C_5H_5(CO)_3MNa \longrightarrow NaBr + Me_3SnM(CO)_3C_5H_5 \quad (ref.24)$$
$$(M = W, Cr)$$

$$Ph_2SnCl_2 + 2C_5H_5(CO)_2FeNa \longrightarrow 2NaCl + Ph_2Sn[Fe(CO)_2C_5H_5]_2 \, (ref. 82)$$

$$PhSnCl_3 + 3C_5H_5(CO)_2FeNa \longrightarrow 3NaCl + PhSn[Fe(CO)_2C_5H_5]_3 \quad (ref. 82)$$

$$PhSnCl_3 + 2C_5H_5(CO)_3MNa \longrightarrow 2NaCl + ClPhSn[M(CO)_3C_5H_5]_2$$
$$(M = Mo, W) \hspace{6cm} (ref. 81)$$

A related synthetic procedure is to replace the alkali metal in a
stannyl-lithium or stannyl-sodium compound with a transition metal
by using a suitably substituted transition metal halide. Thus
triphenylstannylsodium reacted with cyclopentadienyltitanium
chlorides in tetrahydrofuran to give $(C_5H_5)_2TiSnPh_3 \cdot THF$,
$(C_5H_5)_2ClTiSnPh_3$ and $(C_5H_5)_2Ti(SnPh_3)_2$[32]. Stannyl-zirconium and
-hafnium derivatives were made in a similar manner[71].

$$Ph_3SnLi + (C_5H_5)_2MCl_2 \longrightarrow (C_5H_5)_2ClMSnPh_3 + LiCl$$
$$(M = Zr, Hf)$$

Organotin halides react with other transition metal species to form
tin–metal bonds.

$$Ph_3SnCl + HCo(D_2H_2)py \longrightarrow Ph_3SnCo(D_2H_2)py + HCl \quad (ref. 102)$$
$$(py = pyridine, D_2H_2 = dimethylglyoxime)$$

$$MeSnCl_3 + bipyMo(CO)_4 \longrightarrow bipy(CO)_3ClMoSnMeCl_2 + CO \quad (ref. 74)$$
$$(bipy = 2,2'\text{-bipyridyl})$$

$$RSnBr_3 + C_5H_5Co(CO)_2 \longrightarrow C_5H_5COBrCoSnRBr_2 + CO \quad (ref. 74)$$
$$(R = Me, Ph)$$

Tributyltin chloride reacts with iron pentacarbonyl under various
conditions[30, 31] to give $[Bu_2SnFe(CO)_4]_2$ (X), $Bu_4Sn_3[Fe(CO)_4]_4$ (XI)
and $Sn[Fe(CO)_4]_4$ (XII). The same three compounds are produced
from reaction between Bu_3SnCl and $Fe_3(CO)_{12}$ and their structures
are considered to be as follows[31].

$$X \hspace{4cm} XI \hspace{4cm} XII$$

Monocyclic compounds such as X can be made by other methods[69] which include reaction between dialkynyltin compounds, $Bu_2Sn(C \equiv CR)_2$ and $Fe_3(CO)_{12}$[67, 68] and also from organotin hydrides and iron pentacarbonyl[74].

$$2Me_2SnH_2 + 2Fe(CO)_5 \longrightarrow [Me_2SnFe(CO)_4]_2 + 2H_2 + 2CO$$

Stannous halides can be inserted into Co—Co bonds[12]:

$$(CO)_3(Bu_3P)Co—Co(Bu_3P)(CO)_3 + SnCl_2 \longrightarrow$$
$$(CO)_3(Bu_3P)CoSnCl_2Co(Bu_3P)(CO)_3$$

and the products converted to the corresponding organotin compounds by use of Grignard reagents.

$$[(CO)_3(Bu_3P)Co]_2SnCl_2 + 2EtMgBr \longrightarrow$$
$$[(CO)_3(Bu_3P)Co]_2SnEt_2 + MgCl_2 + MgBr_2$$

Two further examples of the replacement of tin-bound halogen by an organic group in compounds containing transition metal–tin bonds are:

$$[(C_5H_5)(CO)_2Fe]_2SnI_2 + 2MeLi \longrightarrow [(C_5H_5)(CO)_2Fe]_2SnMe_2 + 2LiI \quad \text{(ref. 55)}$$
$$(CO)_5MnSnPh_2Cl + C_6F_5Li \longrightarrow (CO)_5MnSnPh_2C_6F_5 + LiCl. \quad \text{(ref. 126)}$$

In the last reaction the compound $(CO)_5MnSnPh_2SnPh_2Mn(CO)_5$, having a chain of 4 metal atoms was also formed. Polynuclear species containing 3 different metal atoms have been prepared.

$$(CO)_5MnSnPh_2Cl + NaMo(CO)_3C_5H_5 \longrightarrow (CO)_5MnSnPh_2Mo(CO)_3C_5H_5 + NaCl$$
$$\text{(ref. 83)}$$

$$(CO)_5MnSnPh_2Cl + NaRe(CO)_5 \longrightarrow (CO)_5MnSnPh_2Re(CO)_5 + NaCl$$
$$\text{(ref. 83)}$$

$$(CO)_5MnSnMe_2Cl + NaMo(CO)_3C_5H_5 \longrightarrow (CO)_5MnSnMe_2Mo(CO)_3C_5H_5 + NaCl$$
$$\text{(ref. 95)}$$

The formation of Sn—Sn bonds by reaction between organostannylamines and organotin hydrides has been discussed earlier in this chapter. Where the appropriate hydrides are available the method can be extended to form transition metal–tin bonds.

$$C_5H_5(CO)_3MH + Me_3SnNMe_2 \longrightarrow C_5H_5(CO)_3MSnMe_3 + Me_2NH \quad \text{(ref. 24)}$$
$$\text{(M = Cr, W)}$$

Distannyl metal tetracarbonyl compounds have been prepared by reaction between trialkyltin hydrides and the appropriate metal carbonyls [31a, 71a].

$$6R_3SnH + M_3(CO)_{12} \xrightarrow[\text{light}]{\text{heat or}} 3(R_3Sn)_2M(CO)_4 + 3H_2$$

$$(M = Fe, Ru, Os)$$

Dessy and his coworkers have examined the factors involved in the preparation of compounds containing metal–metal bonds by metal exchange reactions which include the following types [42, 43].

$$R_3Sn^{\ominus} + L_nMM'L_m \rightleftharpoons R_3SnML_n + L_mM'^{\ominus}$$

$$L_nM^{\ominus} + R_3SnSnR_3 \rightleftharpoons R_3SnML_n + R_3Sn^{\ominus}$$

$$L_nM^{\ominus} + R_3SnM'L_m \rightleftharpoons L_nMSnR_3 + L_mM'^{\ominus}$$

$$(L = \text{ligand})$$

These reactions will be successful if the most stable anion is produced in the forward reaction as in the following equations [42].

$$Ph_3Sn^{\ominus} + C_5H_5(CO)_2FeFe(CO)_2C_5H_5 \longrightarrow C_5H_5(CO)_2FeSnPh_3 + C_5H_5(CO)_2Fe^{\ominus}$$

$$C_5H_5(CO)_2Fe^{\ominus} + Ph_3SnMo(CO)_3C_5H_5 \longrightarrow$$

$$C_5H_5(CO)_2FeSnPh_3 + C_5H_5(CO)_3Mo^{\ominus}$$

Crystal structure determinations have been reported for several compounds containing transition metal–tin bonds [9, 10, 22a, 77, 93, 136a]. For example it has been shown [137] that, in $Ph_3SnMn(CO)_5$, the configuration is essentially tetrahedral at tin and octahedral at manganese with the mean bond angles Mn—Sn—C = 112·7° and C—Sn—C = 106·0°. The mean bond lengths are Sn—Mn = 2·674 Å and Sn—C = 2·154 Å; these are almost identical with the bond lengths in $Me_3SnMn(CO)_5$ [16]. The geometry about the tin atom in $(CO)_4CoSnPh_2Mn(CO)_5$ is similar [22, 23] with $\angle$ Mn—Sn—Co = 114°, $\angle$ C—Sn—C = 107·5° and an Sn—C bond distance of 2·18 Å. The structure of bipy(CO)$_3$-ClMoSnMeCl$_2$ is such that the coordination numbers are Sn = 5 and Mo = 7 since, in addition to the Mo—Sn bond, there is also a chlorine bridge between these atoms [52]. There are two molecules with different conformations in the unit cell of $Ph_3SnFe(CO)_2C_5H_5$ crystals, the Fe—Sn distances of 2·533 and 2·540 Å indicate that these bonds are covalent [17]. The configuration about the tin atom in triphenyltin tetracarbonyltriphenylphosphinemanganese is somewhat distorted from Td symmetry, this distortion is ascribed to lattice effects rather than intra-molecular steric hindrance [17].

Although alternative views have been expressed [54a] the transition metal–tin bond lengths indicate that $d_\pi - d_\pi$ interactions occur in many of these compounds.

The structure of the product of reaction of $Ph_3SnMn(CO)_5$ and tetracyclone which was originally thought to be *XIII* [62] has now been shown to be *XIV* [18, 63].

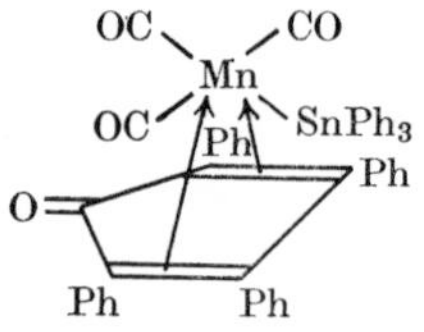

XIII 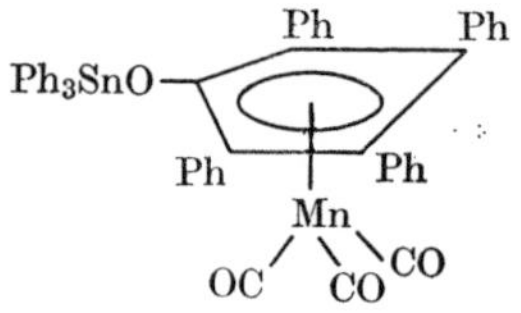   *XIV*

Clark and Tsai [28, 29] have shown that, at 50° in the presence of ultraviolet light, $Me_3SnMn(CO)_5$ and tetrafluoroethylene react to give 55% of trimethyltin fluoride and five carbonylmanganese compounds, one of which was the crystalline insertion product $Me_3SnCF_2CF_2Mn(CO)_5$ (14%). There was no evidence that similar insertion reactions could be effected with $CF_2 = CFH$ or $CF_2 = CFCl$ and ethylene gave the unstable product $Me_3SnMn(CO)_4C_2H_4$ [29]. Tetrafluoroethylene reacted with $R_3SnCo(CO)_4$ under irradiation with ultraviolet light to give, primarily, insertion products but, when the reactants were heated in acetone, an ionic reaction occurred to give R_3SnF and cobalt fluorides [8]. An X-ray crystal analysis of the product of reaction of $[C_5H_5Fe(CO)_2]_2SnPh_2$ with sulphur dioxide showed that insertion into the Sn—C bonds had occurred giving $[C_5H_5Fe(CO)_2]_2Sn(SO_2Ph)_2$ which contained Sn—OS(O)—C units [17a].

Carbonyl groups are displaced from $Ph_3SnRe(CO)_5$ by Ph_3M (M = P, As, Sb) [84]. The nature of the fission products obtained by the action of reagents such as halogen, Ph_2PCl and Me_2AsI on $Me_3SnCo(CO)_4$ and $Me_3SnMn(CO)_5$ has been examined [2].

The infrared spectra of compounds containing transition metal–tin bonds have been studied by a number of workers [1, 40, 70, 97, 128, 129]; interest is centred on the assignment of the $\nu(CO)$ modes and the re-relationship between the positions of these bands and the nature of the metal–metal bonds.

TABLE III

*Melting points of some compounds in which tin
is bonded to a transition metal*

Compound	m.p.
$Me_3SnCo(CO)_4$	74·5°
$Ph_3SnCo(CO)_4$	123°
$Ph_2Sn[Co(CO)_4]_2$	139–141° decomp.
$(Ph_3Sn)_2Fe(CO)_4$	145–150° decomp.
$Ph_3SnMn(CO)_4PPh_3$	232–234°
$Ph_2Sn[Mn(CO)_5]_2$	137–139°
$Ph_2ClSnMn(CO)_5$	97–98°
$PhCl_2SnMn(CO)_5$	83–85°
$Ph_3SnRe(CO)_5$	142–143°
$Ph_2Sn[Re(CO)_5]_2$	139°
$PhSn[Re(CO)_5]_3$	189°

References

1. Abel, E. W., Crow, J. P. and Illingworth, S. M., *Chem. Comm.*, 1968, 817.
2. Abel, E. W. and Hutson, G. V., *J. Inorg. Nucl. Chem.*, 1968, **30**, 2339.
3. Aleksandrov, Yu. A., Fomin, V. M. and Spiridonova, M. N., *Zh. Obshch. Khim.*, 1968, **38**, 1410 (*Chem. Abs.*, 1968, **69**, 76213).
4. Aleksandrov, Yu. A. and Radbil, B. A., *Zh. Obshch. Khim.*, 1966, **36**, 543.
5. Amberger, E. and Muehlhofer, E., *J. Organometallic Chem.*, 1968, **12**, 55.
6. Amberger, E. and Salazar, R. W., *J. Organometallic Chem.*, 1967, **8**, 111.
7. Balashova, L. D., Bruker, A. B. and Soborovskii, L. Z., *Zh. Obshch. Khim.*, 1965, **35**, 2207.
8. Beveridge, V. A. D. and Clark, H. C., *J. Organometallic Chem.*, 1968, **11**, 601.
9. Biryukov, B. P., Ainisimov, K. N., Struchkov, Yu. T., Kolobova, N. E. and Skripkin, V. V., *Zh. Strukt. Khim.*, 1967, **8**, 556 (*Chem. Abs.*, 1967, **67**, 94823d).
10. Biryukov, B. P., Solodova, O. P. and Struchkov, Yu. T., *Zh. Strukt. Khim.*, 1968, **9**, 228 (*Chem. Abs.*, 1968, **69**, 46843).
11. Boeseken, J. and Rutgers, J. J., *Rec. Trav. Chim.*, 1923, **42**, 1017.
12. Bonati, F., Cenini, S., Morelli, D. and Ugo, R., *J. Chem. Soc. (A)*, 1966, 1052.
13. Boue, S., Gielen, M. and Nasielski, J., *Bull. Soc. Chim. Belges.*, 1964, **73**, 864.
14. Breitschaft, S. and Basolo, F., *J. Amer. Chem. Soc.*, 1966, **88**, 2702.
15. Bruker, A. B., Balashova, L. D. and Soborovskii, L. Z., *Zh. Obshch. Khim.*, 1966, **36**, 75.
16. Bryan, R. F., *Chem. Comm.*, 1967, 355; *J. Chem Soc. (A)*, 1968, 696.
17. Bryan, R. F., *J. Chem. Soc. (A)*, 1967, 172; 192.
17a. Bryan, R. F. and Manning, A. R., *Chem. Comm.*, 1968, 1220.

18. Bryan, R. F. and Weber, H. P., *J. Chem. Soc. (A)*, 1967, 843.

19. Bulten, E. J. and Noltes, J. G., *Tetrahedron Letters*, 1966, 4389.

20. Burg, A. B. and Spielman, J. R., *J. Amer. Chem. Soc.*, 1961 **83**, 2667.

21. Bychkov, V. T. and Vyazankin, N. S., *Zh. Obshch. Khim.*, 1965, **35**, 687.

22. Bir'yukov, B. P., Anisimov, K. N., Sruchkov, Yu. T., Kolobova, N. E., Osipova, O. P. and Zakharova, M. Ya., *Zh. Strukt. Khim.*, 1967, **8**, 554 (*Chem. Abs.*, 1967, **67**, 94822c).

22a. Bir'yukov, B. P., Struchkov, Yu. T., Anisimov, K. N., Kolobova, N. E. and Skripkin, V. V., *Chem. Comm.*, 1968, 1193.

23. Bir'yukov, B. P., Struchkov, Yu. T., Anisimov, K. N., Kolobova, N. E., Osipova, O. P. and Zakharova, M. Ya., *Chem. Comm.*, 1967, 749.

23a. Campbell, I. G. M., Fowles, G. W. A. and Nixon, L. A., *J. Chem. Soc.*, 1964, 1389; 3026.

24. Cardin, D. J., Keppie, S. A. and Lappert, M. F., *Inorg. Nucl. Chem. Letters*, 1968, **4**, 365.

25. Cardin, D. J. and Lappert, M. F., *Chem. Comm.*, 1966, 506.

26. Carey, N. A. D. and Clark, H. C., *Can. J. Chem.*, 1968, **46**, 643.

27. Clark, H. C., Cotton, J. D. and Tsai, J. H., *Can J.. Chem.*, 1966, **44**, 903.

28. Clark, H. C. and Tsai, J. H., *Chem. Comm.*, 1965, 111.

29. Clark, H. C. and Tsai, J. H., *Inorg. Chem.*, 1966, **5**, 1407.

30. Cotton, J. D., Duckworth, J., Knox, S. A. R., Lindley, P. F., Paul, I., Stone, F. G. A. and Woodward, P., *Chem. Comm.*, 1966, 253.

31. Cotton, J. D., Knox, S. A. R., Paul, I. and Stone, F. G. A., *J. Chem. Soc. (A)*, 1967, 264.

31a. Cotton, J. D., Knox, S. A. R. and Stone, F. G. A., *J. Chem. Soc. (A)*, 1968, 2758.

32. Coutts, R. S. P. and Wailes, P. C., *Chem. Comm.*, 1968, 260.

33. Creemers, H. M. J. C., Ph.D. Thesis, University of Utrecht, 1967.

34. Creemers, H. M. J. C. and Noltes, J. G., *J. Organometallic Chem.*, 1967, **7**, 237.

35. Creemers, H. M. J. C. and Noltes, J. G., *Rec. Trav. Chim.*, 1965, **84**, 382; 590.

36. Creemers, H. M. J. C. and Noltes, J. G., *Rec. Trav. Chim.*, 1965, **84**, 1589.

36a. Creemers, H. M. J. C., Noltes, J. G. and van der Kerk, G. J. M., *J. Organometallic Chem.*, 1968, **14**, 217.

37. Creemers, H. M. J. C., Verbeek, F. and Noltes, J. G., *J. Organometallic Chem.*, 1967, **8**, 469.

38. Cullen, W. R., Dawson, D. S. and Styan, G. E., *J. Organometallic Chem.*, 1965, **3**, 405.

39. Cullen, W. R. and Styan, G. E., *Can. J. Chem.*, 1966, **44**, 1225.

40. Dalton, J., Paul, I., Smith, J. G. and Stone, F. G. A., *J. Chem. Soc. (A)*, 1968, 1195.

41. Dessy, R. E., Kitching, W. and Chivers, T., *J. Amer. Chem. Soc.*, 1966, **88**, 453.

42. Dessy, R. E. and Weissman, P. M., *J. Amer. Chem. Soc.*, 1966, **88**, 5124; 5129.

43. Dessy, R. E., Weissman, P. M. and Pohl, R. L., *J. Amer. Chem. Soc.*, 1966, **88**, 5117.

44. Des Tombe, F. J. A., van der Kerk, G. J. M., Creemers, H. M. J. C. and Noltes, J. G., *Chem. Comm.*, 1966, 914.
45. Des Tombe, F. J. A., van der Kerk, G. J. M. and Noltes, J. G., *J. Organometallic Chem.*, 1968, **13**, P9.
46. Devaud, M., *Compt. Rend.*, 1966, **263**, 1269.
47. Devaud, M., *J. Chim. Phys.*, 1967, **64**, 791.
48. Devaud, M. and Souchay, P., *J. Chim. Phys.*, 1967, **64**, 1778.
49. Devaud, M., Souchay, P. and Person, M., *J. Chim. Phys.*, 1967, **64**, 646.
50. Doretti, L. and Tagliavini, G., *J. Organometallic Chem.*, 1968, **12**, 203.
51. Eaborn, C., Thompson, A. R. and Walton, D. R. M., *Chem. Comm.*, 1968, 1051.
52. Elder, M., Graham, W. A. G., Hall, D. and Kummer, R., *J. Amer. Chem. Soc.*, 1968, **90**, 2189.
53. Farrar, W. V. and Skinner, H. A., *J. Organometallic Chem.*, 1964, **1**, 434.
54. Fenton, D. E., Massey, A. G. and Urch, D. S., *J. Organometallic Chem.*, 1966, **6**, 352.
54a. Fenton, D. E. and Zuckerman, J. J., *J. Amer. Chem. Soc*, 1968, **90**, 6226.
55. Flitcroft, N., Harbourne, D. A., Paul, I., Tucker, P. M. and Stone, F. G. A., *J. Chem. Soc. (A)*, 1966, 1130.
56. George, M. V., Talukdar, P. B., and Gilman, H., *J. Organometallic Chem.*, 1966, **5**, 397.
57. Gilman, H., Atwell, W. H. and Cartledge, F. K., *Advances in Organometallic Chemistry*, Academic Press, New York and London, 1966, **4**, 1.
58. Gilman, H. and Cartledge, F. K., *J. Organometallic Chem.*, 1966, **5**, 48.
59. Gilman, H., Cartledge, F. K. and Sim, S. Y., *J. Organometallic Chem.*, 1963, **1**, 8.
60. Gilman, H., Cartledge, F. K. and Sim, S. Y., *J. Organometallic Chem.*, 1965, **4**, 332.
61. Gilman, H. and Rosenberg, S. D., *J. Amer. Chem. Soc.*, 1953, **75**, 2507.
62. Gorsich, R. D., *J. Amer. Chem. Soc.*, 1962, **84**, 2486.
63. Gorsich, R. D., *J. Organometallic Chem.*, 1966, **5**, 105.
64. Hague, D. N. and Prince, R. H., *J. Inorg. Nucl. Chem.*, 1966, **28**, 1039.
65. Hein, F. and Jehn, W., *Annalen*, 1965, **684**, 4.
66. Hengge, E. and Brychcy, U., *Monatsh. Chem.*, 1966, **97**, 1309.
67. Ibekwe, S. D. and Newlands, M. J., *Chem. Comm.*, 1965, 114.
68. Ibekwe, S. D. and Newlands, M. J., *J. Chem. Soc. (A)*, 1967, 1783.
69. Kahn, O. and Bigorgne, M., *Compt. Rend.*, 1965, **261**, 2483.
70. Kahn, O. and Bigorgne, M., *J. Organometallic Chem.*, 1967, **10**, 137.
71. Kingston, B. M. and Lappert, M. F., *Inorg. Nucl. Chem. Letters*, 1968, **4**, 371.
71a. Knox, S. A. R., Mitchell, C. M. and Stone, F. G. A., *J. Organometallic Chem.*, 1969, **16**, P67.
72. Kraus, C. A. and Neal, A. M., *J. Amer. Chem. Soc.*, 1929, **51**, 2403.
73. Kuivila, H. G. and Beumel, O. F., *J. Amer. Chem. Soc.*, 1961, **83**, 1246.
74. Kummer, R. and Graham, W. A. G., *Inorg. Chem.*, 1968, **7**, 310; 523.
75. Lorbeth, J., Krapf, H. and Noeth, H., *Chem. Ber.*, 1967, **100**, 3511.
75a. Mackay, K. M. and Watt, R., *Organometallic Chem. Rev.*, A, 1969, **4**, 137.
76. McWhinnie, W. R., Poller, R. C. and Thevarasa, M., *J. Organometallic Chem.*, 1968, **11**, 499.

77. Mannan, Kh. A. I. F. M., *Acta Crystallogr. Sect. B.*, 1968, **24**, 603.
78. Morris, M. D., *Anal Chem.*, 1967, **39**, 476.
79. Morris, M. D., *J. Electroanal. Chem. Interfacial Electrochem.*, 1968, **16**, 569 (*Chem. Abs.*, 1968, **68**, 65134).
80. Nefedov, O. M. and Manakov, M. N., *Angew. Chem. Intern. Ed. Engl.*, 1966, **5**, 1021.
81. Nesmeyanov, A. M., Anisimov, K. N., Kolobova, N. E. and Khandozhko, V. N., *Izv. Akad. Nauk S.S.S.R. Ser. Khim.*, 1967, 1395.
82. Nesmeyanov, A. N., Anisimov, K. N., Kolobova, N. E. and Skripkin, V. V., *Izv. Akad. Nauk S.S.S.R. Ser. Khim.*, 1966, 1292.
83. Nesmeyanov, A. N., Anisimov, K. N., Kolobova, N. E. and Zakharova, M. Ya., *Izv. Akad. Nauk S.S.S.R. Ser. Khim.*, 1965, 1122.
84. Nesmeyanov, A. N., Kolobova, N. E., Anisimov, K. N. and Khandozhko, V. N., *Izv. Akad. Nauk S.S.S.R. Ser. Khim.*, 1966, 163.
85. Neumann, W. P. and Koenig, K., *Angew. Chem.*, 1964, **76**, 892.
86. Neumann, W. P. and Koenig, K., *Annalen*, 1964, **677**, 1; 12.
87. Neumann, W. P., Koenig, K. and Burkhardt, G., *Annalen*, 1964, **677**, 18.
88. Neumann, W. P., Pedain, J. and Sommer, R., *Annalen*, 1966, **694**, 9.
89. Neumann, W. P., Petersen, E. and Sommer, R., *Angew. Chem.*, 1965, **77**, 622.
90. Neumann, W. P. and Schneider, B., *Angew. Chem.*, 1964, **76**, 891.
91. Neumann, W. P., Schneider, B. and Sommer, R., *Annalen*, 1966, **692**, 1.
92. Noltes, J. G., *Rec. Trav. Chim.*, 1964, **83**, 515.
93. O'Connor, J. E. and Corey, E. R., *Inorg. Chem.*, 1967, **6**, 968.
94. Patil, H. R. H. and Graham, W. A. G., *Inorg. Chem.*, 1966, **5**, 1401.
95. Patil, H. R. H. and Graham, W. A. G., *J. Amer. Chem. Soc.*, 1965, **87**, 673.
96. Patmore, D. J. and Graham, W. A. G., *Inorg. Chem.*, 1966, **5**, 2222.
97. Patmore, D. J. and Graham, W. A. G., *Inorg. Chem.*, 1967, **6**, 981.
98. Patmore, D. J. and Graham, W. A. G., *Inorg. Chem.*, 1968, **7**, 771.
99. Rundle, R. E. and Olson, D. H., *Inorg. Chem.*, 1964, **3**, 596.
100. Saville, B., *Angew. Chem. Intern. Ed. Engl.*, 1967, **6**, 928.
101. Sawyer, A. K., *J. Amer. Chem. Soc.*, 1965, **87**, 537.
102. Schrauzer, G. N. and Kratel, G., *Angew. Chem.*, 1965, **77**, 130.
103. Schumann, H. and Jutzi, P., *Chem. Ber.*, 1968, **101**, 24.
104. Schumann, H., Jutzi, P., Roth, A., Schwabe, P. and Schauer, E., *J. Organometallic Chem.*, 1967, **10**, 71.
105. Schumann, H., Jutzi, P. and Schmidt, M., *Angew. Chem.*, 1965, **77**, 912.
106. Schumann, H., Jutzi, P. and Schmidt, M., *Angew. Chem. Intern. Ed. Engl.*, 1965, **4**, 787.
107. Schumann, H., Koepf, H. and Schmidt, M., *J. Organometallic Chem.*, 1964, **2**, 159.
108. Schumann, H., Oestermann, T. and Schmidt, M., *Chem. Ber.*, 1966, **99**, 2057.
109. Schumann, H., Oestermann, T. and Schmidt, M., *J. Organometallic Chem.*, 1967, **8**, 105.
110. Schumann, H. and Ronecker, S., *Z. Naturforsch.*, 1967, **B22**, 542.
111. Schumann, H. and Roth, A., *J. Organometallic Chem.*, 1968, **11**, 125.

112. Schumann, H. and Schmidt, M., *Angew. Chem. Intern. Ed. Engl.*, 1965, **4**, 1007.
113. Schumann, H. and Stelzer, O., *Angew. Chem. Intern. Ed. Engl.*, 1968, **7**, 300.
114. Seyferth, D. and Armbrecht, F. M., *J. Amer. Chem. Soc.*, 1967, **89**, 2790.
115. Sisido, K., Kozima, S. and Isibasi, T., *J. Organometallic Chem.*, 1967, **10**, 439.
116. Sisido, K., Miyanisi, T., Nabika, K. and Kozima, S., *J. Organometallic Chem.*, 1968, **11**, 281.
117. Sommer, R., Neumann, W. P. and Schneider, B., *Tetrahedron Letters*, 1964, 3875.
118. Sommer, R., Schneider, B. and Neumann, W. P., *Annalen*, 1966, **692**, 12.
119. Tagliavini, G., *Anal. Chim. Acta*, 1966, **34**, 24.
120. Tagliavini, G. and Doretti, L., *Chem. Comm.*, 1966, 562.
121. Tagliavini, G., Faleschini, S. and Genero, E., *Ric. Sci.*, 1966, **36**, 717 (*Chem. Abs.*, 1967, **67**, 53417f).
122. Tagliavini, G., Faleschini, S., Pilloni, G. and Plazzogna, G., *J. Organometallic Chem.*, 1966, **5**, 136.
123. Tagliavini, G., Pilloni, G. and Plazzogna, G., *Ric. Sci.*, 1966, **36**, 114 (*Chem. Abs.*, 1966, **65**, 7023e).
124. Tamborski, C., Ford, F. E. and Soloski, E. J., *J. Org. Chem.*, 1963, **28**, 181; 237.
125. Tamborski, C. and Soloski, E. J., *J. Amer. Chem. Soc.*, 1961, **83**, 3734.
126. Thompson, J. A. J. and Graham, W. A. G., *Inorg. Chem.*, 1967, **6**, 1365; 1875.
127. Tsu Tzu Tsai, and Lehn, W. L., *J. Org. Chem.*, 1966, **31**, 2981.
128. Ugo, R., Cariati, F., Bonati, F., Cenini, S. and Morelli, D., *Ric. Sci.*, 1966, **36**, 253 (*Chem. Abs.*, 1966, **65**, 6512e).
129. Ugo, R., Cenini, S. and Bonati, F., *Inorg. Chem. Acta*, 1967, **1**, 451.
130. Vyazankin, N. S. and Bychkov, V. T., *Zh. Obshch. Khim.*, 1966, **36**, 1684.
131. Vyazankin, N. S., Gladyshev, E. N. and Korneva, S. P., *Zh. Obshch. Khim.*, 1967, **37**, 1736.
132. Vyazankin, N. S., Gladyshev, E. N., Razuvaev, G. A. and Korneva, S. P., *Zh. Obshch. Khim.*, 1966, **36**, 952.
133. Vyazankin, N. S., Kalinina, G. S., Kruglaya, O. A. and Razuvaev, G. A., *Zh. Obshch. Khim.*, 1968, **38**, 906 (*Chem. Abs.*, 1968, **69**, 77387).
134. Vyazankin, N. S., Razuvaev, G. A. and Bychkov, V. T., *Izv. Akad. Nauk S.S.S.R. Ser. Khim.*, 1965, 1665.
135. Vyazankin, N. S., Razuvaev, G. A. and Kruglaya, O. A., *Organometallic Chem. Rev. Sec. A*, 1968, **3**, 323.
136. Vyazankin, N. S., Razuvaev, G. A., Kruglaya, O. A. and Semchikova, G. S., *J. Organometallic Chem.*, 1966, **6**, 474.
136a. Watkins, S. F., *J. Chem. Soc.* (A), 1969, 1552.
137. Weber, H. P. and Bryan, R. F., *Chem. Comm.*, 1966, 443.
138. Wells, W. L. and Brown, T. L., *J. Organometallic Chem.*, 1968, **11**, 271.
139. Wiberg, E., Amberger, E. and Cambensi, H., *Z. Anorg. Allg. Chem.*, 1967, **351**, 164.
140. Willemsens, L. C. and van der Kerk, G. J. M., *J. Organometallic Chem.*, 1964, **2**, 260.

141. Netherlands Patent 6,508,201, Dec. 27, 1965 (*Chem. Abs.*, 1966, **64**,17640e).
142. U.S. Patent 3,211,769, Oct. 12, 1965 (*Chem. Abs.*, 1965, **63**, 18152d).
143. U.S. Patent 3,322,801, May 30, 1967 (*Chem. Abs.*, 1967, **67**, 22001m).
144. U.S. Patent 3,347,889, Oct. 17, 1967 (*Chem. Abs.*, 1968, **68**, 49779).
145. U.S.S.R. Patent 170,976, May 11, 1965 (*Chem. Abs.*, 1965, **63**, 9985d).
146. U.S.S.R. Patent 170,977, May 11, 1965 (*Chem. Abs.*, 1965, **63**, 9985d).

10

ORGANOTIN ESTERS

Organotin hydroxides and protic acids react in a manner which is formally analogous to esterification.

$$R_3SnOH + HA \rightleftharpoons R_3SnA + H_2O$$

Since the Sn—A bonds in the products are, essentially, covalent it is more appropriate to call these compounds 'esters' rather than 'salts' though neither term is entirely satisfactory. The choice of 'A' groups dealt with in this chapter is arbitrary (for example, esters of hydrochloric acid are discussed in Chapter 4) and the main emphasis is on the organotin carboxylates.

10.1 ORGANOTIN CARBOXYLATES

The widespread use of compounds of the type $R_2Sn(OCOR')_2$ for the stabilisation of poly(vinyl chloride) has provided a great stimulus to the study of organotin esters of carboxylic acids and, in 1964, over 400 esters were known[54]. These compounds are readily prepared by a number of methods one of the most important being reaction between organotin oxides (or hydroxides) and carboxylic acids or anhydrides[10, 18, 52, 60, 63, 67, 72, 75].

$$R_3SnOSnR_3 + 2R'COOH \longrightarrow 2R_3SnOCOR' + H_2O$$
$$R_3SnOSnR_3 + (R'CO)_2O \longrightarrow 2R_3SnOCOR'$$
$$R_2SnO + 2R'COOH \longrightarrow R_2Sn(OCOR')_2 + H_2O$$

The water produced in these reactions is usually removed azeotropically, for example a number of carboxylates containing oxime residues have been made for biological testing by distilling a benzene solution of the reactants[65, 66].

$$(Bu_3Sn)_2O + RR'C{=}NOCH_2COOH \longrightarrow 2Bu_3SnOCOCH_2ON{=}CRR' + H_2O$$

Alternatively the reactants are heated until there is no further evolution of water[58].

$$(Bu_3Sn)_2O + 2o\text{-}HOC_6H_4COOH \xrightarrow{100-110°} 2Bu_3SnOCOC_6H_4OH\text{-}o + H_2O$$

Another major preparative method is reaction between organotin halides and alkali metal or, less commonly, silver salts of carboxylic acids[22].

$$R_3SnCl + R'COOM \longrightarrow R_3SnOCOR' + MCl \quad (M = Na, K, Ag)$$

By heating organosilicon carboxylates with diphenyltin oxide, carboxylate groups are readily transferred from silicon to tin[28, 50].

$$Et_2Si(OCOR)_2 + Ph_2SnO \longrightarrow Ph_2Sn(OCOR)_2 + Et_2SiO$$
$$(R = Me, Et)$$

Although rarely used as a preparative method esters can be made by the cleavage of organic groups (usually vinyl or phenyl) from tin by carboxylic acids[25].

$$R_4Sn + R'COOH \longrightarrow R_3SnOCOR' + RH$$

Organotin hydrides react with carboxylic acids with the evolution of hydrogen.

$$R_3SnH + R'COOH \longrightarrow R_3SnOCOR' + H_2$$

When dihydrides react with carboxylic acids the initially formed dicarboxylate equilibrates with unreacted dihydride as follows[32].

$$R_2SnH_2 + 2R'COOH \longrightarrow R_2Sn(OCOR')_2 + 2H_2$$
$$R_2Sn(OCOR')_2 + R_2SnH_2 \rightleftharpoons 2R_2Sn(H)OCOR'$$
$$I$$

Dibutyltin hydride in the presence of excess acid gives the dicarboxylate, $Bu_2Sn(OCOR')_2$ but, with a deficiency of acid, the hydride carboxylates (I, $R = Bu$) are formed which then slowly decompose to give hydrogen and the distannane dicarboxylates (II, $R = Bu$). When

$$2R_2Sn(H)OCOR' \longrightarrow R_2Sn(OCOR')Sn(OCOR')R_2 + H_2$$
$$II$$

diphenyltin dihydride reacts with carboxylic acids the distannane dicarboxylates (II, $R = Ph$) are, effectively, the only products irrespective of the conditions used[32].

Organotin halocarboxylates, $R_2Sn(X)OCOR'$, are most conveniently made by heating together, in an inert solvent, equimolar proportions of a dihalide and a dicarboxylate[2, 43].

$$Bu_2SnBr_2 + Bu_2Sn(OCOMe)_2 \longrightarrow 2Bu_2Sn(Br)OCOMe$$

Other methods include reaction between a dihalide and a metal carboxylate[35, 37]:

$$Bu_2SnCl_2 + KOCOR' \longrightarrow Bu_2Sn(Cl)OCOR' + KCl$$

or from a dihalodistannoxane and a carboxylic acid or anhydride[53].

$$ClR_2SnOSnR_2Cl + 2R'COOH \longrightarrow 2R_2Sn(Cl)OCOR' + H_2O$$

Various dialkyltin derivatives of maleic acid are used in the stabilisation of poly(vinyl chloride) including the dibutyltin *bis*(alkyl maleates) (*III*) which are prepared by reaction between dibutyltin oxide and an alkyl hydrogen maleate[35, 57]. Dibutyltin maleate was isolated as a monohydrate (*IV*) and also in anhydrous trimeric and

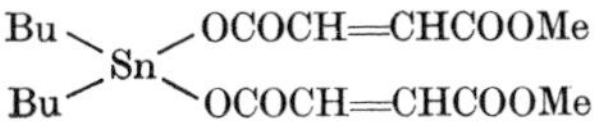

III

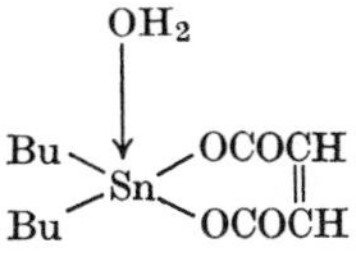

IV

tetrameric forms[35] by the reaction of dibutyltin oxide with maleic acid or maleic anhydride[35, 71]. From the method of manufacture and its physical properties (see for example ref. 71) it appears that the dibutyltin maleate made commercially for the stabilisation of poly(vinyl chloride) is a mixture of oligomeric and probably polymeric forms.

The structures of the organotin carboxylates are discussed in Chapter 12 where it is shown that the trialkyltin compounds $R_3SnOCOR'$ have an associated structure and hence low solubilities in the common organic solvents. When, however, trimethyltin-formate and acetate are heated in a sealed tube at 90° with cyclo-hexane, they are converted to soluble forms and measurement of such properties as infrared spectra can be made on solutions of these compounds whereas the normal forms are too insoluble[45]. When trimethyltin cations, generated electrochemically from tetramethyl-tin, combine with $RCOO^\ominus$ ions the soluble forms of the trimethyltin carboxylates are obtained[37a].

The organotin maleates undergo Diels-Alder reactions with dienes, with cyclopentadiene the products are crystalline solids[35].

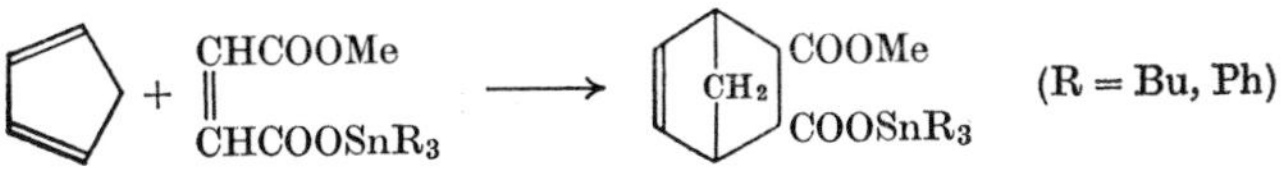

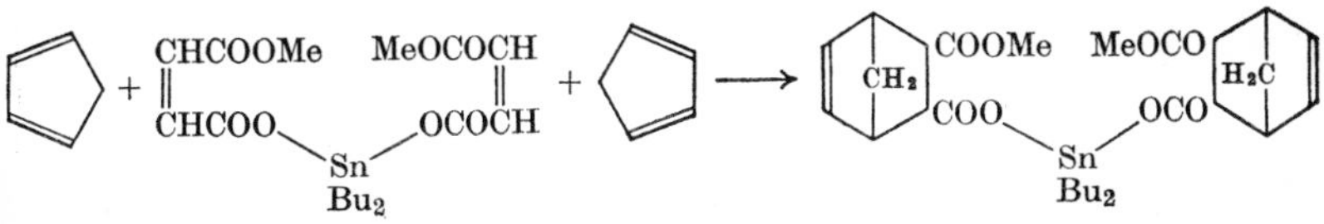

There is evidence that, when a mixture of tributylstannyl trichloroacetate, triphenylphosphine and benzaldehyde were warmed at 40° in benzene some of the β-lactone of 3-hydroxy-3-phenyl-2,2-dichloropropanoic acid was formed.

$$Bu_3SnOCOCCl_3 + Ph_3P + PhCHO \longrightarrow [PhCHCCl_2CO] + Ph_3PO + Bu_3SnCl$$
$$\underset{O}{\underline{\quad\quad\quad}}$$

The lactone was not isolated but the crude product (a) gave a small quantity of $PhCH=CCl_2$ on thermal decomposition and (b) on hydrolysis formed the hydroxyacid $PhCH(OH)CCl_2COOH$[36]. When trimethyltin o-bromobenzoate was heated at 210° the products were tetramethyltin and $o\text{-}BrC_6H_4COOSnMe_2OSn(OCOC_6H_4Br\text{-}o)Me_2$, it appears that a disproportionation reaction occurred followed by hydrolysis of the dicarboxylate[44].

A number of bis[trialkyl(aryl)tin] esters of dicarboxylic acids $R_3SnOCOXCOOSnR_3$ ($X = (CH_2)_n$, $CH=CH$, etc.) have been described[18, 35] and also various polymeric dialkyl(aryl)tin esters, $[-SnR_2OCOXCOO-]_x$, for example poly(dibutyltin adipate), poly-(dibutyltin fumarate) and poly(dibutyltin terephthalate)[8, 20]. An entirely different type of polymer can be made by the polymerisation of stannyl esters of unsaturated acids, thus tributyltin methacrylate can be polymerised in solution, or as an emulsion, to give a tough rubbery solid[26].

$$xBu_3SnOCOC(Me)=CH_2 \longrightarrow [-MeC(COOSnBu_3)CH_2-]_x$$

Copolymerisation of tributyltin methacrylate with dibutyltin methacrylate produces a very clear elastomer with potentially useful properties[26]. Film-forming polymers were obtained by the copolymerisation of an organotin maleate with maleic anhydride and styrene, the reaction was formulated as follows[42].

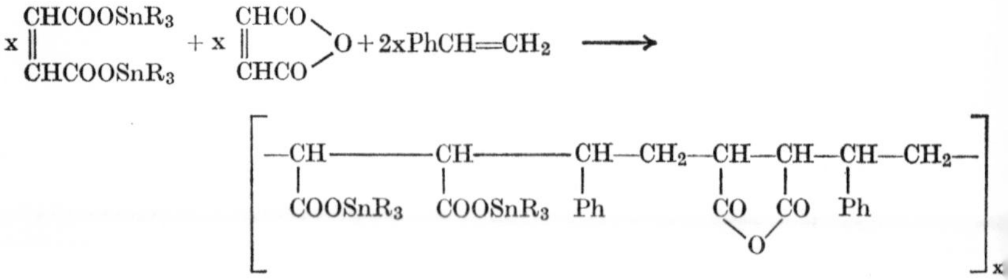

Similar products were formed by reaction between an organotin hydroxide and a copolymer of styrene and maleic anhydride[11, 29]. By the use of torsional vibrational measurements it is possible to distinguish true copolymers of methyl methacrylate and tripropyltin methacrylate from simple mixtures of the polymers[40].

Polymers containing both silicon and tin atoms in the backbone were prepared by reaction between an organotin dichloride and di(carboxyphenyl)silanes[34].

$$x R_2SnCl_2 + x(p\text{-}HOCOC_6H_4)_2SiMe_2 \longrightarrow$$
$$[-SnR_2OCOC_6H_4SiMe_2C_6H_4COO-]_x + 2x\ HCl$$

Reaction between dimethyltin dichloride and sodium sodioacetate gave a product with a molecular weight range 2300–2500[73].

$$x Me_2SnCl_2 + x NaCH_2COONa \longrightarrow [-SnMe_2CH_2COO-]_x + 2x NaCl$$

Organotin esters of α-aminoacids are readily prepared and it is possible that they may be used in peptide synthesis since they have

TABLE I

Physical properties of some organotin carboxylates

Compound	m.p.	b.p.	n_D^{20}
$Me_3SnOCOMe$	196–197°	—	—
$Bu_3SnOCOH$	—	120–125°/0·7 mm	1.4942
$Bu_3SnOCOMe$	85°	—	—
$Bu_3SnOCOCH(NH_2)Me$	130–132°	—	—
$Bu_3SnOCOCCl_3$	80·5°	—	—
$Bu_3SnOCOC_{11}H_{23}$	14–15°	180–185°/2 mm	1·4825
$(Bu_3SnOCO)_2CH_2$	86–87°	—	—
$Bu_2Sn(OCOMe)_2$	8·5–10°	144·5–145·5/10 mm	1·4706
$Bu_2Sn\genfrac{}{}{0pt}{}{/OCOCH}{\backslash OCOCH}\ \|\ \cdot H_2O$	69–70°	—	—
$Bu_2Sn(OCOMe)OSn(OCOMe)Bu_2$	58–60°	—	—
$Bu_2Sn(Cl)OCOMe$	63–65°	—	—
$Bu_2Sn(Cl)OCOCH=CHCOOMe$	41–43°	—	—
$Bu_2Sn(OCOC_{11}H_{23})_2$	22–24°	—	1·4683
$BuSn(OCOMe)_3{}^a$	46°	117–119°/1 mm	1·476
$Ph_3SnOCOMe$	122–124°	—	—
$Ph_3SnOCOCH_2Cl$	182–183°	—	—
$Ph_3SnOCOPh$	84–85·5°	—	—

$^a d^{20} = 1\cdot474.$

an advantage over the corresponding alkyl esters in that the masking R_3Sn group is readily removed by treatment with dilute acid or base[21, 69]. Using tributylstannyl oxide in boiling benzene, 6-amino-penicillin was converted to tributylstannyl 6-aminopenicillinate (V) which can be used for the preparation, in high yields, of pure methyl-penicillins[4].

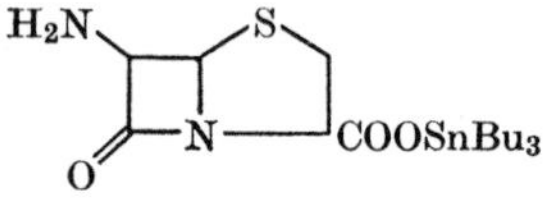

$$V$$

Carboxyl groups in $R_3SnOCOR'$ and $R_2Sn(OCOR')_2$ compounds are conveniently determined by non-aqueous titrations using sodium methoxide in pyridine[24].

10.2. ORGANOTIN PHOSPHATES AND RELATED ESTERS

A large number of esters of phosphorus-containing acids have been prepared, mainly for testing for biological activity. Some illustrative examples are as follows.

$(RO)_2P(O)ONa + R_3'SnCl \longrightarrow (RO)_2P(O)OSnR_3' + NaCl$ (ref. 30)

$(RO)_2P(S)SNa + R_3'SnCl \longrightarrow (RO)_2P(S)SSnR_3' + NaCl$

(refs. 30, 62, 77)

$2(RO)_2P(S)SH + (R_3'Sn)_2O \longrightarrow 2(RO)_2P(S)SSnR_3' + H_2O$ (ref. 64)

$(R_2N)_2P(O)ONa + R_3'SnCl \longrightarrow (R_2N)_2P(O)OSnR_3' + NaCl$ (ref. 30)

$R_2P(O)OH + R_3'SnOH \longrightarrow R_2P(O)OSnR_3' + H_2O$ (ref. 74)

$R_2P(S)SNa + R_3'SnCl \longrightarrow R_2P(S)SSnR_3' + NaCl$ (refs. 6, 7)

$2R_2P(S)SNa + R_2'SnCl_2 \longrightarrow [R_2P(S)S]_2SnR_2' + 2NaCl$ (ref. 31)

$F_2P(O)OP(O)F_2 + (Bu_3Sn)_2O \longrightarrow 2F_2P(O)OSnBu_3$ (ref. 41)

$R_2P(O)OH + R'Sn_3H \longrightarrow R_2P(O)OSnR_3' + H_2$ (ref.40a)

Secondary phosphine oxides react with trialkyl(aryl)tin oxides to give stannyl phosphinates and organotin hydrides.

$R_2P(O)H + (R_3'Sn)_2O \longrightarrow R_2P(O)OSnR_3' + R_3SnH$

This reaction is considered to proceed by a two-stage mechanism.

(i) $R_2P(O)H + (R_3'Sn)_2O \longrightarrow$ [...] $\longrightarrow R_2POSnR_3' + R_3'SnOH$

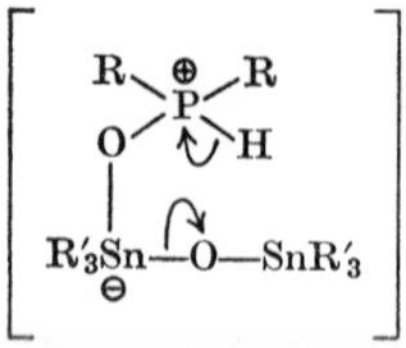

(ii) $R_2POSnR_3' + R_3'SnOH \longrightarrow R_2P(O)OSnR_3' + R_3'SnH$

The disproportionation reaction (ii) was shown, in separate experiments, to occur readily at room temperature[27]. When alkyl esters of phosphorous acid are treated with trialkyltin halides stannyl phosphonates, containing P—O—Sn links are formed, possible mechanisms for this reaction have been discussed[39]. On the other hand treatment of dialkyl phosphites with sodium ethoxide and trialkyltin chlorides gave products with P—Sn bonds[61].

$$(EtO)_2P(H)O + NaOEt + Bu_3SnCl \longrightarrow (EtO)_2P(O)SnBu_3 + NaCl + EtOH$$

Dialkyltin phosphates have been prepared by direct reaction between tin and alkyl phosphates using multicomponent catalyst systems such as alkyl iodide-magnesium-butanol[68].

10.3. ORGANOTIN NITRATES

Esters of nitric acid can be made by a number of methods as indicated by the following three reactions for the preparation of triaryltin nitrates.

$$Ar_3SnX + AgNO_3 \xrightarrow[MeCN]{C_6H_6 \text{ or}} Ar_3SnNO_3 + AgX \quad (X = Cl, I) \quad (refs. 46, 47, 49)$$

$$(Ar_3Sn)_2O + 2HNO_3(dilute) \xrightarrow{EtOH} 2Ar_3SnNO_3 + H_2O \quad (refs. 47, 49)$$

$$Ph_3SnSnPh_3 + 2AgNO_3 \xrightarrow[0°]{acetone} 2Ph_3SnNO_3 + 2Ag \quad (ref. 51)$$

Triphenyltin nitrate is stable for several weeks at room temperature[49]; when heated at 180° in o-dichlorobenzene it disproportionates to give tetraphenyltin and diphenyltin oxide[51].

Dialkyltin oxides react with nitric acid to give the dinitrates, $R_2Sn(NO_3)_2$[55] which can also be prepared from the dichlorides and silver nitrate[19, 23]; dimethyltin dinitrate has been obtained by treating tetramethyltin with dinitrogen tetroxide[1]. The dinitrates are unstable in moist air and also slowly decompose in sealed tubes at room temperature[23]; they do, however, form stable adducts with 1,10-phenanthroline[19, 23] and dimethylsulphoxide[1].

The organotin dinitrates resemble the dihalides in that treatment with carefully controlled amounts of alkali allows the isolation of the partially hydrolysed products $R_2Sn(NO_3)OSn(NO_3)R_2$, $R_2Sn(NO_3)OSn(OH)R_2$ and $R_2Sn(OH)NO_3$; the same products are obtained when the oxide, R_2SnO, is allowed to react with limited quantities of nitric acid[55]. Some organotin nitrates form stable hydrates, for example, $Pr_2Sn(OH)NO_3 \cdot H_2O$[55] and

$Me_3SnNO_3 . H_2O$ [14, 56], vacuum sublimation of the latter gives anhydrous trimethyltin nitrate [56].

Triphenyltin hyponitrite has been made from the iodide and silver hyponitrite [5].

$$2Ph_3SnI + Ag_2N_2O_2 \longrightarrow 2AgI + (Ph_3Sn)_2N_2O_2$$

10.4. MISCELLANEOUS ORGANOTIN ESTERS

There is a great diversity of compounds which could be discussed under this heading and only a very selective treatment of the remaining classes of organotin ester will be attempted. Many of these compounds were synthesised for structural studies and these will be referred to again in Chapter 12.

A very versatile method for the preparation of esters is the reaction of silver salts with organotin halides.

$$R_{4-n}SnCl_n + nAgA \longrightarrow R_{4-n}SnA_n + nAgCl$$

Examples of compounds which have been prepared by this method are $(Ph_3Sn)_3PO_4$ [48], $(Ph_3Sn)_3AsO_4$ [48], $(Ph_3Sn)_2SO_4$ [48], Me_2SnSO_4 [12], $Ph_3SnO_3SC_6H_4Me\text{-}p$ [38], Me_3SnAsF_6 [13], Me_3SnSbF_6 [13], Me_2SnCO_3 [12], and Me_3SnClO_4 [13]. Ethyltin triperchlorate, which decomposes violently at $\sim 50°$ was prepared by reaction between the trichloride and silver perchlorate in ethanol [17].

Certain acids such as the sulphonic acids [3], boric acid [70] and carbonic acid react with organotin oxides and hydroxides to give good yields of the corresponding esters. Carbonate formation occurs so readily that lower organotin hydroxides (other than trimethyltin hydroxide) must be protected from the atmosphere [33].

$$2R_3SnOH + CO_2 \longrightarrow (R_3Sn)_2CO_3 + H_2O$$

Tetrakis(triphenylstannyl) titanate was prepared by evaporating a benzene solution of triphenyltin hydroxide and tetrabutyl titanate [15].

$$4Ph_3SnOH + (BuO)_4Ti \longrightarrow (Ph_3SnO)_4Ti + 4BuOH$$

Similar compounds have been made from alkyl titanates and organotin carboxylates [59]. Reaction between triethanolamine and tetraisopropyl titanate gave the ester *VI* from which the remaining isopropoxyl group can be displaced by treatment with triphenyltin hydroxide to give *VII* (R = Ph) [16]. Attempts to form *VII* (R = Bu)

i-PrOTi$\begin{smallmatrix} \diagup OCH_2CH_2 \diagdown \\ -OCH_2CH_2-N \\ \diagdown OCH_2CH_2 \diagup \end{smallmatrix}$ + R_3SnOH $\longrightarrow$ R_3SnOTi$\begin{smallmatrix} \diagup OCH_2CH_2 \diagdown \\ -OCH_2CH_2-N \\ \diagdown OCH_2CH_2 \diagup \end{smallmatrix}$ + i-PrOH

            VI   *VII*

were unsuccessful. Dibutyltin oxide reacted with tetraisopropyl titanate to give dibutylisopropoxystannyl triisopropyl titanate[76].

$$Bu_2SnO + (i\text{-}PrO)_4Ti \longrightarrow Bu_2(i\text{-}PrO)SnOTi(OPr\text{-}i)_3$$

In many cases stannyl esters can be made by the acid cleavage of tin–carbon bonds. While protic acids are usually employed cleavage can also be effected by Lewis acids as in the preparation of trimethyltin tetrafluoroborate from reaction between trimethylperfluorophenyltin and boron trifluoride in carbon tetrachloride[9].

$$Me_3SnC_6F_5 + 2BF_3 \longrightarrow Me_3SnBF_4 + C_6F_5BF_2$$

TABLE II

Physical properties of some organotin esters (other than carboxylates)

Compound	m.p.	b.p.
$Et_2P(S)SSnEt_3$	—	190°/2 mm
$Ph_2P(O)OSnBu_3$	209°	—
$(Et_2N)_2P(O)OSnBu_3$	151–153°	—
$(EtO)_2P(O)OSnPh_3$	193°	—
$(Ph_3SnO)_3PO$	117–118°	—
$(EtO)_2P(S)SSnPh_3$	105°	—
$[EtP(S)S]_2SnPh_2$	149·5°	—
Me_3SnNO_3	140°	—
$Me_3SnNO_3 . H_2O$	98–99°	—
Ph_3SnNO_3	181–182°	—
$(Ph_3Sn)_2N_2O_2$	128° decomp.	—
$(Bu_3SnO)_3B$	—	230–235°/0·001 mm
$Et_2Sn(OSO_2Me)_2$	334°	—
$Ph_3SnOSO_2C_6H_4Me\text{-}p$	229–231°	—
$(Ph_3Sn)_2SO_4$	> 290°	—
$(Ph_3Sn)_3AsO_4$	195–196°	—

References

1. Addison, C. C., Simpson, W. B. and Walker, A., *J. Chem. Soc.*, 1964, 2360.
2. Alleston, D. L. and Davies, A. G., *J. Chem. Soc.*, 1962, 2050.
3. Anderson, H. H., *Inorg. Chem.*, 1964, **3**, 108.
4. Bamber, P., Ekstrom, B. and Sjoberg, B., *Acta Chem. Scand.*, 1968, **22**, 367.
5. Beck, W., Engelmann, H. and Smedal, H. S., *Z. Anorg. Allg. Chem.*, 1968, **357**, 134.
6. Bonati, F., Cenini, S. and Ugo, R., *J. Organometallic Chem.*, 1967, **9**, 395.
7. Bonati, F., Cenini, S. and Ugo, R., *Rend. Ist. Lombardo Sci. Lettere*, 1965, **A99**, 825 (*Chem. Abs.*, 1966, **65**, 1749c).

8. Bruck, S. D., *J. Polymer Sci. Pt. A*-1, 1967, **5**, 2458.
9. Chambers, R. D. and Chivers, T., *Proc. Chem. Soc.*, 1963, 208.
10. Jen-Hsi Cho, Han-Sheng Hsu, Hsu-Yu Ch'eng, Ch'ang-Lieh Fan, Yuan-Lieh Muo and Chih-Fu Yu, *Hua Hseveh Hseuch Pao*, 1966, **32**, 196 (*Chem. Abs.*, 1966, **65**, 13753e).
11. Chuveleva, E. A., Rzaev, Z. M., Nazarov, P. P., Kochkin, D. A., Chmutov, K. V. and Zubov, P. I., *Zh. Fiz. Khim.*, 1967, **41**, 1389 (*Chem. Abs.*, 1968, **68**, 59917).
12. Clark, H. C. and Goel, R. G., *J. Organometallic Chem.*, 1967, **7**, 263.
13. Clark, H. C. and O'Brien, R. J., *Proc. Chem. Soc.*, 1963, 113; *Inorg. Chem.*, 1963, **2**, 740.
14. Clark, H. C., O'Brien, R. J. and Pickard, A. L., *J. Organometallic Chem.*, 1965, **4**, 43.
15. Cohen, H. J., *J. Org. Chem.*, 1960, **25**, 154.
16. Cohen, H. J., *J. Organometallic Chem.*, 1967, **9**, 177.
17. Devaud, M., *Rev. Chim. Miner.*, 1967, **4**, 921 (*Chem. Abs.*, 1968, **69**, 87122).
18. Dunn, P. and Norris, T., Report No. 269 Australian Defence Scientific Service, Defence Standards Laboratories, 1964.
19. Fenster, A. N. and Becker, E. I., *J. Organometallic Chem.*, 1968, **11**, 549.
20. Frankel, M., Gertner, D., Wagner, D. and Zilkha, A., *J. Appl. Polymer. Sci.*, 1965, **9**, 3383.
21. Frankel, M., Gertner, D., Wagner, D. and Zilkha, A., *J. Org. Chem.*, 1965, **30**, 1596.
22. Frankel, M., Gertner, D., Wagner, D. and Zilkha, A., *J. Organometallic Chem.*, 1967, **9**, 83.
23. Gormley, J. J. and Rees, R. G., *J. Organometallic Chem.*, 1966, **5**, 291.
24. Groagova, A. and Pribyl, M., *Annalen*, 1968, **234**, 423.
25. Henderson, A. and Holliday, A. K., *J. Organometallic Chem.*, 1965, **4**, 377.
26. Ingham, R. K. and Gilman, H., *Inorganic Polymers*, Eds. Stone, F. G. A. and Graham, W. A. G., Academic Press, New York and London, 1962, 386.
27. Issleib, K. and Walter, B., *J. Organometallic Chem.*, 1967, **10**, 177.
28. Kocheshkov, K. A., Gol'der, V. G., Panov, E. M., Zemlyanskii, N. N. and Syutkina, O. P., *Izv. Akad. Nauk S.S.S.R. Ser. Khim.*, 1967, 1171.
29. Kochin, D. A., Rzaev, Z. M., Sukhareva, L. A. and Zubov, P. I., *Vysokomol. Soedin. Ser. A.*, 1967, **9**, 2208 (*Chem. Abs.*, 1968, **68**, 3216).
30. Kubo, H., *Agr. Biol. Chem.* (*Tokyo*), 1965, **29**. 43.
31. Kuchen, W., Jadat, A. and Metten, J., *Chem. Ber.*, 1965, **98**, 3981.
32. Kuivila, H. G., *Advances in Organometallic Chemistry*, Academic Press, New York and London, 1964, **1**, 82.
33. Lohmann, D. H., *J. Organometallic Chem.*, 1965, **4**, 382.
34. Migdal, S., Gertner, D. and Zilkha, A., *J. Organometallic Chem.*, 1968, **11**, 441.
35. Mufti, A. S. and Poller, R. C., *J. Chem. Soc.* (*C*), 1967, 1362; 1767.
36. Ohara, M., Okada, T. and Okawara, R., *Tetrahedron Letters*, 1968, 3489.
37. Okawara, R. and Rochow, E. G., *J. Amer. Chem. Soc.*, 1960, **82**, 3285.
37a. Peruzzo, V., Plazzogna, G. and Tagliavini, G., *J. Organometallic Chem.*, 1969, **18**, 89.
38. Poller, R. C., *J. Inorg. Nucl. Chem.*, 1962, **24**, 593.
39. Pudovik, A. N. and Muritova, A. A., *Dokl. Akad. Nauk S.S.S.R.*, 1964, **158**, 419.

40. Rehage, G. and Schaefer, E. E., *Annalen*, 1968, **235**, 137.
40a. Ridenour, R. E. and Flagg, E. E., *J. Organometallic Chem.*, 1969, **16**, 393.
41. Roesky, H. W., *Chem. Ber.*, 1967, **100**, 2147.
42. Rzaev, Z. M., Kochkin, D. A. and Zubov, P. I., *Dokl. Akad. Nauk S.S.S.R.*, 1967, **172**, 364.
43. Sawyer, A. K. and Kuivila, H. G., *Chem. and Ind.*, 1961, 260.
44. Seyferth, D. and Burlitch, J. M., *Z. Naturforsch*, 1967, **22B**, 1358.
45. Simons, P. B. and Graham, W. A. G., *J. Organometallic Chem.*, 1967, **8**, 479.
46. Simpson, W. B., *Chem. and Ind.*, 1966, 854.
47. Srivastava, T. N. and Bhattacharya, S. N., *Indian J. Chem.*, 1966, **4**, 474.
48. Srivastava, T. N. and Tandon, S. K., *Indian J. Appl. Chem.*, 1963, **26**, 171.
49. Srivastava, T. N. and Tandon, S. K., *Indian J. Chem.*, 1965, **3**, 535.
50. Syutkina, O. P., Panov, E. M., Zemlyanskii, N. N. and Kocheshkov, K. A., *Dokl. Akad. Nauk S.S.S.R.*, 1967, **177**, 615.
51. Tsu Tzu Tsai, Cutler, A. and Lehn, W. L., *J. Org. Chem.*, 1965, **30**, 3049.
52. Vilarem, M. and Maire, J. C., *Compt. Rend.*, 1966, **262**, 480.
53. Wada, M., Shindo, M. and Okawara, R., *J. Organometallic Chem.*, 1963, **1**, 95.
54. Weiss, R. W., Ed. *Organometallic Compounds*, **2**, 2nd Edn., Springer-Verlag, New York, 1967
55. Yasuda, K., Matsumoto, H. and Okawara, R., *J. Organometallic Chem.*, 1966, **6**, 528.
56. Yasuda, K. and Okawara, R., *J. Organometallic Chem.*, 1965, **3**, 76.
57. Belgium Patent 661,479, July 16, 1965 (*Chem. Abs.*, 1966, **64**, 19911a).
58. French Patent 1,386,350, Jan. 22, 1965 (*Chem. Abs.*, 1965, **62**, 16296a).
59. French Patent 1,392,648, March 19, 1965 (*Chem. Abs.*, 1965, **63**, 1816d).
60. Japanese Patent 3532, March 1, 1966 (*Chem. Abs.*, 1966, **65**, 2298g).
61. Japanese Patent 4575, March 14, 1966 (*Chem. Abs.*, 1966, **65**, 2299a).
62. Japanese Patent 4576, March 14, 1966 (*Chem. Abs.*, 1966, **65**, 2298h).
63. Japanese Patent 8856, May 11, 1966 (*Chem. Abs.*, 1966, **65**, 12240e).
64. Japanese Patent 15,290, Aug. 29, 1966 (*Chem. Abs.*, 1966, **65**, 20164g).
65. Japanese Patent 16,296, Sept. 4, 1967 (*Chem. Abs.*, 1968, **68**, 105366).
66. Japanese Patent 21,334, Oct. 21, 1967 (*Chem. Abs.*, 1968, **69**, 27527).
67. Japanese Patent 07941, March 26, 1968 (*Chem. Abs.*, 1968, **69**, 87186).
68. Netherlands Patent 296,687, May 25, 1965 (*Chem. Abs.*, 1965, **63**, 18152b).
69. Netherlands Patent 6,505,425, July 26, 1965 (*Chem. Abs.*, 1966, **64**, 2159e).
70. Netherlands Patent 6,510,469, Feb. 14, 1966 (*Chem. Abs.*, 1966, **65**, 7218d).
71. Netherlands Patent 6,604,233, Oct. 17, 1966 (*Chem. Abs.*, 1967, **66**, 76156p).
72. Polish Patent 51,771, Oct. 15, 1966 (*Chem. Abs.*, 1968, **68**, 49776).
73. U.S. Patent 3,161,664, Dec. 15, 1964 (*Chem. Abs.*, 1965, **62**, 7890b).
74. U.S. Patent 3,179,676, April 20, 1965 (*Chem. Abs.*, 1965, **63**, 2999g).
75. U.S. Patent 3,257,194, June 21, 1966 (*Chem. Abs.*, 1966, **65**, 12240b).
76. U.S. Patent, 3,307,973, March 7, 1967 (*Chem. Abs.*, 1967, **66**, 95179q).
77. U.S.S.R. Patent 181,103, April 15, 1966 (*Chem. Abs.*, 1966, **65**, 8962e).

11

ORGANOTIN COMPOUNDS AS LEWIS ACIDS

The coordination chemistry of organotin compounds was reviewed by the author in 1965[63] and by Gielen and Sprecher in 1966[33]. The concept of organotin compounds as Lewis acids is of fundamental importance to an understanding of many problems of structure and reactivity.

Tin differs from the lighter Group IVb elements in that its d-orbitals are of sufficiently low energy for them to be frequently used in bonding so that tin can readily expand its coordination number above 4. Two main consequences follow from this. (a) Many organotin compounds form stable adducts with Lewis bases. (b) Reactions involving nucleophilic attack at tin are facilitated by the formation of coordinated intermediates which, however unstable, lower the energy of the transition state (see Section 11.5). Both of these aspects of the Lewis acidity of tin are examined here but discussion of the effects of covalency expansion of tin on the structure of organotin compounds, is deferred until the following chapter.

11.1. ABILITY OF ORGANOTIN COMPOUNDS TO FORM COORDINATION COMPLEXES

The readiness with which the stannic halides form thermodynamically stable adducts of the type $SnX_4 . 2L$ with Lewis bases is well known[7]. The ability of tin compounds to form complexes is, to a large extent, dependent upon the electronegativity of the substituents, thus the acceptor strengths of the halides are in the order $SnCl_4 \gg SnBr_4 > SnI_4$[33]. When the chlorine atoms of stannic chloride are replaced by relatively electropositive organic groups the acceptor strength of tin declines as indicated by the sequences $SnCl_4 > RSnCl_3 > R_2SnCl_2 > R_3SnCl$[34], $Ph_3SnCl > Bu_3SnCl$[33] and $PhSnCl_3 > MeSnCl_3 > BuSnCl_3$[83]. (These sequences may show minor variations according to the nature of the donor molecule.)

Organotin trihalides and other species of the type $RSnX_3$ readily form stable complexes, the lowering in the acceptor strength of tin only becomes apparent when weak donors are involved such as dialkyl sulphides which give adducts with $SnCl_4$ but not with $PhSnCl_3$ [7, 35]. With monodentate ligands the adducts usually have the composition $RSnX_3.2L$. A spectroscopic study of the interaction between some aryltin trichlorides and substituted anilines indicated that only 1:1 adducts were formed but the measurements were complicated by the presence of the donor solvent, ether[83].

Stable 1:2 adducts are obtained from the organotin dihalides and complexes such as $Et_2SnCl_2.2pyridine$ were included in Werner's classical studies of coordination chemistry[84]. The acceptor properties of dialkyl(aryl)tin compounds, R_2SnX_2, are no longer a simple function of the electronegativities of the X groups, thus stable complexes are known for compounds in which $X = I$ but not when $X = C$ or S. There are few well-authenticated reports of adducts formed from dialkyltin diacetates though the 2,2'-bipyridyl and 1,10-phenanthroline derivatives of $Bu_2Sn(OAc)_2$ have been isolated[38].

The organotin monohalides differ from the di- and tri-halides in that 1:1 adducts, $R_3SnX.L$, are usually formed. A calorimetric method of calculating the enthalpy of formation of adducts of trimethyltin chloride with Lewis bases has been reported[10]. There was good agreement between calculated and observed enthalpy values, some examples of the latter are $Me_3SnCl.Me_2SO$, $-\Delta H = 8.2$; $Me_3SnCl.MeCONMe_2$, $-\Delta H = 7.9$; $Me_3SnCl.pyridine$, $-\Delta H = 6.5$ kcal. $mole^{-1}$ [10]. Although the halides form 1:1 complexes with monodentate ligands, other trialkyl(aryl)tin compounds form 1:2 adducts, for example $Me_3SnX.2NH_3$ $(X = NO_3, ClO_4)$[22].

No adducts of tetraalkyl(aryl)tin compounds have yet been reported and attempts to detect interactions between tetraethyl- or tetraphenyl-tin and dioxan by dipole moment measurements gave negative results[74].

It is widely accepted that tin(IV) is a Class A acceptor[2] or using the 'hard and soft acids and bases' concept[60], it is a hard acid. The presence of organic groups on tin may confer some softness[70] and organotin compounds sometimes show characteristics of Class B acceptors towards certain ligands[33] but, in general, the stability of organotin complexes seems to indicate that tin retains its class A character. Thus trimethyltin chloride forms more stable adducts with oxygen or nitrogen donors than it does with corresponding sulphur or phosphorus compounds[10].

There is an interesting report[36] of the formation of π-complexes between organotin halides and olefins, the capacity for π-complex formation is given by the sequence $SnCl_4 > PhSnCl_3 > Bu_2SnCl_2 > Ph_2SnCl_2 > Ph_3SnCl > Bu_3SnCl$.

11.2. ADDUCTS FORMED FROM MONODENTATE LIGANDS

For the reasons given above, the most commonly used ligands are those with donor atoms from the first row of the Periodic Table, particularly oxygen and nitrogen.

Dimethylsulphoxide ($=$DMSO) gives stable adducts with organotin compounds and the compositions of these are as expected, i.e. $RSnCl_3.2$DMSO, $R_2SnCl_2.2$DMSO and $R_3SnCl.$DMSO although, by using excess tin compound, the adduct $PhSnCl_3.$DMSO was obtained[46]. The positions of the $\nu(S-O)$ bands in the infrared spectra of sulphoxide complexes shows that, as expected, tin as a hard acid bonds to oxygen rather than sulphur[25, 46]. The preparation of adducts from a large number of sulphoxides $RSOR'$ (R and R' = alkyl or aryl) with triphenyltin chloride and triphenyltin isothiocyanate has been described[90, 96] and complexes have been obtained from reaction of the cyclic ligands 1,4-dithianemonosulphoxide[66] and tetramethylene sulphoxide[50] with a number of organotin halides. The complexes $R_2SnCl_2.2(Me_2SeO)$ contain O—Sn bonds and the donor strength of the selenium ligand appears to be comparable with that of dimethylsulphoxide[77].

A large number of 1:1 adducts of triphenyltin chloride with substituted pyridine-N-oxides are known[88] and the stability constants of complexes $Me_3SnCl.L$ and $Me_2SnCl_2.2L$ (L = a substituted pyridine-N-oxide) have been measured[41]. For a series of complexes of trimethyltin chloride with pyridine-N-oxides carrying substituents in the 4-position, $Me_3SnCl.ONC_5H_4X$-4, there is a linear relationship between the stability constants and the Hammett σ constants for the substituents X[41]. Examples of coordination compounds containing the triphenylphosphine oxide and triphenylarsine oxide ligands are $MeSnI_3.2L$, $R_2SnX_2.2L$ and $R_3SnX.L$ where R = Me, Ph; X = Cl, Br, I and L = Ph_3PO and Ph_3AsO[23, 45]. In addition a series of complexes has been obtained by treating triphenyltin chloride with a number of tertiary phosphine oxides[91].

Compounds of the type $[Me_3SnL_2]^\oplus BPh_4^\ominus$ (L = Me_2SO[44], $RCONMe_2$[44] and H_2O[82]) illustrate the stabilisation of a trialkyltin cation by complex formation; salts containing $(Me_2SnL_4)^{2\oplus}$ cations have also been isolated[44]. Ions containing coordinated water molecules such as $[Me_2Sn(OH_2)_4]^{2\oplus}$ lose protons in alkaline solutions giving, eventually, $[Me_2Sn(OH)_4]^{2\ominus}$ and detailed studies have been reported of the proton transfer equilibria[6, 80].

Complexes formed from a wide range of primary, secondary and tertiary amines are reasonably stable and show the normal acceptor: donor ratio. While early workers favoured pyridine complexes for the characterisation of organotin halides, today the more stable adducts derived from the bidentate ligands 2,2'-bipyridyl or 1,10-phenanthroline are preferred (see below). A recent patent[93] which describes the triethanolamine adducts $R_3SnX.N(CH_2CH_2OH)_3$ (R = Ph, alkyl; X = Cl, AcO) is noteworthy since very few adducts of trialkyltin carboxylates have been reported.

Whereas stannic halides only form 6-coordinate complexes with halide ions, $SnX_6^{2\ominus}$, the reduced acceptor properties of the organotin dihalides allows the formation of both $R_2SnX_3^{\ominus}$ and $R_2SnX_4^{2\ominus}$ species[24]. In accordance with the Class A or hard acid character of organotin compounds the tendency of halide ions to coordinate is in the sequence $F^{\ominus} \gg Cl^{\ominus} > Br^{\ominus} > I^{\ominus}$[14] and salts containing the anions $Me_2SnI_3^{\ominus}$ or $Me_2SnI_4^{2\ominus}$ could not be isolated[24]. A large number of investigations of complex formation between R_3Sn, R_2Sn and RSn species and halide ions have been carried out[13, 15, 16, 18, 19, 21, 75]. The tendency to form anionic chloride and bromide complexes is in the order $Me_2Sn \sim PhSn > MeSn$[14] and the stability constants of some $Me_3SnCl_n^{1-n}$, $Me_2SnCl_n^{2-n}$ and $MeSnCl_n^{3-n}$ ions have been reported[21].

In a number of cases, stable crystalline salts containing these complex anions have been isolated: the stability of salts containing the anion $Me_2SnX_3^{\ominus}$ (X = Cl, Br) depends upon the size of the associated cation and increases in the series $Cs^{\oplus} < NMe_4^{\oplus} < NEt_4^{\oplus}$[24]. Some examples of complex salts which have been isolated are $K_2[Me_2SnF_4]$[86], $NH_4[Me_2SnF_3]$[86], $Ph_4As[R_2SnCl_3]$[76] and $Ph_4As[RSnCl_4]$[76] (R = Et, Bu, Ph). Adducts of the type $Ph_4As[R_{4-n}SnCl_{n+1}]$ are strong 1 to 1 electrolytes. The relative stabilities of the anions are $PhSnCl_4^{\ominus} > EtSnCl_4^{\ominus} > BuSnCl_4^{\ominus} > Ph_2SnCl_3^{\ominus} > Me_2SnCl_3^{\ominus} > Et_2SnCl_3^{\ominus} > Pr_2SnCl_3^{\ominus}$ indicating that it is electronic rather than steric effects of the R groups which determine stability[87]. A large number of quaternary phosphonium salts containing the anions Ph_3SnXY (X and Y are halogens) have been prepared[95], for example:

$$[Ph_3PC_6H_4Cl\text{-}p]Br + Ph_3SnCl \longrightarrow [Ph_3PC_6H_4Cl\text{-}p][Ph_3SnClBr]$$
$$[Ph_3PCH_2Cl]Cl + Ph_3SnCl \longrightarrow [Ph_3PCH_2Cl][Ph_3SnCl_2]$$
$$[Et_3PCH_2Ph]Br + Ph_3SnCl \longrightarrow [Et_3PCH_2Ph][Ph_3SnClBr].$$

The above compounds were prepared as potential biocidal agents, and, for similar purposes, a series of *bis* quaternary ammonium salts of the type $[R_3N(CH_2)_nNR_3][Ph_3SnXY]$ has been made[89].

The formation of thiocyanate complexes of organotin compounds in aqueous media was studied using paper electrophoresis and anion exchange paper chromatography. Depending upon the concentration of sodium thiocyanate present, cationic, neutral and anionic complexes were formed[29]. By treating organotin isothiocyanates with tetraethylammonium thiocyanate the following complex salts were prepared, $[Et_4N][Me_3Sn(NCS)_2]$, $[Et_4N]_2[Me_2Sn(NCS)_4]$ and $[Et_4N]_2[MeSn(NCS)_5]$[20].

11.3. ADDUCTS FORMED FROM BIDENTATE LIGANDS

Since the organotin di- and tri-halides are, in the main, liquids or solids with low melting points the well-defined crystalline adducts formed with 2,2'-bipyridyl and 1,10-phenanthroline are useful for characterisation[5, 8, 30]. These derivatives are very readily made by mixing solutions containing equimolar amounts of the two reactants in an inert solvent such as benzene when the complex precipitates. With organotin dihalides and diisothiocyanates, 1:1 adducts are formed but by treating diphenyltin diisocyanate with 2,2'-bipyridyl the 2:1 adduct $[Ph_2Sn(NCO)_2]_2$. bipy was obtained[53]. Although diphenyltin diisocyanate is very readily hydrolysed by atmospheric moisture the bipyridyl derivative is stable; this stabilisation of a reactive species represents a valuable use of complex formation in organotin chemistry.

Some stability constants have been determined and the order of coordinate bond strengths of 2,2'-bipyridyl to chlorotin compounds is $SnCl_4 > BuSnCl_3 > Me_2SnCl_2 \sim Et_2SnCl_2 > Bu_2SnCl_2$[42, 49]. Equilibrium constants for the interaction between dialkyltin species and substituted 1,10-phenanthrolines in aqueous solution have been reported[1]. Adducts of 2,2'-bipyridyl dissociate in a number of solvents[53, 78]; treatment of Bu_2SnCl_2. bipy with mercuric chloride or ferrous ions gives, respectively, the $HgCl_2$. bipy and $[Fe.3bipy]^{2\oplus}$ complexes[4].

Various other diamines have been used as ligands[37a, 83, 98] and 4,4'-bipyridyl gave 1:1 adducts with triphenyltin chloride, phenyltin trichloride and several organotin dihalides[65]. Pyrazine[65], dipyridylamine[47] and tripyridylamine[47] all, apparently, function as bidentate ligands giving stable 1:1 adducts with organotin dihalides.

With 2,2',6',2"-terpyridyl and organotin halides there was considerable variation in the composition of the adducts formed, for example, $[Me_2SnBr_2]_2 . terpy$, $Me_2SnI_2 . terpy$ and $[BuSnCl_3]_3 . 2terpy$[30]. Organotin derivatives of diphenyltriazine ($=dptH$) have been reported[12], for example, $Me_2Sn(dpt)_2$ and $PhSnCl(dpt)_2$.

The most important class of bidentate oxygen donors are the β-diketones and the organometallic derivatives of these compounds were recently reviewed[11]. The principal methods of making complexes from acetylacetone ($=acacH$) are indicated by the following equations:

$$R_2SnCl_2 + 2acacM \xrightarrow{C_6H_6} R_2Sn(acac)_2 + 2MCl \quad (M = Na \text{ or } Tl)$$
(refs. 11, 55, 57, 81)

$$R_2Sn(OR')_2 + 2acacH \xrightarrow{R'OH} R_2Sn(acac)_2 + 2R'OH \quad \text{(refs. 51, 81)}$$

$$RSnX_3 + 2acacH \xrightarrow{H_2O} RSnX(acac)_2 + 2HX \quad (X = Cl, Br)$$
(ref. 81)

The latter method was unsuitable for the iodides but these could be prepared using sodium acetylacetonate[81]:

$$MeSnI_3 + 2acacNa \xrightarrow{CHCl_3} MeSnI(acac)_2 + 2NaI.$$

Other $R_2Sn(chelate)_2$ compounds prepared include those derived from 1-phenylbutan-1,3-dione[55], 1,3-diphenylpropan-1,3-dione[11, 55] and salicylaldehyde[56]. A number of tropolone complexes are known such as Ph_3SnT, Me_2SnXT, R_2SnT_2, $RSnXT_2$ [R = alkyl or Ph; X = Cl, Br, I; TH = tropolone (I)][43]. Of particular interest are the compounds $PhSnT_3$[52] and $BuSnT_3$[43] since there are few compounds of the type $RSn(chelate)_3$. Tin compounds form complexes with

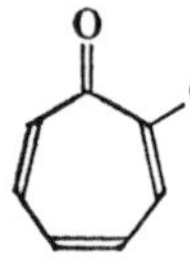

I 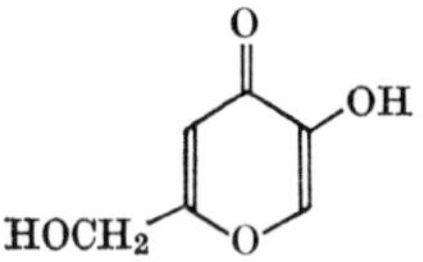 II

kojic acid (II) which include *bis*(kojato)dimethyltin and *bis*(kojato)-halomethyltin compounds[58]. *Cis*-1,4-dithiane disulphoxide (III) and its *trans* isomer (IV) form 1:1 adducts with Ph_2SnCl_2 and $PhSnCl_3$ though there is evidence that the *trans* ligand is bidentate and the *cis* ligand monodentate[66].

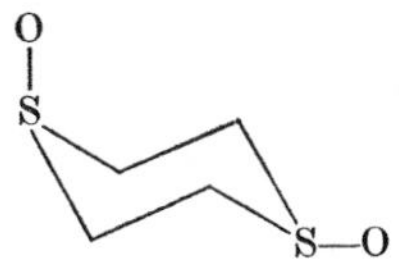

III 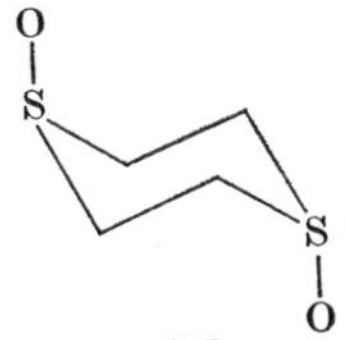 · *IV*

There has been much recent interest in organotin 8-hydroxy-quinolinates and many compounds of the types $R_{4-n}SnOx_n$ ($n = 1, 2$; OxH = 8-hydroxyquinoline) are known[8, 31, 32, 55, 56, 64]. These compounds are prepared either from the organotin halides and sodium oxinate[40]:

$$R_2SnCl_2 + 2NaOx \longrightarrow R_2SnOx_2 + 2NaCl$$

or oxine itself is used, the hydrogen halide formed being removed by treatment with a base such as ammonia[78].

$$R_2SnCl_2 + 2OxH + 2NH_3 \longrightarrow R_2SnOx_2 + 2NH_4Cl$$

Diphenyltin dichloride and oxine react in benzene in the absence of a base to give diphenyltin dioxinate but, when the same compounds are heated above 100° without a solvent, Sn—C bond cleavage occurs and the product is Cl_2SnOx_2[56].

If a dihalide and oxine are allowed to react in a 1:1 molar ratio in the absence of a base the halooxinate is formed[39, 40].

$$R_2SnCl_2 + OxH \longrightarrow R_2SnClOx + HCl$$

These compounds can also be prepared by heating together equimolar proportions of a dihalide and a dioxinate in ethanol or benzene[39, 40, 54, 64, 85]:

$$R_2SnCl_2 + R_2SnOx_2 \longrightarrow 2R_2SnClOx$$

or by reaction between a dioxinate and a silver halide[39].

$$R_2SnOx_2 + AgX \longrightarrow R_2SnXOx + AgOx$$

The halogen in halooxinates can be substituted by other anionic groups.

$$Ph_2SnClOx + KNCS \longrightarrow Ph_2Sn(NCS)Ox + KCl \qquad \text{(ref. 54)}$$
$$2Ph_2SnClOx + Ag_2SO_4 \longrightarrow (Ph_2SnOx)_2SO_4 + 2AgCl \qquad \text{(ref. 39)}$$

In other halogen substitution reactions the product disproportionates[85].

$$R_2SnClOx + I^\ominus \longrightarrow [R_2SnIOx] + Cl^\ominus$$
$$2[R_2SnIOx] \longrightarrow R_2SnI_2 + R_2SnOx_2$$

Evaporation of an ethanol solution of oxine and butyltin trichloride in a $2:1$ molar ratio gave butylchlorotin dioxinate[40] (the correspond-

$$BuSnCl_3 + 2OxH \longrightarrow BuSnClOx_2 + 2HCl$$

ing methyltin compound is also known). A few organotin trioxinates have been prepared[31, 64], good yields were obtained by using either of the following reactions.

$$R_2Sn_2O_3 + 6OxH \xrightarrow[\text{benzene}]{\text{boiling}} 2RSnOx_3 + 3H_2O$$

$$RSnCl_3 + 3NaOx \longrightarrow RSnOx_3 + 3NaCl$$

$$(R = Et, Bu, Ph)$$

Methyltin trioxinate was prepared by prolonged heating of methyltin sesquisulphide with oxine, in a $1:3$ molar ratio, in boiling toluene; similar heating of butyltin sesquisulphide and oxine in a $1:2$ molar ratio gave $(BuSnOx_2)S$ [42a].

When an equimolar mixture of *bis*(8-hydroxy-5-quinolyl)methane ($V = BH_2$) and an organotin dichloride or trichloride was heated in a solvent in the presence of a base the compounds $(RSnCl.B)_n$[67] and $(R_2Sn.B)_n$[67, 71] were isolated, the latter compounds were also formed from direct reaction between V and the oxides R_2SnO. The compound $[(C_8H_{17})_2SnB]_n$ was shown to be anhydrous and trimeric, all the other compounds were hydrated and polymeric[67].

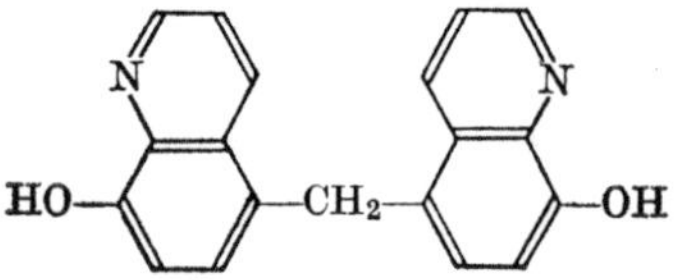

$$V$$

11.4. ORGANOTIN BASES

It is convenient to mention briefly at this point some organotin bases which have been used to form adducts with compounds of tin and other metals. There is evidence that the organotin azides R_3SnN_3 ($R = Me$, Ph) form adducts with stannic chloride[79]. Tetrakis(2-cyanoethyl)tin forms $1:1$ complexes with stannic chloride and bromide, $(NCCH_2CH_2)_4Sn.SnX_4$, regardless of the proportions of the reactants; no adducts were formed when stannic fluoride, iodide, sulphate or oxide were used[69, 97]. The complex VI has been reported[27]: triphenyl[2-(4'-pyridyl)ethyl]tin(VII) forms stable adducts with organotin and other halides e.g. $Ph_2SnCl_2.2L$, $Ph_3SnCl.L$, $ZnCl_2.2L$, $CoCl_2.4L$, $NiCl_2.4L$ and $CuCl.L$ of the same stoichiometry as would be expected with pyridine itself[68].

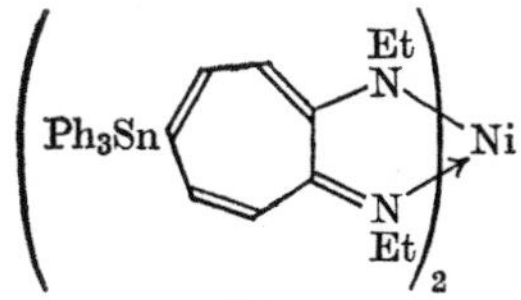

$$\left(\text{Ph}_3\text{Sn}\underset{\underset{\text{Et}}{\overset{\text{N}}{\underset{\text{N}}{}}}{\overset{\overset{\text{Et}}{\text{N}}}{}}\text{Ni}\right)_2$$

$$\text{Ph}_3\text{SnCH}_2\text{CH}_2\overset{}{\underset{}{\longrightarrow}}\text{N}$$

      VI 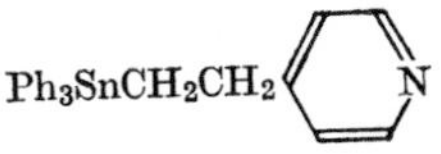   *VII*

11.5. LEWIS ACIDITY AND REACTIVITY

This subject is discussed by Gielen and Sprecher in their review article[33] under the heading 'Intervention of complexes on the course of heterolytic reactions'. It is clear that, in a nucleophilic substitution at tin of the type:

$$\text{>SnX} + \text{N} \longrightarrow \text{>SnN} + \text{X}$$

a pentacoordinate intermediate >SnX.N may be formed. In an excellent discussion of analogous substitutions at a silicon atom, Eaborn[26] has pointed out that, although the rate of reaction is governed by the free energy difference between the separated reactants and the transition state, the energy of the latter may be substantially decreased by the formation of an intermediate. Since complex formation is much more marked with tin than with silicon the facilitation of nucleophilic substitution reactions by this means is correspondingly more important. It should be stressed that this facilitation will occur even if the intermediate is too unstable to be detected[26].

Lewis acidity influences are not confined to nucleophilic substitution reactions and, as was seen in Chapter 3, tin–carbon bond cleavage involves, not only electrophilic attack at carbon, but also nucleophilic assistance at the tin atom. Hence in the halogen cleavage of an Sn—C bond the presence of a donor solvent, such as an alcohol, even though it cannot form a stable adduct with the tin compound can affect the rate of halogen cleavage by rendering nucleophilic assistance (see p. 38). The abnormally ready hydrolysis of a fluorocarbon–tin bond has been partly ascribed to the increased susceptibility of tin to nucleophilic attack[22]. The hydrolysis of perfluorophenyltrimethyltin is

$$\text{C}_6\text{F}_5\text{SnMe}_3 + \text{H}_2\text{O} \longrightarrow \text{C}_6\text{F}_5\text{H} + \text{Me}_3\text{SnOH}$$

catalysed by halide ions and a mechanism which postulates co-ordination of halide ion to tin has been proposed[22].

11.6. THE USE OF ORGANOTIN COMPLEXES IN ANALYSIS AND PURIFICATION

The tendency of certain organotin compounds to form complexes has been exploited in analytical chemistry, either in the spectrophotometric determination of compounds which form coloured complexes or, in the removal of interfering organotin species. Reagents which have been used for the colorimetric determination of organotin compounds include dithizone[3], diphenylcarbazone[73], pyrocatechol violet[28] and alizarin red-S[17]. In general, these reagents form stable complexes with mono- and di-alkyl(aryl)tin species which are, in consequence, readily determined. In addition the complexes formed by dialkyl- and diaryl-tin compounds with 1-(2-pyridylazo)-2-naphthol were shown, by spectroscopic measurements in aqueous dioxan, to have high stabilities[61]. Diethyltin compounds at concentrations in the range 2×10^{-6} to 10^{-4} M have been determined by spectrophotometric estimation of the 1:1 complexes formed with 4-(2-pyridylazo)resorcinol[62].

Mixtures of different types of organotin compounds have been analysed by using complex formation to remove $RSnX_3$ and R_2SnX_2 species leaving the R_3SnX compounds which can then be determined gravimetrically by conversion to stannic oxide[9]. Dialkyltin compounds have been estimated in amounts of 5–100 μg by separating them from other organotin species by thin-layer chromatography followed by treatment with dithizone, elution and photometric determination[37]. It is not usually easy to detect, for example on t.l.c. plates, R_4Sn or R_3SnX (R = alkyl or aryl) compounds by the usual colorimetric reagents since complexes are either not formed or are too unstable. This difficulty is overcome by exposing the plates to bromine vapour[59] so that tin–carbon bond cleavage occurs and the product has sufficient Lewis acidity to form a stable complex with the reagent.

Technical grade dibutyltin dichloride which contained 20% tributyltin chloride was treated with a solution of ammonia in acetone to give the pure adduct $Bu_2SnCl_2 . 2NH_3$ which can be converted directly to the dicarboxylate[94].

$$Bu_2SnCl_2 . 2NH_3 + 2RCOOH \longrightarrow Bu_2Sn(OCOR)_2 + 2NH_4Cl$$

In a similar manner the amount of dibutyltin dichloride which contaminated a specimen of tributyltin chloride was reduced from 5·7 to < 0·01% by treatment with aqueous ammonia[92].

TABLE I

Melting points of some organotin complexes

Compound	m.p.
$MeSnCl_3 \cdot 2Me_2SO$	175°
$Me_2SnCl_2 \cdot 2Me_2SO$	113°
$Me_3SnCl \cdot Me_2SO$	49°
$PhSnCl_3 \cdot 2Me_2SO$	152°
$Ph_2SnCl_2 \cdot 2Me_2SO$	135°
$Ph_3SnCl \cdot Me_2SO$	114–115°
$Ph_3SnNCO \cdot Me_2SO$	147–150°
$Ph_3SnCl \cdot$ pyridine-N-oxide	131–133°
$[Ph_4P][Ph_3SnCl_2]$	203–205°
$[Et_4N][Me_3Sn(NCS)_2]$	120–121°
$Me_2SnCl_2 \cdot 2$pyridine	163°
$Me_3SnCl \cdot$ pyridine	40–41°
$BuSnCl_3 \cdot 2$pyridine	119°
$Pr_2SnI_2 \cdot 2PhNEt_2$	63–64°
$Et_2SnCl_2 \cdot$ bipy[a]	200–201°
$Bu_2SnCl_2 \cdot$ bipy	180°
$Bu_2SnI_2 \cdot$ bipy	163°
$Bu_2Sn(NCS)_2 \cdot$ bipy	153°
$Bu_2Sn(OAc)_2 \cdot$ bipy	50° decomp.
$Me_3SnCl \cdot$ bipy	70°
$BuSnCl_3 \cdot$ bipy	227–228°
$Ph_2SnCl_2 \cdot$ bipy	243–245°
$Et_2SnCl_2 \cdot$ phen[b]	235–236°
$Bu_2SnCl_2 \cdot$ phen	200°
$Bu_2Sn(OAc)_2 \cdot$ phen	118° decomp.
$Ph_2SnCl_2 \cdot$ phen	235° decomp.
$Me_2Sn(acac)_2$[c]	177–178°
$MeClSn(acac)_2$	135–136°
$Et_2Sn(acac)_2$	86·5–87°
$Ph_2Sn(acac)_2$	125–126°
Me_2SnOx_2[d]	235–237°
Bu_2SnOx_2	150–151°
Ph_2SnOx_2	251°
Ph_3SnOx	145–146°
$Et_2SnClOx$	120°
$Pr_2SnClOx$	90–91°
$BuSnClOx_2$	182°
$BuSnOx_3$	229°
$[Me_2SnB \cdot H_2O]_x$[e]	410° decomp.

[a] bipy = 2,2′-bipyridyl.
[b] phen = 1,10-phenanthroline.
[c] acacH = acetylacetone.
[d] OxH = 8-hydroxyquinoline.
[e] BH_2 = *bis*(8-hydroxy-5-quinolyl)methane.

References

1. Affolter, J., Jocot-Guilarmod, A. and Bernauer, K., *Helv. Chim. Acta*, 1968, **51**, 293.
2. Ahrland, S., Chatt, J. and Davies, N. R., *Quart. Rev.*, 1958, **12**, 265.
3. Aldridge, W. N. and Cremer, J. E., *Analyst*, 1957, **82**, 37.
4. Alleston, D. L. and Davies, A. G., *Chem. and Ind.*, 1961, 551.
5. Alleston, D. L. and Davies, A. G., *J. Chem. Soc.*, 1962, 2050.
6. Asso, M. and Carpeni, G., *Omagiu Raluca Ripan*, 1966, 81 (*Chem. Abs.*, 1967, **67**, 76725u).
7. Beattie, I. R., *Quart. Rev.*, 1963, **17**, 382.
8. Blake, D., Coates, G. E. and Tate, J. M., *J. Chem. Soc.*, 1961, 756.
9. Bock, R., Gorbach, S. and Oeser, H., *Angew. Chem.*, 1958, **70**, 272.
10. Bolles, T. F. and Drago, R. S., *J. Amer. Chem. Soc.*, 1965, **87**, 5015; 1966, **88**, 3921; 5730.
11. Bonati, F., *Organometallic Chem. Rev.*, 1966, **1**, 379.
12. Brinckman, F. E. and Haiss, H. S., *Chem. and Ind.*, 1963, 1124.
13. Cassol, A., *Gazz. Chim. Ital.*, 1966, **96**, 1764 (*Chem. Abs.*, 1967, **67**, 15434c).
14. Cassol, A. and Barbieri, R., *Ann. Chim. Rome*, 1965, **55**, 606 (*Chem. Abs.*, 1965, **63**, 14358c).
15. Cassol, A. and Magon, L., *Gazz. Chim. Ital.*, 1966, **96**, 1724 (*Chem. Abs.*, 1967, **67**, 15432z).
16. Cassol, A. and Magon, L., *Gazz. Chim. Ital.*, 1966, **96**, 1752 (*Chem. Abs.*, 1967, **67**, 15434b).
17. Cassol, A. and Magon, L., *J. Inorg. Nucl. Chem.*, 1965, **27**, 1297.
18. Cassol, A., Magon, L. and Barbieri, R., *J. Chromatog.*, 1965, **19**, 57.
19. Cassol, A. and Portanova, R., *Gazz. Chim. Ital.*, 1966, **96**, 1734 (*Chem. Abs.*, 1967, **67**, 15433a).
20. Cassol, A., Portanova, R. and Barbieri, R., *J. Inorg. Nucl. Chem.*, 1965, **27**, 2275.
21. Cassol, A., Portanova, R. and Magon, L., *Ric. Sci.*, 1966, **36**, 1180 (*Chem. Abs.*, 1967, **66**, 119421r).
22. Chambers, R. D. and Chivers, T., *Organometallic Chem. Rev.*, 1966, **1**, 279.
22a. Clark, H. C., O'Brien, R. J. and Pickard, A. L., *J. Organometallic Chem.*, 1965, **4**, 43.
23. Clark, J. P., Langford, V. M. and Wilkins, C. J., *J. Chem. Soc.* (*A*), 1967, 792.
24. Clark, J. P. and Wilkins, C. J., *J. Chem. Soc.* (*A*), 1966, 871.
25. Cotton, F. A., Francis, R. and Horrocks, W. D., *J. Phys. Chem.*, 1960, **64**, 1534.
26. Eaborn, C., *Organosilicon Compounds*, Butterworths, London, 1960, 103.
27. Eaton, D. R. and McClellan, W. R., *Inorg. Chem.*, 1967, **6**, 2134.
28. Efer, J., Quaas, D. and Spichale, W., *Z. Chem.*, 1965, **5**, 390.
29. Faraglia, G., Cassol, A. and Barbieri, R., *Chromatogr. Methods Immed. Separ.*, *Proc. Meeting, Athens*, 1965, **2**, 271 (*Chem. Abs.*, 1968, **69**, 46300).
30. Fergusson, J. E., Roper, W. R. and Wilkins, C. J., *J. Chem. Soc.*, 1965, 3716.
31. Foldesi, I. and Straner, G., *Acta Chim. Acad. Sci. Hung.*, 1965, **45**, 313 (*Chem. Abs.*, 1966, **64**, 3591h).
32. Gerrard, W., Mooney, E. F. and Rees, R. G., *J. Chem. Soc.*, 1964, 740.
33. Gielen, M. and Sprecher, N., *Organometallic Chem. Rev.*, 1966, **1**, 455.

34. Gol'dshtein, I. P., Gur'yanova, E. N., Deleneskaya, E. D. and Kocheshkov, K. A., *Dokl. Akad. Nauk S.S.S.R.*, 1961, **136**, 1079.
35. Gol'dshtein, I. P., Gur'yanova, E. N. and Kocheshkov, K. A., *Dokl. Akad. Nauk S.S.S.R.*, 1961, **138**, 1099.
36. Gol'dshtein, I. P., Gur'yanova, E. N. and Kocheshkov, K. A., *Sintez i Svoistva Monomerov, Akad. Nauk S.S.S.R., Inst. Neftekhim. Sinteza, Sb. Rabot 12-oi (Dvenadtsatoi) Konf. po Vysokomolekul. Soedin*, 1962, 109 (*Chem. Abs.*, 1965, **62**, 6501b).
37. Helberg, D., *Deut. Lebensm. Rundsch.*, 1967, **63**, 69 (*Chem. Abs.*, 1968, **68**, 92801).
37a. Hendricker, D. G., *Inorg. Chem.*, 1969, **8**, 2328.
38. Huber, F., Enders, M. and Kaiser, R., *Z. Naturforsch.*, 1966, **21B**, 83.
39. Huber, F. and Kaiser, R., *J. Organometallic Chem.*, 1966, **6**, 126.
40. Kawakami, K. and Okawara, R., *J. Organometallic Chem.*, 1966, **6**, 249.
41. Kawasaki, Y., Hori, M. and Uenaka, K., *Bull. Chem. Soc. Japan*, 1967, **40**, 2463.
42. Komura, M., Kawasaki, Y., Tanaka, T. and Okawara, R., *J. Organometallic Chem.*, 1965, **4**, 308.
42a. Komura, M. and Okawara, R., *Inorg. Nucl. Chem. Letters*, 1966, **2**, 93.
43. Komura, M., Tanaka, T. and Okawara, R., *Inorg. Chim. Acta*, 1968, **2**, 321.
44. Kumar Das, V. G. and Kitching, W., *J. Organometallic Chem.*, 1967, **10**, 59.
45. Kumar Das, V. G. and Kitching, W., *J. Organometallic Chem.*, 1968, **13**, 523.
46. Langer, H. G. and Blut, A. H., *J. Organometallic Chem.*, 1966, **5**, 288.
47. McWhinnie, W. R., Poller, R. C., Ruddick, J. N. R. and Thevarasa, M., *J. Chem. Soc. (A)*, 1969, 2327.
48. Maggio, F., Bosco, R., Cefalu, R. and Barbieri, R., *Inorg. Nucl. Chem. Letters*, 1968, **4**, 389.
49. Matsubayashi, G., Kawasaki, Y., Tanaka, T. and Okawara, R., *J. Inorg. Nucl. Chem.*, 1966, **28**, 2937.
50. Matwiyoff, N. A. and Drago, R. S., *Inorg. Chem.*, 1964, **3**, 337.
51. Mehrotra, R. C. and Gupta, V. G., *J. Organometallic Chem.*, 1965, **4**, 237.
52. Muetterties, E. L. and Wright, C. M., *J. Amer. Chem. Soc.*, 1964, **86**, 5132; 1965, **87**, 4706.
53. Mufti, A. S. and Poller, R. C., *J. Chem. Soc.*, 1965, 5055.
54. Mullins, M. A. and Curran, C., *Inorg. Chem.*, 1968, **7**, 2584.
55. Nelson, W. H. and Martin, D. F., *J. Inorg. Nucl. Chem.*, 1965, **27**, 89.
56. Nelson, W. H. and Martin, D. F., *J. Organometallic Chem.*, 1965, **4**, 67.
57. Nelson, W. H., Randall, W. R. and Martin, D. F., *Inorg. Synth.*, 1967, **9**, 52.
58. Otera, J., Kawasaki, Y. and Tanaka, T., *Inorg. Chim. Acta*, 1967, **1**, 294.
59. Otto, P. Ph. H. L., Creemers, H. M. J. C. and Luijten, J. G. A., *J. Labelled Compounds*, 1966, **2**, 339.
60. Pearson, R. G., *J. Amer. Chem. Soc.*, 1963, **85**, 3553.
61. Pilloni, G., *Anal. Chim. Acta*, 1967, **37**, 497.
62. Pilloni, G. and Plazzogna, G., *Anal. Chim. Acta*, 1966, **35**, 325.
63. Poller, R. C., *J. Organometallic Chem.*, 1965, **3**, 321.
64. Poller, R. C. and Ruddick, J. N. R., *J. Chem. Soc. (A)*, 1969, 2273.
65. Poller, R. C. and Toley, D. L. B., *J. Chem. Soc. (A)*, 1967, 1578.

66. Poller, R. C. and Toley, D. L. B., *J. Chem. Soc. (A)*, 1967, 2035.
67. Poller, R. C. and Toley, D. L. B., *J. Inorg. Nucl. Chem.*, 1969, 2973.
68. Poller, R. C. and Toley, D. L. B., *J. Organometallic Chem.*, 1968, **14**, 453.
69. Reifenberg, G. H. and Considine, W. J., *J. Organometallic Chem.*, 1967, **10**, 285.
70. Saville, B., *Angew. Chem. Intern. Ed. Engl.*, 1967, **6**, 928.
71. Sen, D. N. and Umpathy, P., *Indian J. Chem.*, 1967, **5**, 209.
72. Shindo, M., Matsumura, Y. and Okawara, R., *J. Organometallic Chem.*, 1968, **11**, 299.
73. Skeel, R. T. and Bricker, C. E., *Anal. Chem.*, 1961, **33**, 428.
74. Strohmeier, W. and Miltenberger, K., *Z. Physik. Chem.*, 1958, **17**, 274.
75. Tagliavini, G. and Zanella, P., *Anal. Chim. Acta*, 1968, **40**, 33.
76. Tagliavini, G. and Zanella, P., *J. Organometallic Chem.*, 1966, **5**, 299.
77. Tanaka, T. and Kamitani, T., *Inorg. Chim. Acta*, 1968, **2**, 175.
78. Tanaka, T., Komura, M., Kawasaki, Y. and Okawara, R., *J. Organometallic Chem.*, 1964, **1**, 484.
79. Thayer, J. S. and West, R., *Inorg. Chem.*, 1965, **4**, 114.
80. Tobias, R. S. and Freidline, C. E., *Inorg. Chem.*, 1965, **4**, 215.
81. Ueeda, R., Kawasaki, Y., Tanaka, T. and Okawara, R., *J. Organometallic Chem.*, 1966, **5**, 194.
82. Wada, M. and Okawara, R., *J. Organometallic Chem.*, 1965, **4**, 487.
83. Wardell, J. L., *J. Organometallic Chem.*, 1967, **9**, 89; 1967, **10**, 53.
84. Werner, A., *Z. Anorg. Chem.*, 1898, **17**, 82.
85. Westlake, A. H. and Martin, D. F., *J. Inorg. Nucl. Chem.*, 1965, **27**, 1579.
86. Wilkins, C. J. and Haendler, H. M., *J. Chem. Soc.*, 1965, 3174.
87. Zanella, P. and Tagliavini, G., *J. Organometallic Chem.*, 1968, **12**, 355.
88. French Patent 1,389,821, Feb. 19, 1965 (*Chem. Abs.*, 1965, **63**, 1816e).
89. German Patent 1,204,226, Nov. 4, 1965 (*Chem. Abs.*, 1966, **64**, 2128f).
90. German Patent 1,215,709, May 5, 1966 (*Chem. Abs.*, 1966, **65**, 5489f).
91. German Patent 1,216,300, May 12, 1966 (*Chem. Abs.*, 1966, **65**, 5490b).
92. German Patent 1,271,113, June 27, 1968 (*Chem. Abs.*, 1968, **69**, 77504).
93. Japanese Patent 10,102, May 30, 1966 (*Chem. Abs.*, 1966, **65**, 12240c).
94. Netherlands Patent, 109,491, Sept. 15, 1964 (*Chem. Abs.*, 1965, **62**, 9173g).
95. Netherlands Patent 6,414,757, June 21, 1965 (*Chem. Abs.*, 1966, **64**, 757g).
96. Netherlands Patent 6,510,858, Feb. 21, 1966 (*Chem. Abs.*, 1966, **65**, 17004).
97. Netherlands Patent 6,700,012, July 4, 1967 (*Chem. Abs.*, 1968, **68**, 114745).
98. U.S. Patent 3,242,105, March 22, 1966 (*Chem. Abs.*, 1966, **64**, 17825h).

12

THE STRUCTURE OF ORGANOTIN COMPOUNDS

A review by Gielen and Sprecher[26] includes a discussion of organotin structures in which the coordination number of tin is greater than 4; the same topic is treated in an article by Okawara and Wada[64].

The increasing use of spectroscopic methods is the main reason for the very rapid growth now taking place in this absorbing field of study. Infrared and Raman spectroscopy, proton- and to a less extent ^{119}Sn–nmr studies, and Mössbauer spectroscopy have all contributed. Fortunately, during the last year or so, there has been an acceleration in the publication of X-ray crystal structure determinations providing an increased number of structural reference points between which the less informative spectroscopic techniques can be applied.

It is usual to vizualise regular trigonal bipyramids and octahedra for 5- and 6-coordinate tin compounds but, as will be seen, molecules can have such highly distorted structures that they bear only slight resemblance to the regular geometrical models. For this reason any structure assignments based solely on infrared spectra measurements and symmetry arguments must be received with caution. This is particularly true when infrared measurements are made on compounds in the solid state where, for example, crystal field effects may increase the complexity of the spectra by resolving degenerate bands.

Although the bonds formed between tin and other elements may have considerable ionic character they are, with a few exceptions, treated here as directed covalent bonds. Tobias[22, 89] has shown that it is often illuminating to discuss the solid state structures of, for example, R_3SnX compounds in terms of $R_3Sn^{\oplus}$ and $X^{\ominus}$ ions with varying degrees of covalent interaction between these ions. Although the two methods of describing Sn—X bonds differ only in semantics they represent two different and complementary view-points which should both be kept in mind.

In the sections which follow unequivocal structure assignments, made on the basis of X-ray, electron diffraction or microwave measurements, are discussed in some detail followed by a briefer treatment of structures which have been assigned by less certain methods.

12.1. COMPOUNDS WITH FOUR Sn—C BONDS

In the symmetrically substituted R_4Sn compounds (R = alkyl or aryl) the four Sn—C bonds are tetrahedrally disposed making use of sp^3 hybrid orbitals at tin. An electron diffraction study[10] of tetramethyltin in the vapour state showed Td symmetry with an Sn—C bond distance of $2 \cdot 18 \pm 0 \cdot 03$ Å. It is likely that, when the four organic groups attached to tin are not identical some distortion from a regular tetrahedron would occur but no precise information is available.

Mössbauer-, infrared- and mass-spectra measurements gave good evidence for intramolecular O → Sn coordination in R_3SnCH_2COR' compounds[42]. The nmr spectra showed an abnormally slow inversion at the nitrogen atom in $Et_3SnCH_2\overline{NCH_2CH_2}$ indicative of inter- or intra-molecular coordination from nitrogen to tin but this was not confirmed by Mössbauer measurements[44].

The nature of the bonding in di(cyclopentadienyl)tin [prepared from stannous chloride and $(C_5H_5)_2Mg$[74]] is not known but simple σ-bonds are unlikely since electron diffraction measurements showed that, in the vapour state, the two cyclopentadienyl rings are not parallel, the angle between the planes being about 55°[4].

12.2. COMPOUNDS WITH THREE Sn—C BONDS

Electron diffraction studies[85] of trimethyltin chloride, bromide and iodide showed tetrahedral molecules in the vapour phase. The optical resolution of ethylmethylpropyltin iodide was reported in 1900[73] but this work has never been repeated. A suggestion[9] that the original reports are in error has now been confirmed since nmr measurements have shown that an asymmetrically substituted organotin halide RR′R″SnX undergoes rapid configurational inversion in solution[68].

An X-ray crystal study of 2,3,5,6-tetrachloro-1,4-*bis*(triethylstannoxy)benzene showed the structure I in which the *trans* conformation is maintained by steric inhibition of free rotation about the C—O bonds[95].

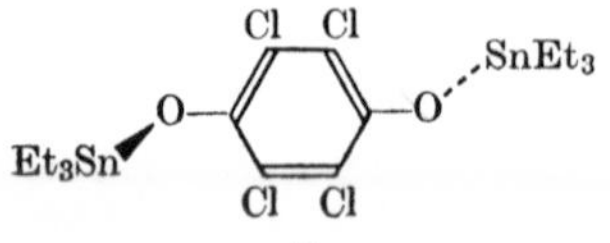

I 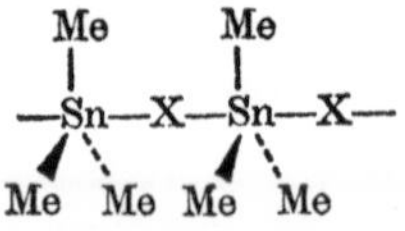   II

Recent X-ray diffraction studies of crystalline Me_3SnX compounds have shown the prevalence of polymeric structures with bridging X groups and 5-coordinate tin (II). Only in trimethyltin cyanide $(II, X = CN)$ is the configuration at tin unequivocally that of a regular trigonal bipyramid with equatorial methyl groups and cyanide groups which are equidistant from neighbouring tin atoms[79]. Interpretation of the X-ray studies of trimethyltin fluoride[15, 98] and hydroxide[38] is somewhat less certain and, although both of these compounds are polymeric with structures corresponding to II $(X = F$ or $OH)$ the precise geometry at the tin atoms is not known. Preliminary X-ray results[19a] and Mössbauer spectra measurements[67a] suggest that polymeric structures may be general for R_3SnX $(X = halogen)$ compounds in the solid state although the configuration at the two tin atoms in $Ph_2ISn(CH_2)_4SnIPh_2$ is close to tetrahedral[17a]. X-ray crystal structure determinations of tribenzyltin acetate[1a] and trimethyltin isothiocyanate[24a] show that these compounds have structures similar to II with, respectively, bridging acetate and isothiocyanate groups though, in both cases, there are distortions from a regular trigonal bipyramid.

Nuclear magnetic resonance measurements on

$$Me_2Sn(X)C(Ph){=}C(Ph)C(Ph){=}CPhX \qquad (X = Cl, Br, I)$$

showed that the two methyl groups were non-equivalent with $\delta = 0{\cdot}48$ and $0{\cdot}98$ at $36°$ in carbon tetrachloride, the bands merged to a single peak at $87°$. An X-ray crystal structure determination of the compound where $X = Br$ showed that the molecule existed in the preferred dissymetric conformation III[5]. This appears to be the first example of a compound in which a carbon-bound halogen atom coordinates to tin; the $Br \rightarrow Sn$ bond, however, is weak and the contribution it makes to maintaining the preferred conformation may, in fact, be very small[25]. The bond distances were Sn—Br $2{\cdot}507 \pm 0{\cdot}007$ Å, Sn......Br $3{\cdot}769 \pm 0{\cdot}007$ Å.

A partial X-ray study of triphenyltin chloride has been reported[6].

Polymeric structures of the type shown in Fig. II have been assigned to many R_3SnX compounds in the solid state on the basis of their infrared spectra and the arguments used may be illustrated by reference to the trimethyltin carboxylates $Me_3SnOCOR$[69]. The infrared spectra of these compounds in the solid and molten states show two C—O stretching bands at ~1570 and ~1410 cm^{-1} indicating a symmetrical (C_{2v}) COO group. The absence of a νs(Sn—C) band

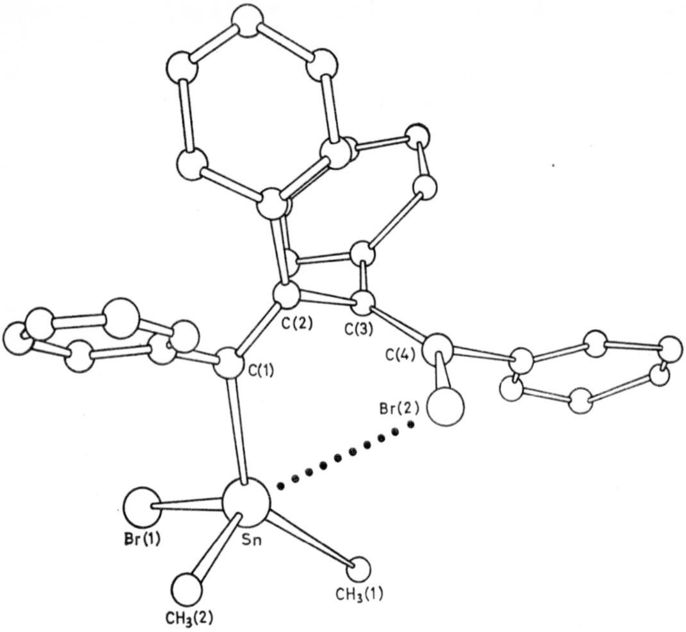

III

[Reproduced (with permission) from Boer, Flynn, Freedman, McKinley and Sandel, *J. Amer. Chem. Soc.*, 1967, **89**, *5068*.]

suggests that the Me$_3$Sn group is planar and all the infrared evidence is compatible with structures *IV* or *V*. The polymeric structure *V*

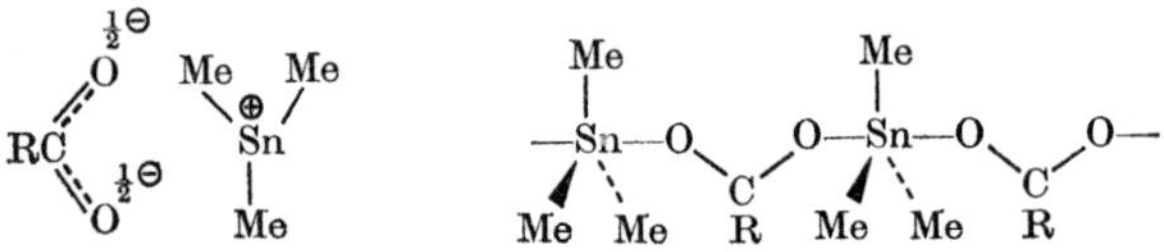

IV   *V*

was shown to be the correct one from molecular weight measurements and because, in carbon tetrachloride solution, the carboxylate absorption bands were shifted to ~1650 and ~1300 cm^{-1} and both the νas(Sn—C) and νs(Sn—C) bands were observed. This indicated that the polymeric structure *V* was broken down in solution to give ester-type carboxyl groups and pyramidal Me$_3$Sn units corresponding to *VI*. Infrared evidence has been cited to show that triethyl- and tripropyl-tin formate exist as low polymers in organic solvents, i.e.

$$\begin{array}{c} \text{Me} \\ \text{Me} \rightarrow \text{Sn---O} \\ \text{Me} \end{array} \begin{array}{c} \text{O} \\ \diagdown \text{C} \diagup \\ | \\ \text{R} \end{array}$$

VI

only partial break-down of the polymeric structure has occurred[63]. Splitting of the carboxylate bands in the infrared spectra of tri-methyltin mono- and di-chloroacetates has been shown to be caused by conformational isomerism in solution[82].

With the unequivocal demonstration of the structure of tribenzyltin acetate[1a] noted earlier there is now no doubt that the majority of $R_3SnOCOR'$ compounds have polymeric structures in the crystalline state but this may not be the case for compounds in which R or R' are bulky groups. Thus X-ray studies show that tricyclohexyltin acetate molecules exist as discrete tetrahedra in the solid phase[1b].

Trimethyltin formate and acetate are rather insoluble but can be converted to soluble forms by heating with cyclohexane in a sealed tube at $100°$. The soluble formate is monomeric in ethanol but has a degree of association of between 3 and 4 in carbon tetrachloride, it is suggested that these soluble forms may be cyclic oligomers[82].

Similar arguments have been used to assign polymeric structures of the type shown in Fig. *II* to a large number of R_3SnX compounds in the solid state including examples where $X = ClO_4$[11, 62], BF_4[32], AsF_6[13], SbF_6[13], $AlCl_4$[60], N_3[88] and a number of heterocyclic ring systems containing 1,3-nitrogen atoms such as imidazole[37]. Since both mono- and bi-dentate NO_3 groups have C_{2v} symmetry the infrared spectrum of the nitrate group in R_3SnNO_3 compounds is not diagnostic and there has been some disagreement concerning the structure of these compounds which are now considered to be as in Fig. *II* with bridging NO_3 groups and 5-coordinate tin atoms[12, 14, 83, 100].

The infrared spectra of $(R_3Sn)_2SO_4$ compounds indicate Td symmetry for the SO_4 group and this has been interpreted in terms of an ionic structure[47]. In an alternative interpretation $(Me_3Sn)_2SO_4$ and $(Me_3Sn)_2CrO_4$ are both considered to have polymeric structures in which every oxygen atom of the SO_4 and CrO_4 groups is coordinated to a tin atom[11, 12]. The trialkyltin carbonates $(R_3Sn)_2CO_3$ (R = Me, Et) appear to be monomeric with two 4-coordinate tin atoms linked by a carbonate group[47, 77].

The infrared spectrum of triethyltin oxide was interpreted in terms of the ribbon structure *VII* and it has been suggested that triethyltin hydroxide is a hydrated oxide[47]. The properties of triethyl-

$$
\begin{array}{ccccccc}
\text{Et}_3 & & \text{Et}_3 & & \text{Et}_3 & \\
\text{Sn} & & \text{Sn} & & \text{Sn} & \\
& \text{O} & & \text{O} & & \text{O} & \\
\text{Sn} & & \text{Sn} & & \text{Sn} & \\
\text{Et}_3 & & \text{Et}_3 & & \text{Et}_3 &
\end{array}
$$

VII

and higher trialkyl-tin hydroxides are somewhat different from those of Ph_3SnOH and Me_3SnOH which contain well defined Sn—OH groups. The solid-state structure of trimethyltin hydroxide has already been discussed (p. 201) and it has been suggested that this compound has the dimeric structure *VIII* in solution[66]; a similar 4-membered ring structure *IX* has been proposed for Ph_3SnSLi in benzene[81].

$$
\begin{array}{cc}
\text{H} & \text{Li} \\
\text{O} & \text{S} \\
\text{Me}_3\text{Sn} \diagdown \diagup \text{SnMe}_3 & \text{Ph}_3\text{Sn} \diagdown \diagup \text{SnPh}_3 \\
\text{O} & \text{S} \\
\text{H} & \text{Li} \\
\textit{VIII} & \textit{IX}
\end{array}
$$

Although there is only limited evidence for the structure of the organotin (iso)cyanates it was shown[56, 84] that an *N*-bonded structure R_3SnNCO is most likely.

The Mössbauer parameters of organotin nitramines such as $Me_3SnN(Me)NO_2$ indicate that tin has a coordination number of 5 or 6 and the absence of resonance absorption at room temperature suggests a monomeric structure. A solution of the compound in nitrobenzene is non-conducting and the infrared spectrum shows the presence of pyramidal Me_3Sn groups so that a cyclic structure such as *X* is indicated.

$$
\begin{array}{c}
\text{Me} \\
| \\
\text{Me} \diagdown \\
\qquad \text{Sn—O} \\
\text{Me} \diagup \quad | \quad | \\
\qquad \text{N——N} \rightarrow \text{O} \\
\qquad | \\
\qquad \text{Me}
\end{array}
$$

X

12.2.1. Adducts of trialkyl(aryl)tin compounds

An X-ray crystal structure determination of Me_3SnCl.pyridine revealed a regular trigonal bipyramid with equatorial methyl groups[36]. Although this is the only complete structure determination there is evidence that other 1:1 adducts from trialkyltin halides and monodentate ligands have similar structures[7, 50]. From a study of the $J(^{119}Sn$—C—H) coupling constants in the nmr spectra of the adducts and the enthalpy of formation it was concluded that the distortion of the tetrahedral Me_3SnCl molecule on complex formation with a ligand L was dependent upon the strength of the Sn—L bond. Little distortion occurs when the tin–ligand bond is weak but, with a strong Sn—L bond, the rehybridisation at tin is more complete and an essentially trigonal bipyramidal structure results[7].

In those R_3SnX compounds in which X is a group not usually found in the coordination sphere of complexes 1:2 adducts are formed with the ionic structure $[R_3SnL_2]^{\oplus}X^{\ominus}$[14]. Although this structure is based mainly on infrared evidence it is reasonably well established so that, despite the change in stoichiometry, tin remains pentacoordinate. Examples of this type of compound are $[Me_3Sn.2pyridine]^{\oplus}ClO_4^{\ominus}$ [14], $[Ph_3Sn.2(Me_2SO)]^{\oplus}NO_3^{\ominus}$ [83] and $[Me_3SnL_2]^{\oplus}Ph_4B^{\ominus}$ (L = Me_2SO, Me_2NCOH and Me_2NCOMe)[46].

Unstable 2,2'-bipyridyl adducts are known of composition Me_3SnX.bipy (X = Cl, Br, I) and it is possible that these contain 6-coordinate tin atoms but the only structural information available is that the adducts have low electrical conductivities and are therefore probably non-ionic[23]. It has been suggested that the tin atoms in trimethyltin nitrate monohydrate are 6-coordinate and that the structure is polymeric with $Me_3Sn.OH_2$ units linked by bridging NO_3 groups[100].

The structure of the oxinates R_3SnOx (OxH = 8-hydroxyquinoline) is not settled. Ultraviolet absorption measurements on solutions of triphenyltin oxinate in protic solvents are complicated by solvolysis reactions[75] though the spectrum in cyclohexane suggests that the oxine group is chelating and that tin is pentacoordinate[91]. The stability of triphenyltin oxinate to moisture indicates that the oxine group is chelating in the solid state[69] and there is some infrared evidence that Me_3SnOx contains a coordinating oxine group in the pure, liquid, state[40].

12.3. COMPOUNDS WITH TWO Sn—C BONDS

There are, essentially, three structural possibilities for compounds of the type R_2SnX_2: tetrahedral with 4-coordinate tin (XI), octahedral monomeric with 6-coordinate tin (XII) and octahedral polymeric also with 6-coordinate tin (XIII). Electron diffraction measurements on the dimethyltin dihalides Me_2SnX_2 (X = Cl, Br, I) show that these are tetrahedral molecules corresponding to XI in the

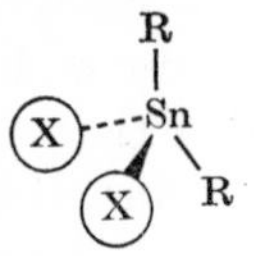

XI	XII	XIII

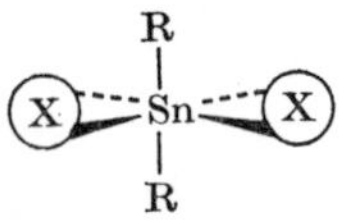

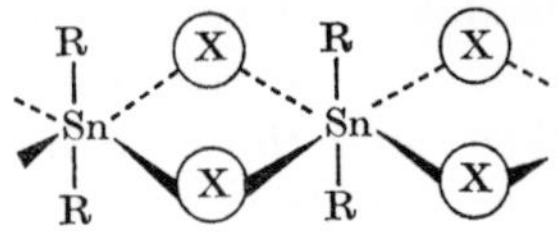

vapour state[85]. Turning to solid state structures, a comparison of physical properties such as melting points (Me_2SnI_2 m.p. 44°, Me_2SnBr_2 m.p. 75–77°, Me_2SnCl_2 m.p. 107–8°, Me_2SnF_2 m.p. ~400° decomp.) indicates that dimethyltin difluoride differs from the other dihalides. An X-ray crystal study showed structure XIII (R = Me, X = F)[80].

The infrared spectrum of $Me_2Sn(NO_3)_2$ was interpreted in terms of a tetrahedral configuration at tin with covalent monodentate nitrate groups[1]. However, this is not confirmed by the Mössbauer spectrum[76] which indicates that hexacoordinate tin atoms are probably present.

In a preliminary account[33] of the results of an X-ray crystal study of the diphenyltin sulphide trimer the molecule was shown to have a boat conformation XIV with $\angle Sn—S—Sn = 104\cdot0 + 1\cdot6°$ and $\angle S—Sn—S = 111\cdot9 \pm 0\cdot3°$. From dipole moment measurements it was suggested that the 6-membered ring in $(Me_2SnS)_3$ also has a boat conformation[27].

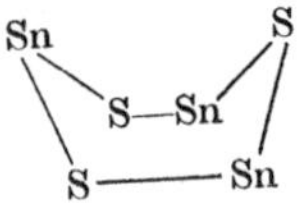

XIV

A number of dialkyl- and diaryl-tin *bis*(chelate) compounds, $R_2Sn(Ch)_2$, are known and the structure of dimethyltin dioxinate, which was determined by X-ray diffraction[78], is shown in Fig. XV. The molecule has chelating oxine groups and a highly distorted octahedral configuration with C—Sn—C bond angles close to tetrahedral at 110·7°. These bond angles indicate some uncertainty in the

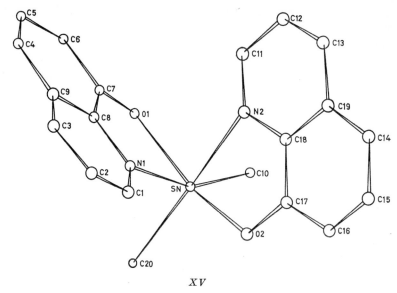

XV

[Reproduced (with permission) from Schlemper, *Inorg. Chem.*, 1967, **6**, 2012.]

extent to which the tin $5d$ orbitals are involved in the bonding and it has been suggested[78] that tin sp^3 orbitals are used, participating in normal covalent bonds to the methyl groups and three-centre bonds to the oxinate groups. It seems more likely that the bonds from tin are somewhere between the two extremes represented by the sp^3 and sp^3d^2 hybridisation states though there are grounds for supposing that the extent of d-orbital participation in the bonding in 6-coordinate complexes is less than was previously supposed[35]. There has been considerable interest in the use of spectroscopic methods to assign the configurations of 6-coordinate complexes of the type *XII* and it is disconcerting to note that one nmr study[40] of Me_2SnOx_2 indicated *trans* methyl groups although another interpretation[48] of $J(^{117,\ 119}Sn—C—H)$ coupling constants correctly suggested that the C—Sn—C angle might be nearly tetrahedral. It appears that the most reliable spectroscopic method for determining the stereochemistry of compounds of this type is Mössbauer spectroscopy[24] (see Chapter 13).

An X-ray crystal structure determination[20] of the complex $[Me_2SnCl_2]_2 . 2,2',6',2''$-terpyridyl confirmed the formulation $[Me_2SnCl . terpy]^{\oplus}[Me_2SnCl_3]^{\ominus}$ made earlier[23] on the basis of electrical conductivity measurements. The cation is a highly distorted octahedron with *trans* methyl groups and all three terpyridyl nitrogen atoms coordinated to tin (the structure of the anion is discussed below).

The compound $Br_2Sn[CH(COOEt)CH_2COOEt]_2$ exists in two forms and an X-ray crystal study[101] of the form with the lower melting point showed the octahedral configuration XVI. The geometry is irregular with $\angle O^2—Sn—Br^1 = 166°$, $\angle C^1—Sn—C^2 = 148°$ and $\angle O^1—Sn—C^2 = 80°$.

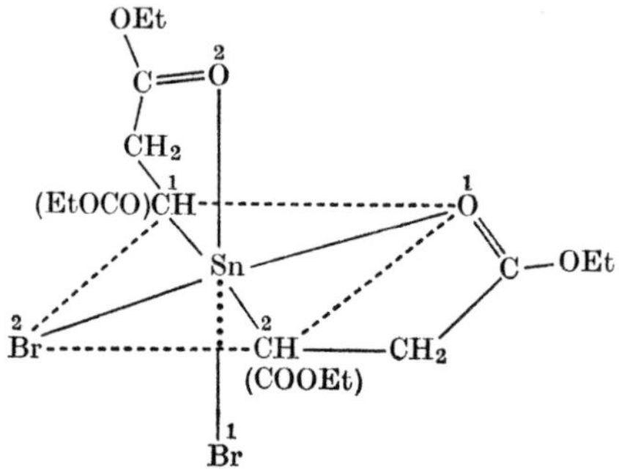

XVI

There is evidence that the chlorine atoms in the chloromethyl groups form intramolecular coordinate bonds to the tin atom in compounds such as $Cl_2Sn(CH_2Cl)_2$ but the extent of this interaction is uncertain[45].

Some compounds of the types R_2SnX_2 and $R_2SnX_2 \cdot 2L$ which have been assigned octahedral structures are listed in Table I.

TABLE I

Some structures containing hexacoordinate tin atoms

Compound	Phase	Method	Remarks
Me_2SnF_2	Solid	X-ray[80]	Polymeric, *trans* Me groups
Me_2SnSO_4	Solid	IR[11, 12]	Polymeric, tetradentate SO_4 groups
$R_2Sn(NCS)_2^{a,b}$	Solid	Mössbauer[58]	R = Bu, Ph: polymeric, *trans* R groups
$R_2Sn(OCOMe)_2$	Solid or neat liquid	IR[49]	Polymeric

TABLE I—cont.

Compound	Phase	Method	Remarks
$R_2Sn(OCOMe)_2$	Solution	IR [49]	Monomeric
$Me_2Sn(OH)NO_3$	Solid	IR [100]	Polymeric
Me_2SnOx_2[c]	Solid	X-ray[78], Mössbauer [24]	Distorted octahedron, monomeric, cis Me groups
R_2SnOx_2	Solid	Mössbauer [24, 70]	Monomeric, cis R groups
$Me_2Sn(acac)_2$[d]	Solid	Mössbauer [24] ⎫	Monomeric, trans Me
$Me_2Sn(acac)_2$	Solution	nmr [40, 48] ⎬	groups
$Ph_2Sn(acac)_2$	Solid	Mössbauer [24]	Monomeric, cis Ph groups
$Ph_2Sn(acac)_2$	Solution	Dipole moment [59]	Stereochemistry uncertain;[59] probably cis Ph groups [8]
Me_2SnK_2[e]	Solution	nmr [67]	Distorted octahedron, monomeric, trans Me groups
$Ph_2SnX_2 \cdot 2pyridine$	Solid	Mössbauer and IR [48a]	X = Cl, Br; monomeric, trans Ph and cis X groups
$Ph_2SnX_2 \cdot dipyam$[f]	Solid ⎫	Mössbauer and IR [48a]	X = Cl, Br; distorted octahedra, monomeric,
$Ph_2SnX_2 \cdot tripyam$[g]	Solid ⎭		trans Ph groups, cis halogens
$Me_2SnCl_2 \cdot bipy$[h]	Solid	IR [16, 87] Mössbauer [24]	Monomeric, trans Me groups, cis Cl atoms
$Bu_2SnX_2 \cdot phen$[i]	Solid	Mössbauer [24, 57] ⎫	X = Cl, Br, I;
$Bu_2SnX_2 \cdot phen$	Solution	dipole moment [57] ⎬	Monomeric, trans Bu groups
$Bu_2SnX_2 \cdot bipy$	Solid	Mössbauer [24, 57]	X = Cl, Br, I; monomeric, trans Bu groups
$R_2Sn(NCS)_2 \cdot bipy$ ⎱	Solid	Mössbauer [58]	Monomeric, trans R groups when R = Bu and cis R groups when R = Ph
$R_2Sn(NCS)_2 \cdot phen$ ⎰			
$Me_2SnCl_2 \cdot 2Me_2SO$	Solid	X-ray[36a], IR [12, 86]	Monomeric; trans Me groups and cis Cl atoms
$Me_2SnCl_2 \cdot 2pyO$[j]	Solid	X-ray[4a]	Monomeric; all trans
$Me_2SnBr_2 \cdot 2Me_2SO$	Solid	IR [86]	Monomeric, trans Me groups, trans Br atoms
$[Me_2SnF_4]^{2\ominus}$	Solid	IR [96]	Monomeric, trans Me groups

TABLE I—*cont.*

Compound	Phase	Method	Remarks
[Me$_2$SnCl$_4$]$^{2\ominus}$	Solid	IR [16]	Monomeric, *trans* Me groups
[Me$_2$Sn(OH)$_4$]$^{2\ominus}$	Solution	Raman [90]	Monomeric, *trans* Me groups
R$_2$SnCl$_2$·L^k	Solution Solid	nmr [33a] ⎫ IR [33a] ⎬	Monomeric; *trans* R groups

[a] Dipole moment measurements indicate that these compounds are tetrahedral monomers in benzene[58].

[b] Infrared spectra were interpreted in terms of tetrahedral unassociated molecules in the solid state[92].

[c] OxH = 8-hydroxyquinoline.

[d] acacH = acetylacetone.

[e] KH = kojic acid.

[f] dipyam = 2,2'-dipyridylamine.

[g] tripyam = 2,2',2''-tripyridylamine.

[h] bipy = 2,2'-bipyridyl.

[i] phen = 1,10-phenanthroline.

[j] pyO = pyridine-N-oxide.

[k] L = 2,7,dimethyl-1,8-naphthyridine.

The crystal study[20] of [Me$_2$SnCl$_2$]$_2$.terpy referred to above showed that the [Me$_2$SnCl$_3$]$^{\ominus}$ anion is a trigonal bipyramid with the two methyl groups and a chlorine atom occupying equatorial positions, the equatorial Sn—Cl distance is 2·35 Å and the axial Sn—Cl distance is 2·54 Å. This configuration conforms with the generalisation that, in trigonal bipyramidal structures, the more electronegative atoms occupy the axial sites[54]. Similarly the infrared spectra of a number of chlorides and bromides of structure [NR$_4$]$^{\oplus}$[Me$_2$SnX$_3$]$^{\ominus}$ (R = H, Me, Et) indicate that the axial positions in the pentacoordinate anions are occupied by halogen atoms[16]. An apparent exception to the rule is [NH$_4$]$^{\oplus}$[Me$_2$SnF$_3$]$^{\ominus}$ where the presence of single ν(Sn—F) and ν(Sn—C) bands in the infrared spectrum was interpreted in terms of a trigonal bipyramidal arrangement at tin with equatorial fluorine atoms and axial methyl groups[96].

The positions of the νas(COO) bands in the infrared spectra of dialkylchlorotin acetates, R$_2$Sn(Cl)OCOMe, indicate that the carboxylate groups are bidentate. When solid-state and solution spectra are compared the shifts in the positions of the νas(COO) bands are relatively small so that a 5-coordinate monomeric structure with chelating carboxyl groups was proposed[94].

From studies of the Mössbauer spectra of the oxinates $R_2Sn(X)Ox$ (R = Ph; X = NCS)[58], (R = Me, Et, Pr, Bu, Oct, Ph; X = Cl, Br, I, NCS)[70] it was shown that the oxine group is chelating and the tin atom pentacoordinate. An exception is $Bu_2Sn(NCS)Ox$ in which the NCS group is considered to be bridging and a dimeric structure containing hexacoordinate tin atoms was proposed for the solid[58]. Similarly $Ph_2Sn(NO_3)Ox$ is considered to be polymeric with bridging NO_3 groups[34].

When organotin dialkoxides and dihalides are mixed reaction occurs and the products may be formulated as $R_2Sn(OR')X$ or as the complexes $R_2Sn(OR')_2 \cdot R_2SnX_2$. Kocheshkov and his coworkers[27, 28, 29, 102] have obtained calorimetric, dipole moment, and other evidence in favour of the $R_2Sn(OR')_2 \cdot R_2SnX_2$ structure. However, Davies and Harrison[19] have shown that, when the product from dibutyltin dimethoxide and diethyltin dichloride is treated with 2,2'-bipyridyl both $Bu_2SnCl_2 \cdot$ bipy and $Et_2SnCl_2 \cdot$ bipy are isolated; they therefore propose an equilibrium similar to that exhibited by the Grignard reagent:

$$R_2Sn\underset{O}{\overset{O}{<}}SnR_2X_2 \rightleftharpoons R_2Sn(OMe)_2 + R_2SnX_2 \rightleftharpoons 2R_2Sn\overset{X}{\underset{OMe}{<}} \rightleftharpoons XR_2Sn\underset{O}{\overset{O}{<}}SnR_2X$$

12.3.1. Hydrolysis products of organotin dihalides

It was noted in Chapter 5 that, by controlled hydrolysis of the organotin dihalides, a number of products can be isolated. All of the

$$R_2SnX_2 \longrightarrow \underset{XVII}{R_2Sn(OH)X} \longrightarrow \underset{XVIII}{R_2Sn(X)OSn(X)R_2} \longrightarrow$$

$$\underset{XIX}{R_2Sn(X)OSn(OH)R_2} \longrightarrow \underset{XX}{R_2Sn(OH)OSn(OH)R_2} \longrightarrow \underset{XXI}{R_2SnO}$$

intermediates XVII–XX and the final oxide, XXI, represent interesting structural problems which still lack unequivocal solutions.

Little information is available concerning the structure of the unstable hydroxyhalides XVII but the pseudohalide $Ph_2Sn(OH)NCO$ was shown to be dimeric and infrared and chemical evidence indicate structure $XVIIa$ with bridging hydroxyl groups[56].

$$Ph_2(OCN)Sn\underset{O}{\overset{\overset{H}{|}}{\underset{\underset{H}{|}}{<}}}Sn(NCO)Ph_2$$

$$XVIIa$$

Compounds with the composition XVIII are well known but their structure remains uncertain. The key pieces of evidence are summarised in a paper by Davies et al.[3]; most of the early work in this field was carried out by Harada who has recently re-examined some of the arguments used in assigning structures[31]. These compounds are dimeric in benzene[3] though some of the evidence here is confusing particularly the assertion[31] that the apparent molecular weights increase with decreasing concentration. For the compounds XVIII R = Bu; X = Cl, Br[3] and also XVIII, R = Me, X = OSiMe$_3$ [18], the ^{119}Sn magnetic resonance spectra indicated two non-equivalent tin atoms in solution. From these observations and because there is some X-ray evidence for the presence of a 4-membered ring containing two oxygen and two tin atoms in XVIII R = Me, X = OSiMe$_3$ [61], the structures $XVIIIa$ and $XVIIIb$ have been proposed[3, 31]. In the solid

$$
\begin{array}{ccc}
\text{R}_2\text{SnX} & \text{R}_2\text{SnX}_2 & \text{SnR}_2\text{X} \\
| & \uparrow & \\
\text{XR}_2\text{Sn} \overset{O}{\underset{O}{\diagup \diagdown}} \text{SnR}_2\text{X} & \text{R}_2\text{Sn} \overset{O}{\underset{O}{\diagup \diagdown}} \text{SnR}_2 & \text{R}_2\text{Sn} \overset{O}{\underset{O}{\diagup \diagdown}} \text{SnR}_2 \\
| & \downarrow & \\
\text{R}_2\text{SnX} & \text{R}_2\text{SnX}_2 & \text{XR}_2\text{Sn} \\
\end{array}
$$

$XVIIIa$ $XVIIIb$ $XVIIIc$

state, however, Mössbauer spectra measurements [76] indicate only one type of tin atom; unless the Mössbauer parameters for the two types of tin atom are identical this suggests different structures in the solid and solution states. When compounds of this type were treated with hydrogen sulphide, either a mixture of the compounds R$_2$SnS and R$_2$SnX$_2$[71] or the compound XR$_2$SnSSnR$_2$X[52] were isolated depending upon conditions so that it is still not possible to choose between structures $XVIIIa$ and $XVIIIb$. A compromise between these two structures is $XVIIIc$ which allows all the tin atoms to have coordination number 5 and which accounts for the mobility of the X groups. This has been proposed for compounds where X = NO$_3$[99] and NCS[93].

Using similar arguments for compounds XIX the structures $XIXa$ and $XIXb$ were proposed[2]. Since these compounds can be made by the partial hydrolysis of XVIII it has been suggested that attack by a water molecule or OH$^{\ominus}$ ion is more likely to occur at the 4-coordinate atoms making $XIXa$ the preferred structure[69] but the contrary view which favours $XIXb$ has also been proposed[65]. The possibility that,

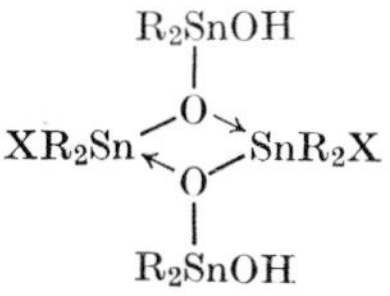

$$
\begin{array}{ccc}
\text{R}_2\text{SnOH} & \text{R}_2\text{SnX(OH)} & \text{SnR}_2\text{X} \\
| & \uparrow & \text{HO} \quad | \\
\text{XR}_2\text{Sn} \underset{\text{O}}{\overset{\text{O}}{<}} \text{SnR}_2\text{X} & \text{R}_2\text{Sn} \underset{\text{O}}{\overset{\text{O}}{<}} \text{SnR}_2 & \text{R}_2\text{Sn} \underset{\text{O}}{\overset{|}{<}} \text{SnR}_2 \\
| & \downarrow & | \quad \text{OH} \\
\text{R}_2\text{SnOH} & \text{R}_2\text{SnX(OH)} & \text{XR}_2\text{Sn} \\
XIXa & XIXb & XIXc
\end{array}
$$

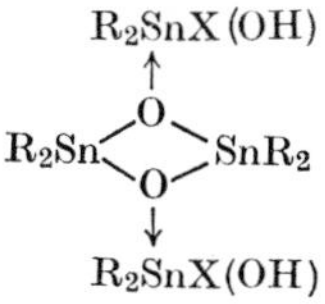

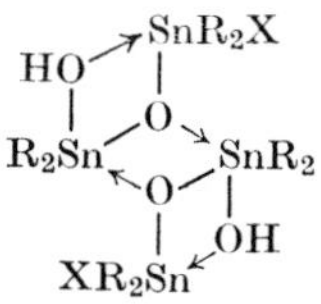

in the solid state, the OH groups may be intramolecularly coordinated as in $XIXc$ has also been discussed[65, 93]. There is some evidence for structure $XIXc$ since the OH rocking mode at $\sim$900 cm^{-1} is absent from the infrared spectra of these compounds. This observation has been quoted[53] as evidence that there is no OH group present at all in these compounds and that, for example, [ClBu$_2$SnOSn(OH)-Bu$_2$]$_2$ is, in fact, ClBu$_2$SnOSnBu$_2$OSnBu$_2$OSnBu$_2$Cl though this seems unlikely.

Few examples of the hydroxyoxides R$_2$Sn(OH)OSn(OH)R$_2$ (XX) are known though two forms of the compound XX R = o-phenoxyphenyl were isolated[68a] and both appeared to be dimeric[2].

Mössbauer spectra measurements indicate that all the tin atoms in the organotin oxides (XXI) have coordination number 5 and the structure $XXIa$ was indicated[30]. This structure has received some support[2] although the infrared spectra of the oxides have been interpreted in terms of polymeric chains of alternating tin and oxygen atoms held together by weak cross links $XXIb$[47].

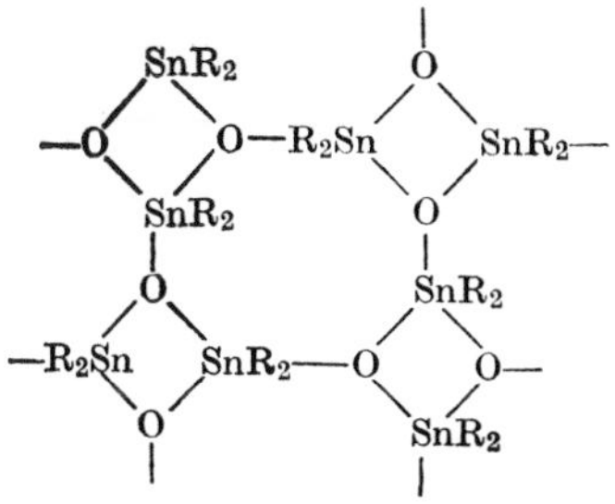

XXIa

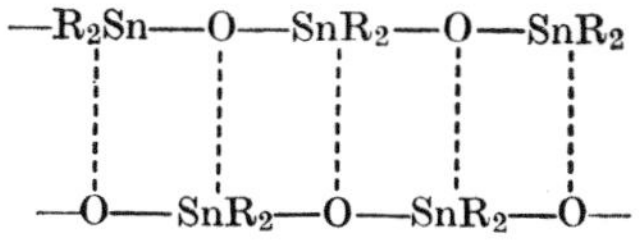

XXIb

12.4. COMPOUNDS WITH ONE Sn—C BOND

Electron diffraction measurements indicate that MeSnX$_3$ (X = Cl, Br, I) molecules are tetrahedral in the gas phase[85] and a similar conclusion regarding methyltin trihydride was reached from the microwave spectrum[73]. When the organotin trihalides are hydrolysed some well-defined intermediates can be isolated (see Chapter 5) but no

TABLE II

Some tin–element bond lengths

Compound	Bond	Length (Å)	Method	Reference
Me$_4$Sn	Sn—C	2·18 ± 0·03	Electron diff[n]	10
Me$_3$SnCl	Sn—C	2·19 ± 0·03 ⎱	Electron diff[n]	85
	Sn—Cl	2·37 ± 0·03 ⎰		
Me$_3$SnBr	Sn—Br	2·49 ± 0·03	Electron diff[n]	85
Me$_3$SnI	Sn—I	2·72 ± 0·03	Electron diff[n]	85
Me$_3$SnCN	Sn—C(methyl)	2·16 ± 0·03	X-ray diff[n]	79
Me$_2$Sn(Br)C(Ph)=C(Ph)C(Ph)=C(Ph)Br	Sn—C(methyl)	2·24 ± 0·04 ⎱	X-ray diff[n]	5
	Sn—Br	2·507 ± 0·007 ⎰		
Cl Cl / OSnEt$_3$ / Cl Cl / Et$_3$SnO	Sn—O	2·08 ± 0·06 ⎱	X-ray diff[n]	95
	Sn—C	2·15 ± 0·06 ⎰		
Me$_2$SnF$_2$	Sn—C	2·07 ± 0·01	X-ray diff[n]	80
Me$_2$SnCl$_2$	Sn—Cl	2·34 ± 0·03	Electron diff[n]	85
Me$_2$SnBr$_2$	Sn—C	2·17 ± 0·05 ⎱	Electron diff[n]	85
	Sn—Br	2·48 ± 0·03 ⎰		
Me$_2$SnI$_2$	Sn—I	2·69 ± 0·03	Electron diff[n]	85
	Sn—C	2·19		
[Me$_2$SnCl$_3$]$^\ominus$	Sn—Cl(equatorial)	2·35 ⎱	X-ray diff[n]	20
	Sn—Cl(axial)	2·54 ⎰		
[Me$_2$SnCl.terpy]$^\oplus$[a]	Sn—Cl	2·63		
Me$_2$SnOx$_2$[b]	Sn—C	2·16 ± 0·02 ⎱	X-ray diff[n]	78
	Sn—O	2·11 ± 0·01		
	Sn—N	2·35 ± 0·04 ⎰		

Compound	Bond	Value	Method	Ref.
(Ph$_2$SnS)$_3$	Sn—C	2·24 ± 0·06	X-ray diff^n	33
	Sn—S	2·42 ± 0·02		
Br$_2$Sn[CH(COOEt)CH$_2$COOEt]$_2$	Sn—C	2·26	X-ray diff^n	101
	Sn—Br	2·516, 2·588		
	Sn—O	244, 2·46		
MeSnCl$_3$	Sn—C	2·19 ± 0·05	Electron diff^n	85
	Sn—Cl	2·32 ± 0·03		
MeSnBr$_3$	Sn—C	2·17 ± 0·10	Electron diff^n	85
	Sn—Br	2·45 ± 0·02		
MeSnI$_3$	Sn—I	2·68 ± 0·02	Electron diff^n	85
MeSnH$_3$	Sn—C	2·143 ± 0·002	Microwave Spectra	85
	Sn—H	1·700 ± 0·015		
Me$_2$SnCl$_2$·2pyO^c	Sn—C	2·225	X-ray	4a
	Sn—Cl	2·584		
	Sn—O	2·251		
Ph$_2$ISn(CH$_2$)$_4$SnIPh$_2$	Sn—C(phenyl)	2·15 ± 0·02	X-ray	17a
	Sn—I	2·729 ± 0·003		
	Sn—C	2·17 ± 0·02		
(PhCH$_2$)$_3$SnOCOMe	Sn—O (intramol.)	2·14 ± 0·02	X-ray	1a
	Sn—O (intermol.)	2·65 ± 0·02		
(C$_6$H$_{11}$)$_3$SnOCOMe^d	Sn—C	2·19 ± 0·03	X-ray	1b
	Sn—O	2·12 ± 0·03		
	Sn—C	2·13 ± 0·03		
Me$_3$SnNCS	Sn—N	2·15 ± 0·02	X-ray	24a
	Sn—S	3·13 ± 0·02		

^{a} terpy = 2,2',6',2"-terpyridyl.
^{b} OxH = 8-hydroxyquinoline.
^{c} pyO = pyridine-N-oxide.
^{d} C$_6$H$_{11}$ = cyclohexyl.

structural information is available. The final products of hydrolysis are the stannonic acids $RSnO_2H$, or their dehydrated forms $R_2Sn_2O_3$, in both cases these are amorphous polymers of uncertain structure.

The corresponding sesquisulphides $R_2Sn_2S_3$ are usually obtained as powders which decompose rather than melt when heated. However $Bu_2Sn_2S_3$ has been obtained in a crystalline form and it exists in benzene as a tetramer[43]. The suggestion[17] that the four tin atoms are at the apices of a tetrahedron, linked by sulphur atoms along the tetrahedron edges, has been shown to be correct by an X-ray study[19b].

There is evidence from ultraviolet-[21] and Mössbauer-spectra[70] that both oxine groups in $RSn(X)Ox_2$ (X = halogen) are chelating so that the compounds contain hexacoordinate tin atoms. The nmr spectra of $PhXSn(acac)_2$ (X = Cl, Br) compounds indicate an octahedral configuration at tin with phenyl and halogen above and below the plane of the acetylacetonate rings[41].

Solutions containing $PhSn(tropolonate)_3$ are nonconducting and, in the absence of further evidence, a structure containing heptacoordinate tin seems likely[55]. Although $BuSnOx_3$ is rather sensitive to hydrolysis with solution measurements being complicated by the loss of an 8-hydroxyquinoline group, ultraviolet-[39], nmr-[39] and Mössbauer[70]-spectra suggest that this compound may also contain 7-coordinate tin.

References

1. Addison, C. C., Simpson, W. B. and Walker, A., *J. Chem. Soc.*, 1964, 2360.
1a. Alcock, N. W. and Timms, R. E., *J. Chem. Soc.* (A), 1968, 1873.
1b. Alcock, N. W. and Timms, R. E., *J. Chem. Soc.* (A), 1968, 1876.
2. Alleston, D. L., Davies, A. G. and Hancock, M., *J. Chem. Soc.*, 1964, 5744.
3. Alleston, D. L., Davies, A. G., Hancock, M. and White, R. F. M., *J. Chem. Soc.*, 1963, 5469.
4. Almenningen, A., Haaland, A. and Motzfeldt, T., *J. Organometallic Chem.*, 1967, **7**, 97.
4a. Blom, E. A., Penfold, B. R. and Robinson, W. T., *J. Chem. Soc.* (A), 1969, 913.
5. Boer, F. P., Flynn, J. J., Freedman, H. H., McKinley, S. V. and Sandel, V. R., *J. Amer. Chem. Soc.*, 1967, **89**, 5068.
6. Bokii, N. G., Avoyan, R. L., Zakharova, G. N., Minasyan, M.Kh., Akopyan, Z. A. and Struchkov, Yu. T., *Zh. Strukt. Khim.*, 1965, **6**, 795 (*Chem. Abs.*, 1966, **64**, 10502b).
7. Bolles, T. F. and Drago, R. S., *J. Amer. Chem. Soc.*, 1965, **87**, 5015; 1966, **88**, 5730.
8. Bonati, F., *Organometallic Chem. Rev.*, 1966, **1**, 379.
9. Bott, R. W., Eaborn, C. and Redl, G., unpublished work (quoted in Bott, R. W., Eaborn, C. and Swaddle, T. W., *J. Organometallic Chem.*, 1966, **5**, 233).

10. Brockway, L. O. and Jenkins, H. O., *J. Amer. Chem. Soc.*, 1936, **58**, 2036.
11. Clark, H. C. and Goel, R. G., *Inorg. Chem.*, 1965, **4**, 1428.
12. Clark, H. C. and Goel, R. G., *J. Organometallic Chem.*, 1967, **7**, 263.
13. Clark, H. C. and O'Brien, R. J., *Inorg. Chem.*, 1963, **2**, 1020.
14. Clark, H. C., O'Brien, R. J. and Pickard, A. L., *J. Organometallic Chem.*, 1965, **4**, 43.
15. Clark, H. C., O'Brien, R. J. and Trotter, J., *J. Chem. Soc.*, 1964, 2332.
16. Clark, J. P. and Wilkins, C. J., *J. Chem. Soc. (A)*, 1966, 871.
17. Coates, G. E., Green, M. L. H., Wade, K., *Organometallic Compounds*, Methuen, London, 3rd Ed., 1967, **1**, 451.
17a. Cody, V. and Cory, E. R., *J. Organometallic Chem.*, 1969, **19**, 359.
18. Considine, W. J., Baum, G. A. and Jones, R. C., *J. Organometallic Chem.*, 1965, **3**, 308.
19. Davies, A. G. and Harrison, P. G., *J. Chem. Soc. (C)*, 1967, 298.
19a. Davies, A. G., Milledge, H. J., and Puxley, D. C., unpublished observations.
19b. Doerfelt, C., Janeck, A., Kobelt, D., Paulus, E. F. and Scherer, H., *J. Organometallic Chem.*, 1968, **14**, P22.
20. Einstein, F. W. B. and Penfold, B. R., *Chem. Comm.*, 1966, 780.
21. Faraglia, G., Roncucci, L. and Barbieri, R., *Ric. Sci. Rend. Sez. A*, 1965, **8**, 205.
22. Farrer, H. N., McGrady, M. M. and Tobias, R. S., *J. Amer. Chem. Soc.*, 1965, **87**, 5019.
23. Fergusson, J. E., Roper, W. R. and Wilkins, C. J., *J. Chem. Soc.*, 1965, 3716.
24. Fitzsimmons, B. W., Seeley, N. J. and Smith, A. W., *Chem. Comm.*, 1968, 390; *J. Chem. Soc. (A)*, 1969, 143.
24a. Forder, R. A. and Sheldrick, G. M., *Chem. Comm.*, 1969, 1125.
25. Freedman, H. H., personal communication.
26. Gielen, M. and Sprecher, N., *Organometallic Chem. Rev.*, 1966, **1**, 455.
27. Gol'dshtein, I. P., Gur'yanova, E. N., Zemlyanskii, N. N., Syutkina, O. P., Panov, E. M. and Kocheshkov, K. A., *Dokl. Akad. Nauk S.S.S.R.*, 1967, **175**, 836.
28. Gol'dshtein, I. P., Gur'yanova, E. N., Zemlyanskii, N. N., Syutkina, O. P., Panov, E. M. and Kocheshkov, K. A., *Izv. Akad. Nauk S.S.S.R., Ser. Khim.*, 1967, 2201.
29. Gol'dshtein, I. P., Zemlyanskii, N. N., Shamagina, O. P., Gur'yanova, E. N., Panov, E. M., Slovokhotova, N. A. and Kocheshkov, K. A., *Dokl. Akad. Nauk S.S.S.R.*, 1965, **163**, 880.
30. Gol'danskii, V. I., Makarov, E. F., Stukan, R. A., Trukhtanov, V. A., and Khrapov, V. V., *Dokl. Akad. Nauk S.S.S.R.*, 1963, **151**, 357.
31. Harada, T., *Bull. Chem. Soc. Japan*, 1968, **41**, 737.
32. Hathaway, B. J. and Webster, D. E., *Proc. Chem. Soc.*, 1963, 14.
33. Hellner, E. and Dittmar, G., unpublished work (quoted in Schumann, H., *Z. Anorg. Allg. Chem.*, 1967, **354**, 192).
33a. Hendricker, D. G., *Inorg. Chem.*, 1969, **8**, 2328.
34. Huber, F. and Kaiser, R., *J. Organometallic Chem.*, 1966, **6**, 126.
35. Hudson, R. F., *Angew. Chem. Intern. Ed. Engl.*, 1967, **6**, 749.
36. Hulme, R., *J. Chem. Soc.*, 1963, 1524.
36a. Isaacs, N. W., Kennard, C. H. L. and Kitching, W., *Chem. Comm.*, 1968, 820.

37. Janssen, M. J., Luijten, J. G. A. and van der Kerk, G. J. M., *J. Organometallic Chem.*, 1964, **1**, 286.
38. Kasai, N., Yasuda, K. and Okawara, R., *J. Organometallic Chem.*, 1965, **3**, 172.
39. Kawakami, K., Kawasaki, Y. and Okawara, R., *Bull. Chem. Soc. Japan*, 1967, **40**, 2693.
40. Kawakami, K. and Okawara, R., *J. Organometallic Chem.*, 1966, **6**, 249.
41. Kawasaki, Y. and Tanaka, T., *J. Chem. Phys.*, 1965, **43**, 3396.
42. Khrapov, V. V., Gol'danskii, V. I., Prokof'ev, A. K., Rochev, V. Ya. and Kostyanovskii, R. G., *Izv. Akad. Nauk S.S.S.R. Ser. Khim.*, 1968, 1261 (*Chem. Abs.*, 1968, **69**, 96850).
43. Komura, M. and Okawara, R., *Inorg. Nucl. Chem. Letters*, 1966, **2**, 93.
44. Kostyanovskii, R. G. and Prokof'ev, A. K., *Dokl. Akad. Nauk S.S.S.R.*, 1965, **164**, 1054.
45. Kostyanovskii, R. G. and Prokof'ev, A. K., *Izv. Akad. Nauk S.S.S.R.*, *Ser. Khim.*, 1968, 274 (*Chem. Abs.*, 1968, **69**, 10518).
46. Kumar Das, V. G. and Kitching, W., *J. Organometallic Chem.*, 1967, **10**, 59.
47. Lohmann, D. H., *J. Organometallic Chem.*, 1965, **4**, 382.
48. McGrady, M. M. and Tobias, R. S., *J. Amer. Chem. Soc.*, 1965, **87**, 1909.
48a. McWhinnie, W. R., Poller, R. C., Ruddick, J. N. R. and Thevarasa, M., *J. Chem. Soc. (A)*, 1969, 2327.
49. Maeda, Y. and Okawara, R., *J. Organometallic Chem.*, 1967, **10**, 247.
50. Matwiyoff, N. A. and Drago, R. S., *Inorg. Chem.*, 1964, **3**, 337.
51. Mendelsohn, J., Pommier, J. C. and Valade, J., *Compt. Rend.*, 1966, **263**, 921.
52. Migdal, S., Gertner, D. and Zilkha, A., *Can. J. Chem.*, 1967, **45**, 2987.
53. Migdal, S., Gertner, D. and Zilkha, A., *Can. J. Chem.*, 1968, **46**, 2409.
54. Muetterties, E. L. and Schunn, R. A., *Quart. Rev.*, 1966, **20**, 259.
55. Muetterties, E. L. and Wright, C. M., *J. Amer. Chem. Soc.*, 1964, **86**, 5132; 1965, **87**, 4706.
56. Mufti, A. S. and Poller, R. C., *J. Chem. Soc.*, 1965, 5055.
57. Mullins, M. A. and Curran, C., *Inorg. Chem.*, 1967, **6**, 2017.
58. Mullins, M. A. and Curran, C., *Inorg. Chem.*, 1968, **7**, 2584.
59. Nelson, W. H. and Martin, D. F., *J. Inorg. Nucl. Chem.*, 1965, **27**, 89.
60. Neumann, W. P., Schick, R. and Koester, R., *Angew. Chem. Intern. Ed. Engl.*, 1964, **3**, 385.
61. Okawara, R., *Proc. Chem. Soc.*, 1961, 383.
62. Okawara, R., Hathaway, B. J. and Webster, D. E., *Proc. Chem. Soc.*, 1963, 13.
63. Okawara, R. and Ohara, M., *J. Organometallic Chem.*, 1964, **1**, 360.
64. Okawara, R. and Wada, M., *Advances in Organometallic Chemistry*, Academic Press, New York and London, 1967, **5**, 137.
65. Okawara, R. and Wada, M., *J. Organometallic Chem.*, 1963, **1**, 81.
66. Okawara, R. and Yasuda, K., *J. Organometallic Chem.*, 1964, **1**, 356.
67. Otera, J., Kawasaki, T. and Tanaka, T., *Inorg. Chim. Acta*, 1967, **1**, 294.
67a. Parish, R. V. and Platt, R. H., *Chem. Comm.*, 1968, 1118.
68. Peddle, G. J. D. and Redl, G., *Chem. Comm.*, 1968, 626.
68a. Poller, R. C., *J. Chem. Soc.*, 1963, 706.

69. Poller, R. C., *J. Organometallic Chem.*, 1965, **3**, 321.
70. Poller, R. C. and Ruddick, J. N. R., *J. Chem. Soc. (A)* 1969, 2273.
71. Poller, R. C. and Spillman, J. A., *J. Organometallic Chem.*, 1967, **7**, 259.
72. Pommier, J. C. and Valade, J., *J. Organometallic Chem.*, 1968, **12**, 433.
73. Pope, W. J. and Peachey, S. J., *Proc. Chem. Soc.*, 1900, **16**, 42; 116.
74. Reid, A. F. and Wailes, P. C., *Austral. J. Chem.*, 1966, **19**, 309.
75. Roncucci, L., Faraglia, G. and Barbieri, R., *J. Organometallic Chem.*, 1964, **1**, 427; 1966, **6**, 278.
76. Ruddick, J. N. R., M.Sc. Thesis, University of Newcastle upon Tyne, 1966.
77. Sato, H., *Bull. Chem. Soc. Japan*, 1967, **40**, 410.
78. Schlemper, E. O., *Inorg. Chem.*, 1967, **6**, 2012.
79. Schlemper, E. O. and Britton, D., *Inorg. Chem.*, 1966, **5**, 507.
80. Schlemper, E. O. and Hamilton, W. C., *Inorg. Chem.*, 1966, **5**, 995.
81. Schumann, H., Thom, K. F. and Schmidt, M., *J. Organometallic Chem.*, 1963, **1**, 167.
82. Simons, P. B. and Graham, W. A. G., *J. Organometallic Chem.*, 1967, **8**, 479; **10**, 457.
83. Simpson, W. B., *Chem. and Ind.*, 1966, 854.
84. Srivastava, T. N. and Bhattacharya, S. N., *J. Inorg. Nucl. Chem.*, 1967, **29**, 1873.
85. Sutton, L. E. (Ed.), *Interatomic Distances*. The Chemical Society, London, 1958.
86. Tanaka, T., *Inorg. Chim. Acta*, 1967, **1**, 217.
87. Tanaka, T., Matsumura, Y., Okawara, R. and Musya, Y., *Bull. Chem. Soc. Japan*, 1968, **41**, 1497.
88. Thayer, J. S. and West, R., *Inorg. Chem.*, 1964, **3**, 389.
89. Tobias, R. S., *Organometallic Chem. Rev.*, 1966, **1**, 93.
90. Tobias, R. S. and Freidline, C. E., *Inorg. Chem.*, 1965, 4, 215.
91. Wada, M., Kawakami, K. and Okawara, R., *J. Organometallic Chem.*, 1965, **4**, 159.
92. Wada, M., Nishino, M. and Okawara, R., *J. Organometallic Chem.*, 1965, **3**, 70.
93. Wada, M. and Okawara, R., *J. Organcmetallic Chem.*, 1967, **8**, 261.
94. Wada, M., Shindo, M. and Okawara, R., *J. Organometallic Chem.*, 1963, **1**, 95.
95. Wheatley, P. J., *J. Chem. Soc.*, 1961, 5027.
96. Wilkins, C. J., and Haendler, H. M., *J. Chem. Soc.*, 1965, 3174.
97. Winters, L. J. and Hill, D. T., *Inorg. Chem.*, 1965, **4**, 1433.
98. Yasuda, K., Kawasaki, Y., Kasai, N. and Tanaka, T., *Bull. Chem. Soc. Japan*, 1965, **38**, 1216.
99. Yasuda, K., Matsumoto, H. and Okawara, R., *J. Organometallic Chem.*, 1966, **6**, 528.
100. Yasuda, K. and Okawara, R., *J. Organometallic Chem.*, 1965, **3**, 76.
101. Yoshida, M., Ueki, T., Yasuoka, N., Kasai, N., Kakudo, M., Omae, I., Kikkawa, S. and Matsuda, S., *Bull. Chem. Soc. Japan*, 1968, **41**, 1113.
102. Zemlyanskii, N. N., Gol'dshtein, I. P., Gur'yanova, E. N., Syutkina, O. P., Panov, E. M., Slovokhotova, N. A., and Kocheshkov, K. A., *Izv. Akad. Nauk S.S.S.R., Ser. Khim.*, 1967, 728.

13

PHYSICAL MEASUREMENTS

A powerful array of physical techniques are available for the investigation of organotin compounds. The most important of these are infrared-, proton magnetic resonance-, and Mössbauer-spectroscopy. Accordingly, most of the space in this chapter is devoted to these methods.

It is the practice in most laboratories to record infrared spectra in the sodium chloride region as routine. While these spectra can yield much useful information it must be remembered that, save for the $\nu(\text{Sn}—\text{H})$ mode, almost all fundamental stretching vibrations involving the tin atom occur at frequencies below 650 cm^{-1}. Proton magnetic resonance spectra are being used increasingly for structural studies and also to obtain information about the nature of the bonding in organotin compounds. There is a radioactive precursor of the ^{119}Sn isotope with properties which are appropriate for observing the Mössbauer effect and the use of γ-ray spectroscopy to study organotin compounds is becoming more widespread. This technique is specialised and very much in its infancy but it could develop into one of the most sensitive probes for the examination of organotin structures.

Other physical methods, which are discussed in this chapter, such as the determination of ultraviolet absorption spectra, dipole moments and molar refractivities are given a much briefer treatment.

13.1. INFRARED SPECTROSCOPY

Discussion is limited here to absorptions due to molecular vibrations which directly involve the tin atom and secondary effects, for example the perturbation of the $\nu(\text{C}\equiv\text{C})$ mode which occurs when an acetylenic carbon atom is attached to tin[71, 201, 202, 204] are omitted. Some Raman spectra are also included in this section.

13.1.1. *Tin–carbon stretching frequencies*

Since the aliphatic carbon–tin stretching frequencies are the most commonly quoted infrared bands, a fairly extensive list is given in Table I. This list is, of course, far from complete; another compilation of ν(Sn—C) bands (which omits reference to the physical states of the compounds) was published recently[154]. The di- and tri-alkyltin compounds generally show two bands, the νas(Sn—C) mode causing absorption essentially in the range 500–600 cm^{-1} with the second band at approximately 470–530 cm^{-1} due to νs(Sn—C) vibration. The intensities of these bands vary considerably and, in some cases, only very weak absorption is observed.

Monoalkyltin compounds should exhibit a single Sn—C stretching band. This prediction is borne out in the spectra of the methyltin trihalides but the spectra of compounds with longer alkyl chains are more complex. Thus butyltin trichloride shows two bands in the ν(Sn—C) region which are considered to arise from the conformational isomers *I* and *II* resulting from restricted rotation about the SnCH$_2$—CH$_2$Et bond[39]. When more than one butyl (or larger alkyl)

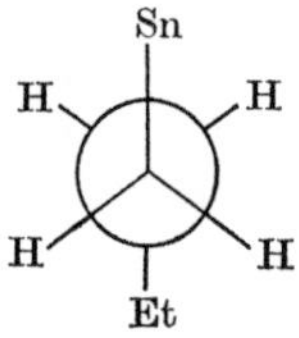

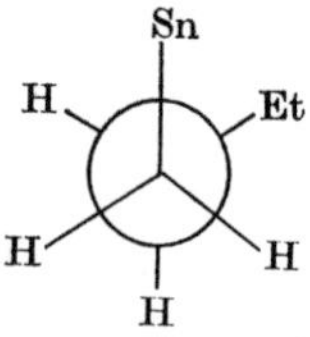

group is attached to tin the number of conformational isomers increases and the ν(Sn—C) region shows more complex absorption for liquid or solution spectra but becomes simplified in the solid state when a particular isomer is usually preferred[39, 41, 72]. In contrast ethyl- and propyl-tin compounds often show more infrared bands in the solid than in liquid state spectra[244, 246], a phenomenon which is commonplace in infrared spectroscopy, though it was claimed that this is evidence for restricted rotation about the Sn—C bonds[244].

Knowledge of the factors which affect the positions of the ν(Sn—C) bands is limited but it has been observed that, for the methyltin halides, electronic effects are dominant. Thus the more electronegative the halogen and the greater the number of halogens the greater is the effective nuclear charge on tin and hence the higher are the ν(Sn—C) frequencies[34].

It can be seen from Table I that the positions of the ν(Sn—C) bands are not particularly sensitive to changes in the coordination number of the tin atom. However, the number of bands observed may be affected, for example, both νas(Sn—C) and νs(Sn—C) bands are observed for tetrahedral $Me_3SnOCOH$ molecules in chloroform solution but, as pyridine is added to the solution, the $\nu s(Sn—C)$ band slowly disappears as the trimethyltin group assumes a planar configuration in the 1:1 pyridine complex[207]. There is considerable variation in the positions of bands assigned to acetylenic C—Sn vibrations and figures in the range 504–340 cm^{-1} have been quoted for different alkynyltin compounds[188, 216].

Recent assignments[194, 247] to the methyltin rocking mode are in the range 770–791 cm^{-1} and bands in the 685–658 cm^{-1} region have been assigned to the ethyltin rocking vibration[136].

TABLE I

Infrared spectra: some Sn—C(*aliphatic*) *stretching frequencies*

Compound	State	ν(Sn—C) (cm^{-1})		Reference
(a) Compounds with 4 Sn—C bonds				
Me_4Sn	Liquid		528s	224
Et_4Sn	Liquid		508s	224
Pr_4Sn	Liquid	590s	500s	224
Bu_4Sn	Soln. in CS$_2$	592s	503s	41
$Me_3SnC{\equiv}CH$	Liquid	538vs	517w	216
(b) Compounds with 3 Sn—C bonds				
Me_3SnCl	Soln. in cyclo-hexane	542s	513w	10
Me_3SnBr	Soln. in cyclo-hexane	539s	511m	34
Me_3SnI	Soln. in cyclo-hexane	536m	508w	34
Me_3SnOH	Solid		540s	165
Me_3SnOH	Soln. in CCl$_4$	571m	504m	165
$Me_3SnOCOH$	Solid		555s	207
$Me_3SnOCOH$	Soln. in CHCl$_3$	550m	513w	207
$Me_3SnMn(CO)_5$	Solid	517s	500s	31
$Me_3SnCF_2CF_2Mn(CO)_5$	Solid	539s	518m	31
Et_3SnCl	Soln. in cyclo-hexane	518vs	489s	223

TABLE I—*cont.*

Compound	State	ν(Sn—C) (cm^{-1})		Reference
Et$_3$SnBr	Soln. in cyclo-hexane	510m	484w	34
Et$_3$SnI	Soln. in cyclo-hexane	506m	482w	34
Et$_3$SnOH	Solid	510vs	485m(sh)	136
(Et$_3$Sn)$_2$O	Liquid	509vs	485s(sh)	136
(Et$_3$Sn)$_2$S	Liquid	505vs	483s	136
Et$_3$SnOCOH	Solid	520		164
Et$_3$SnOCOH	Soln. in heptane	525		164
Et$_3$SnOCOR	Solid	521–517vs	490–487m	136
Et$_3$SnSnEt$_3$	Liquid	496vs	475vs	136
(Et$_3$Sn)$_2$SO$_4$	Solid	524s	490vw	136
(Et$_3$Sn)$_2$CO$_3$	Solid	518vs	491m	136
Bu$_3$SnCl	Liquid	601s	513m	41
Bu$_3$SnBr	Soln. in cyclo-hexane	599s	503m	34
Bu$_3$SnI	Soln. in cyclo-hexane	598s	501s	34
(Bu$_3$Sn)$_2$O	Soln.	648, 595	509	154
Bu$_3$SnOR	Soln.	648 $\pm$ 3, 601 $\pm$ 6	510	154
Me$_3$SnCl.pyridine	Solid	541s	512vw	34
Me$_3$SnCl.bipy^a	Solid	554w	544w	34
Me$_3$SnBr.pyridine	Solid	542s	509vw	34
Me$_3$SnBr.bipy	Solid	551w	543w	34
Me$_3$SnI.pyridine	Solid	541s	504vw	34
Me$_3$SnI.bipy	Solid	567w	538w	34
(c) *Compounds with* 2 Sn—C *bonds*				
Me$_2$SnCl$_2$	Soln. in CS$_2$	560	524	20
Me$_2$SnBr$_2$	Soln. in benzene	554m	518w	34
Me$_2$SnI$_2$	Soln. in cyclo-hexane	542m	511w	34
Me$_2$SnCO$_3$	Solid	576s	523w	30
Me$_2$SnSO$_4$	Solid	600s		30
[Me$_2$Sn]$^{2\oplus}$	Soln. in water	529^b		229

TABLE I—*cont.*

Compound	State	ν(Sn—C) (cm^{-1})		Reference
Et$_2$SnCl$_2$	Soln. in CS$_2$	531	497	20
Et$_2$SnBr$_2$	Soln. in benzene	528m	493m	34
Et$_2$SnI$_2$	Soln. in benzene	520m	490m	34
Et$_2$SnO	Solid	532s	493m	136
(Et$_2$SnS)$_3$	Solid	524m-s	493s	136
Pr$_2$SnCl$_2$	Soln. in CS$_2$	598	512	20
Pr$_2$SnI$_2$	Soln. in CS$_2$	589	501	20
Bu$_2$SnCl$_2$	Soln. in CS$_2$	602	517	20
Bu$_2$SnBr$_2$	Soln. in benzene	600m	511m	34
Bu$_2$SnI$_2$	Soln. in benzene	592m	508m	34
(C$_8$H$_{17}$)$_2$SnCl$_2$	Soln. in CS$_2$	606	518	20
R$_2$Sn(OR′)$_2$	Soln.	~600	521–504	149, 154
Me$_2$SnCl$_2$.2pyridine	Solid	563m		227
Me$_2$SnCl$_2$.bipy	Solid	572m		34
Me$_2$SnCl$_2$.2Me$_2$SO	Solid	575m	507vw	225
[Me$_2$SnCl$_4$]$^{2\ominus}$	Solid	580m		33
Me$_2$SnCl$_2$.phen^c	Solid	578m		34
Me$_2$SnBr$_2$.2pyridine	Solid	563m		227
Me$_2$SnBr$_2$.bipy	Solid	571w		34
Me$_2$SnBr$_2$.phen	Solid	572m	551w	34
[Me$_2$SnBr$_3$]$^{\ominus}$	Solid	566	522	33
Me$_2$SnI$_2$.2pyridine	Solid	550m	514vw	227
Me$_2$SnI$_2$.bipy	Solid	569m		34
Me$_2$SnI$_2$.phen	Solid	560m	554sh	34
[Me$_2$Sn(OH)$_4$]$^{2\ominus}$	Soln. in water	523^b		230
Et$_2$SnCl$_2$.2pyridine	Solid	580w	531m	227
Et$_2$SnCl$_2$.bipy	Solid	529m	481w	34
Et$_2$SnCl$_2$.phen	Solid	525m	470w	34
Et$_2$SnCl$_2$.dipyam^d	Solid	493s		143
Et$_2$SnCl$_2$.tripyam^e	Solid	485s		143
Et$_2$SnBr$_2$.bipy	Solid	523m		34
Et$_2$SnBr$_2$.phen	Solid	520w		34
Et$_2$SnBr$_2$.dipyam	Solid	488s		143

TABLE I—*cont.*

Compound	State	ν(Sn—C) (cm^{-1})		Reference
Et$_2$SnI$_2$.bipy	Solid	514m		34
Et$_2$SnI$_2$.phen	Solid	513w		34
Bu$_2$SnCl$_2$.phen	Solid	624vw	594vw	34
Bu$_2$SnBr$_2$.bipy	Solid	~620sh	587w, br	34
Bu$_2$SnBr$_2$.bipy	Solid	~615vw	588w	34
Bu$_2$SnI$_2$.bipy	Solid	613w	584w	34
Bu$_2$SnI$_2$.phen	Solid	610w	582w	34
(d) Compounds with 1 Sn—C bond				
MeSnCl$_3$	Soln.	551– 546w-m		35, 124
MeSnBr$_3$	Soln.	538m		34, 124
MeSnI$_3$	Soln. in cyclo- hexane	527w		10
EtSnCl$_3$	Soln. in benzene	522w		35
EtSnBr$_3$	Soln. in benzene	511m		34
BuSnCl$_3$	Liquid	596	518	39
BuSnBr$_3$	Soln. in benzene	596w	513w	34
MeSnCl$_3$.bipy	Solid	536w		34
MeSnCl$_3$.phen	Solid	529w		34
MeSnBr$_3$.bipy	Solid	519m		34
MeSnBr$_3$.phen	Solid	515m, 508w		34
MeSnI$_3$.bipy	Solid	497m		34
MeSnI$_3$.phen	Solid	500w, 495w		34
EtSnCl$_3$.bipy	Solid	504w		34
EtSnCl$_3$.phen	Solid	507w		34
EtSnBr$_3$.bipy	Solid	484w		34
EtSnBr$_3$.phen	Solid	496w, 479w		34
BuSnCl$_3$.bipy	Solid	592w		34
BuSnCl$_3$.phen	Solid	608w		34
BuSnBr$_3$.bipy	Solid	595w		34
BuSnBr$_3$.phen	Solid	595w		34

[a] bipy = 2,2'bipyridyl.
[b] Raman spectrum.
[c] phen = 1,10-phenanthroline.
[d] dipyam = 2,2'-dipyridylamine.
[e] tripyam = 2,2',2"-tripyridylamine.

There is some disagreement concerning the positions of the tin-phenyl stretching frequencies. The assignments[122] νas(Sn—Ph) $\sim$ 450 cm^{-1} and νs(Sn—Ph) $\sim$ 210 cm^{-1} for the Ph$_3$Sn group have been accepted by some workers[195, 196]. However the νas(Sn—Ph) band is not expected to occur at such a high frequency and the $\sim$450 cm^{-1} band is, in fact, a benzene ring vibration[176]. The correct assignments for di- and tri-phenyltin compounds are νas(Sn—Ph) $\sim$ 280–260 cm^{-1} and νs(Sn—Ph) $\sim$ 240–225 cm^{-1} with monophenyl compounds showing a single band at $\sim$250–236 cm^{-1}[176, 211, 215]. As with analagous aliphatic compounds the positions of the tin–phenyl stretching frequencies are largely unaffected by changes in the coordination number of tin[143, 179, 180, 225].

The spectra of a number of benzyltin compounds in the region 1800–400 cm^{-1}, together with some assignments have recently been reported[23].

13.1.2. *Tin–halogen stretching frequencies*

Some recent tin–halogen stretching frequency assignments for alkyl- and phenyl-tin halides are given in Table II. Overall ranges are ν(Sn—Cl) 385–318 cm^{-1}, ν(Sn—Br) 264–222 cm^{-1} and ν(Sn—I) 207–170 cm^{-1}. Less information is available for organotin fluorides; these compounds are associated by fluorine bridges (Chapter 12) and hence are structurally dissimilar to the other organotin halides. The highly tentative range obtained from Table II of ν(Sn—F) 372–328 cm^{-1} is lower than would be expected from simple atomic mass considerations and this is probably a consequence of the highly associated structures (see below).

It can be seen from Table III that the tin–halogen stretching frequencies are very sensitive to changes in the coordination of the tin atom. Conversion of a 4-coordinate dichloride to a 6-coordinate adduct with a Lewis base causes a reduction in the frequency of the ν(Sn—Cl) bands by some 100 cm^{-1}. Similar reductions in absorption frequencies are observed for the other halides which form 6-coordinate complexes although the effect is less marked with the iodides. Few results are available concerning the reduction in frequency of tin-halogen stretching bands when organotin monohalides are converted to pentacoordinate complexes R$_3$SnX.L. The single example in Table III suggests that the reduction in frequency may be similar to that occurring in 6-coordinate complexes.

Some tin–pseudohalide stretching frequencies which have been assigned recently include $\nu(\text{Sn}-\text{N}_3)$ 401 cm^{-1} and $\nu(\text{Sn}-\text{NCO})$ 400 cm^{-1} [228].

TABLE II

Infrared spectra: tin-halogen stretching frequencies for organotin halides[a]

Compound	νas(Sn—X) (cm^{-1})	νs(Sn—X) (cm^{-1})	ν(Sn—X) (cm^{-1})	Reference
Bu$_3$SnF[b]	—	—	330	16
Ph$_3$SnF[b]	—	—	350	16
Ph$_3$SnF[b]	—	—	372	176
(C$_6$F$_5$)$_3$SnF[b]	—	—	330	16
R$_3$SnCl	—	—	336–318	35, 72, 136, 223
Ph$_3$SnCl	—	—	346–332	176, 214
R$_3$SnBr	—	—	234–222	34
Ph$_3$SnBr	—	—	256	176
R$_3$SnI	—	—	189–182	34
Ph$_3$SnI	—	—	170[c]	122
Pr$_2$SnF$_2$	330		—	16
R$_2$SnCl$_2$	361–356	356–340	—	20, 72
Ph$_2$SnCl$_2$	364	356, 350	—	176
R$_2$SnBr$_2$	260–248	241–238	—	20, 34
R$_2$SnI$_2$	204–196	186–176	—	20, 34
RSnCl$_3$	384–376	368–358	—	35, 124, 223
BuSnCl$_3$	355[c]		—	72
PhSnCl$_3$	385–364		—	176
RSnBr$_3$	264–256	253–225	—	34, 124, 136
MeSnI$_3$	207	174	—	34

[a] Unless otherwise indicated frequencies refer to liquid or solution spectra. The absorption bands are, generally, of strong intensity.

[b] Solid state spectrum.

[c] Raman spectrum.

13.1.3. *Tin–oxygen stretching frequencies*

Interpretation of the spectra of compounds containing Sn—O bonds has proved rather difficult though, as indicated in Table IV, several problems of assignment have now been solved. The trialkyltin hydroxides show a medium to strong intensity band at ~900 cm^{-1} due to the Sn—OH deformation mode. Why triphenyltin hydroxide shows a doublet in that region (replaced on deuteration by a single band) is not clear. It is suggested that there is Fermi resonance with an out-of-plane γCH vibration[40].

TABLE III

Infrared spectra: tin-halogen stretching frequencies for adducts of organotin halides[a]

Compound		$\nu(Sn—X)$ (cm^{-1})	Reference
Ph$_3$SnCl.phepy[b]		226	179
R$_2$SnCl$_2$.bipy[c]	1 or 2 bands	244–215	35
R$_2$SnCl$_2$.phen[d]	1 or 2 bands	247–220	35
Et$_2$SnCl$_2$.dipyam[e]		284, 275sh	143
Et$_2$SnCl$_2$.tripyam[f]		247br	143
Bu$_2$SnCl$_2$.4,4'-bipy[g]		247–231br	179
Bu$_2$SnCl$_2$.2phepy		225– <200br	179
R$_2$SnCl$_2$.2Me$_2$SO	1 or 2 bands	244–187	225
[Me$_2$SnCl$_3$]$^{\ominus}$		333, 322, 235	33
[Me$_2$SnCl$_4$]$^{2\ominus}$		227	33
Ph$_2$SnCl$_2$.2pyridine		248	143
Ph$_2$SnCl$_2$.bipy		252, 246	143
Ph$_2$SnCl$_2$.dipyam		265, 254sh	143
Ph$_2$SnCl$_2$.tripyam		262, 248	143
Ph$_2$SnCl$_2$.4,4'-bipy		240–228br	179
R$_2$SnBr$_2$.bipy	1 or 2 bands	169–140	34
R$_2$SnBr$_2$.phen	1 or 2 bands	169–149	34
R$_2$SnI$_2$.bipy	2 bands	156–139	34
R$_2$SnI$_2$.phen	2 bands	147–126	34
RSnCl$_3$.bipy	2 bands	294–267	35
RSnCl$_3$.phen	1 or 2 bands	299–270	35
PhSnCl$_3$.4,4'-bipy		330–318, 286, 281	179
RSnBr$_3$.bipy	2 or 3 bands	201–170	34
RSnBr$_3$.phen	2 or 3 bands	200–177	34
MeSnI$_3$.bipy		176, 159, 147	34
MeSnI$_3$.phen		184, 158, 140	34

[a] The spectra were measured with compounds in the solid state. The adsorption bands are generally of strong intensity.

[b] phepy = 4-phenylpyridine.

[c] bipy = 2,2'-bipyridyl.

[d] phen = 1,10-phenanthroline.

[e] dipyam = 2,2'-dipyridylamine.

[f] tripyam = 2,2',2''-tripyridylamine.

[g] 4,4'-bipy = 4,4'-bipyridyl.

An empirical calculation[175] indicated that the Sn—O stretching frequency should occur at ≈ 570 cm^{-1} and a range of 575–550 cm^{-1} has been suggested[230]. Although solutions of trimethyltin hydroxide absorb in this region there is doubt concerning the ν(Sn—O) band in the solid-state spectrum and assignments as different as 370 cm^{-1}[165] and 518 cm^{-1}[123] have been made. While there is no basis for un-equivocal choice the higher frequency assignment appears to be preferable. It is difficult to detect the ν(Sn—O) band in triphenyltin hydroxide though a weak intensity shoulder at 456 cm^{-1} has been assigned[122].

There seems little doubt that absorption due to the νas(SnOSn) mode occurs at ~780–770 cm^{-1} in hexaalkyldistannoxanes and the cor-responding aryl compounds. Weak to medium intensity bands at 415 cm^{-1} in Me$_3$SnOSnMe$_3$[123] and 407 cm^{-1} in Bu$_3$SnOSnBu$_3$[154] have been assigned to the νs(SnOSn) modes but in many compounds these bands are too weak to be detected. This weak intensity of the Sn—O—Sn symmetric stretching vibration has been ascribed to a tendency to linearity of the Sn—O—Sn system[40, 253].

The polymeric dialkyl(aryl)tin oxides also show a single ν(Sn—O) absorption instead of the expected two bands and the frequency of ~570 cm^{-1} observed is much lower than in the monomeric stannoxanes.

It has been suggested that there are characteristic absorption bands associated with the 4-membered $\overline{\text{OSnOSn}}$ ring system which frequently occurs in dimeric tin compounds. Absorptions in the region 531–481 cm^{-1} have been assigned to this ring system in the methoxyl-bridged acetylacetonates III[116].

$$\text{acacXYSn}\underset{\text{O}}{\overset{\text{O}}{\diagdown\!\!\!\diagup}}\text{SnXYacac} \qquad (\text{X} = \text{halogen or alkyl, Y} = \text{halogen})$$

with Me groups above (O) and below (O).

III

The carboxylate groups in compounds such as R$_3$SnOCOR′ and R$_2$Sn(OCOR′)$_2$ are bidentate and absorption bands at ~300 cm^{-1} have been assigned to vibrations of the coordination bond O → Sn[146, 164]. This assignment is tentative, however, and an alternative explanation for the origin of these bands is that they are due to in-plane deformation vibrations of the COO group[102].

All the members of a series of compounds containing the Sn—O—Si group showed an intense broad absorption band centred in the range 996–948 cm^{-1} which has been assigned to the νs(SnOSi) mode[47]. Comparison with related assignments given in Table IV suggests that this may be the antisymmetric stretching vibration.

It is clear that tin–oxygen stretching vibrations occur over an exceptionally wide range of frequencies depending upon the precise environment of the Sn—O group in the molecule. Partial explanations have been advanced in terms of variations in the extent of $d\pi$–$p\pi$ bonding between oxygen and tin[175], variations in the bond angle at oxygen[40] and differences in the ionic character of the Sn—O bond[113].

TABLE IV

Infrared spectra: absorption bands associated with the Sn—O *group*

Compound	State	Absorption bands (cm^{-1}) and assignments	Reference
Me$_3$SnOH	Solid	917m δ(Sn—OH)	123
Et$_3$SnOH	Solid	885s δ(Sn—OH)	136
Ph$_3$SnOH	Solid	897s, 912s δ(Sn—OH)	40, 66, 122, 131, 175
Me$_3$SnOH	Solution	576m ν(Sn—O) 531s ν(Sn—O)	123 165
Ph$_3$SnOSnPh$_3$	Solid or solution	777–770s νas(SnOSn)	40, 66, 122, 131, 175
R$_3$SnOSnR$_3$	Liquid	784–769 νas(SnOSn)	40, 136, 154, 245
Bu$_3$SnOR	Solution	1100–946w-m νas(SnOC) 525–514w νs(SnOC)	42, 150, 154
Ph$_3$SnOP(O)Ph$_2$	Solid	394m ν(Sn—O)	195
R$_2$SnO	Solid	576–561s ν(Sn—O)	40, 136
Ph$_2$SnO	Solid	575–571s ν(Sn—O)	40, 175
[Me$_2$Sn(OH)$_4$]$^{2\ominus}$	Solution	555br^a ν(Sn—O)	230
R$_2$Sn(OR')$_2$	Solution	~600w^b νas(OSnO) 487–466w νs(OSnO)	149, 150, 154
R$_2$Sn(OH)NO$_3$	Solid	2 bands in the range 594–398s ν(Sn—O)	250, 251
RR'Sn(acac)$_2$^{c}	Solid	461–404 ν(Sn—O)	117
(Me$_3$SiO)$_n$SnMe$_{4-n}$ (n = 1–4)	Liquid	1070–938 νas(SiOSn)	194
Me$_n$SnOx$_{4-n}$^{d}(n = 2, 3)	Solid or liquid	528–517 ν(Sn—O)	110, 226
Me$_2$SnX$_2$.2Ph$_3$PO	Solid	320–300 ν(Sn—O)	32
Me$_2$SnX$_2$.2Ph$_3$AsO	Solid	380–370 ν(Sn—O)	32
Me$_2$SnCl$_2$.2PyO^e	Solid	325s,br ν(Sn—O)	113

^{a} Raman spectrum.
^{b} Overlap with ν(CSnC) occurs.
^{c} acacH = acetylacetone.

^{d} OxH = 8-hydroxyquinoline.
^{e} PyO = pyridine-N-oxide.

16

13.1.4. *Some other tin-element stretching frequencies*

Absorption associated with the tin–hydrogen stretching frequency is usually intense and Kuivila[127] lists absorption bands in the range 1880–1790 cm^{-1} for 22 organotin hydrides. The position of the tin–hydrogen stretching frequency depends upon the electronegativities of the groups attached to tin;[57] the ν(Sn—H) band in dialkyltin halide hydrides, R_2SnXH, occurs in the range 1874–1820 cm^{-1} [111, 196]. With the aid of a normal coordinate analysis, complete vibrational assignments have been made for methylstannane[70, 118].

The tin–sulphur stretching vibration occurs at ~350 cm^{-1}, some examples are given in Table V.

TABLE V

Infrared spectra: some tin–sulphur stretching frequencies

Compound	Phase	νas(SnSSn) (cm^{-1})	νs(SnSSn) (cm^{-1})	νs(Sn—S) (cm^{-1})	Reference
Ph$_3$SnSSnPh$_3$	Solid	376s	330s	—	196
Ph$_3$SnSGePh$_3$	Solid	—	—	355m	196
Ph$_3$SnSPbPh$_3$	Solid	—	—	365s	196
bis(ethane-1,2-dithiolato)tin	Solid	327s	388s	—	63, 178
Ph$_3$SnSP(S)Ph$_2$	Solid	—	—	340s	195
(Et$_3$Sn)$_2$S	Liquid	370vs	318s	—	136
(Et$_2$SnS)$_3$	Solid	368vs	330vs	—	136

In symmetrically substituted distannanes, R_3SnSnR_3, the ν(Sn—Sn) mode is infrared inactive but it can be observed in the Raman spectrum or in the infrared spectrum of an unsymmetrically substituted compound. Values for some tin–tin and other tin–metal stretching frequencies are given in Table VI.

The ν(Sn—N) absorption bands, like the ν(Sn—O) bands discussed earlier, are reported to occur over a wide range of frequencies and appear to be very sensitive to changes in the molecular environment of the Sn—N group. The tin–nitrogen stretching frequency in N-trimethylstannylaniline has been shown[183] to occur at 843 cm^{-1}. There is disagreement concerning the bands in (Me$_3$Sn)$_3$N, the most recent assignments being νas(NSn$_3$) 672 cm^{-1} and νs(NSn$_3$) 514 cm^{-1} [9] whereas a band at 728 cm^{-1} has also been assigned to the antisymmetric stretching mode in this compound[209]. Absorption bands of much lower frequency have been assigned to the stretching vibration

TABLE VI

Infrared spectra: some tin–metal stretching frequencies

Compound	Phase	ν(Sn–metal) (cm^{-1})	Reference
Ph$_3$SnSnPh$_3$	Solid	208[a]	69
(Ph$_3$Sn)$_4$Sn	Solid	207[a]	69
Ph$_3$SnMn(CO)$_5$	Solution	174[a]	69
Ph$_3$SnSnEt$_3$	Liquid	208	22
Et$_3$SnSnBu$_3$	Liquid	199	22
Ph$_3$SnSnMe$_3$	Solid	194	22
Me$_3$SnGePh$_3$	Solid	225	22
Et$_3$SnGePh$_3$	Solid	230	22
Ph$_3$SnGeBu$_3$	Liquid	235	22
Me$_3$SnMn(CO)$_5$	Liquid	182	22
Ph$_3$SnMn(CO)$_5$	Solid	174	22
Me$_3$SnMo(CO)$_3$C$_5$H$_5$	Solid	172	22
Ph$_3$SnMo(CO)$_3$C$_5$H$_5$	Solid	169	22
Me$_3$SnFe(CO)$_2$C$_5$H$_5$	Liquid	185	22
Me$_3$SnCo(CO)$_4$	Solid	176	22
Me$_2$ISnMn(CO)$_5$	Liquid	179	22
Me$_2$BrSnMn(CO)$_5$	Liquid	191	22
Me$_2$ClSnMn(CO)$_5$	Liquid	197	22
MeCl$_2$SnMn(CO)$_5$	Liquid	201	22

[a] Raman spectrum.

of N → Sn coordination bands in compounds in which the nitrogen atom is incorporated into an aromatic ring. Thus in the organotin oxinates, bands in the region 406–387 cm^{-1} have been assigned to ν(Sn—N) modes [110, 226] and the tin–nitrogen stretching vibrations in the complexes Me$_2$SnX$_2$.2pyridine are thought to occur at ~200 cm^{-1} [227].

Recent assignments made for compounds in which tin is attached to other Group V elements are as follows: νas(PSn$_3$) 351–347 cm^{-1} and νs(PSn$_3$) 296–284 cm^{-1} in (R$_3$Sn)$_3$P (R = Me[96], Ph[58]); ν(Sn—P) 351 cm^{-1} in Ph$_3$SnPPh$_2$[195]; νas(AsSn$_3$) 233 cm^{-1} and νs(AsSn$_3$) 209 cm^{-1} in (Me$_3$Sn)$_3$As [96].

13.2. NUCLEAR MAGNETIC RESONANCE SPECTROSCOPY

A review article published in 1965[145] dealt with the application of nuclear magnetic resonance spectroscopy to the study of organometallic compounds and includes a section on organotin compounds.

Of the 10 isotopes of tin there are three in which the nuclear spin quantum number $I = \frac{1}{2}$, of these the abundance[167] of the ^{115}Sn isotope (0·34%) is so low that nmr measurements involving this nucleus are rarely possible. Hence the nuclei which are significant in magnetic resonance measurements are those of the ^{117}Sn (7·54% abundance) and ^{119}Sn (8·62% abundance) isotopes. A few direct measurements of ^{119}Sn resonances have been made[19, 101] but there has, so far, been very limited application of this technique to structure investigations[5, 36]. By observing the ^{119}Sn—C—H coupling and using a heteronuclear double resonance technique it is possible to obtain, indirectly, ^{119}Sn chemical shift values[46, 46a]. The greater part of this section, however, is concerned with proton magnetic resonance measurements involving the hydrogen atoms of the organic groups attached to tin. The magnitudes of the indirect tin–proton coupling constants $J(^{117}$Sn—C—H) and $J(^{119}$Sn—C—H) are often informative in structural studies. In proton magnetic resonance measurements on the organotin hydrides direct spin–spin coupling between tin and hydrogen nuclei occurs and the parameters $J(^{117}$Sn—H) and $J(^{119}$Sn—H) can be evaluated. Some results of ^{19}F magnetic resonance measurements on compounds containing fluorocarbon groups attached to tin are also discussed briefly.

The results of proton magnetic resonance measurements on a number of methyltin compounds are given in Table VII. Although there has been some discussion[17, 98, 231, 232] of the factors which influence the values of the chemical shift, most interest centres around the magnitude of the indirect tin–proton spin–spin coupling constants. Holmes and Kaesz[97a] and van der Kelen[235] showed that the coupling constants for methyltin chlorides showed strong dependence on the solvent. For example, $J(^{119}$Sn—C—H) for Me$_3$SnCl is 58·5 cps in carbon tetrachloride and 68·4 cps in water[97a] (similar measurements have been made recently over a very wide range of solvents[152]). It was suggested that these J values are a measure of the percentage s-character in the tin–carbon bond. Thus Me$_3$SnCl in carbon tetrachloride with sp^3 hybridisation at tin has 25% s-character in the Sn—C bonds whereas the same compound exists in water as sp^2 hybrid Me$_3$Sn$^{\oplus}$ ions with $33\frac{1}{3}$% s-character in the metal-carbon bonds. Since it is now known that these ions are hydrated it has to be assumed that negligible d-orbital mixing occurs[229]. Hence, in principle, we have in the determination of coupling constants an important probe for investigating the coordination number at tin.

TABLE VII

Proton magnetic resonance spectra of some methyltin compounds

Compound	Solvent	τ (ppm relative to Me_4Si)	$J(^{119}Sn{-}C{-}H)$ (cps)	$J(^{117}Sn{-}C{-}H)$ (cps)	Reference
(a) Compounds with 4 Sn—C bonds					
Me_4Sn	$CDCl_3$, CCl_4	9·92–9·93	53·4–54·3	51·1–52	2, 105, 138, 235
Me_3SnCCl_3	$CHCl_3$	9·50	59a		48
Me_3SnCBr_3	$CHCl_3$	9·50	57·4	54·6	48
(SnMe₃ / Ph aromatic structure)	CCl_4	10·03	54	52	60
(bis-SnMe₃ aromatic structure)	CCl_4	9·69	53	51	60
(chloro ring with SnMe₃, CCl₂, Ph structure)	CCl_4	9·90	58	56	198
(Sn ring with Ph, Ph, Me, Me structure)	CCl_4	9·38	58·5	55·1	7

TABLE VII—cont.

Compound	Solvent	τ (ppm relative to Me$_4$Si)	J(^{119}Sn—C—H) (cps)	J(^{117}Sn—C—H) (cps)	Reference
(b) Compounds with 3 Sn—C bonds					
Me$_3$SnF	MeOH	9.55	69.0	66.1	138
Me$_3$SnCl	CCl$_4$	9.34–9.39	58.1–58.5	55.7–56	138, 232
Me$_3$SnBr	CCl$_4$	9.20–9.27	57.8–58.5	55.2–56	138, 232
Me$_3$SnI	CCl$_4$	9.03–9.12	57.2–58.5	54.7–56	138, 232
Me$_3$SnSMe	CHCl$_3$, CCl$_4$	9.59–9.61	56.5–56.9	54–54.5	2, 234
Me$_3$SnOCOH	CHCl$_3$, CDCl$_3$	9.44	62.7	59.9	207
Me$_3$SnOCOCH$_2$Cl	CDCl$_3$	9.39	58.9	56.4	208
Me$_3$SnOSnMe$_3$	CCl$_4$	9.77	56.0	53.6	194
Me$_3$SnOSiMe$_3$	CCl$_4$	9.66	57.4	54.9	194
Me$_3$SnNEt$_2$	CHCl$_3$	9.82		56^a	48
Me$_3$SnN(Me)SiMe$_3$	CCl$_4$	9.78	57.5	53.4	193
Me$_3$SnSeSnMe$_3$	CCl$_4$	9.43	56	53	186
Me$_3$SnOx^b	CHCl$_3$	9.54	57.0	54.6	110
[Me$_3$Sn. 2Me$_2$SO]$^{\oplus}$Ph$_4$B$^{\ominus}$	CH$_2$Cl$_2$	9.74, 9.71	62, 56	—	130
Me$_3$SnCo(CO)$_4$	CDCl$_3$	9.37	52.6	50.6	169
Me$_3$SnMo(CO)$_3$C$_5$H$_5$	CHCl$_3$	9.52	49.0	47.0	168
Me$_3$SnW(CO)$_3$C$_5$H$_5$	CHCl$_3$	9.40	48.7	46.4	168
(c) Compounds with 2 Sn—C bonds					
Me$_2$SnF$_2$	MeOH	9.25	84.1	80.2	138
Me$_2$SnCl$_2$	CCl$_4$	8.76–8.85	69.0–70	66.0–67	138, 232
Me$_2$SnBr$_2$	CCl$_4$	8.53–8.67	66–66.3	63.5–63.6	138, 232
Me$_2$SnI$_2$	CCl$_4$	8.28–8.37	62.4–62.5	59.6–60	138, 232
Me$_2$Sn(SMe)$_2$	CHCl$_3$, CCl$_4$	9.28–9.29	60.5–61.2	58.5	2, 234

Me⟨S,S⟩SnMe2	CCl₄	9·06	63	59	249
Me₂Sn(OSiMe₃)₂	CCl₄	9·44	71·4	68·0	194
Me₂Sn[N(Me)SiMe₃]₂	CCl₄	9·67	60·0	57·5	193
Me₂SnOx₂	CDCl₃	9·56	71·2	67·9	142
Me₂SnClOx	CDCl₃	8·96	79·9	76·4	110
Me₂Sn(acac)₂ᶜ	CDCl₃	9·51	99·3	95·0	142
Me₂Sn(pic)₂ᵃ	CDCl₃	9·40	77·6	73·5	142
[Me₂Sn.4Me₂SO]²⊕2Ph₄B⊖	CH₂Cl₂	9·55, 9·25	65, 57	—	130
Me₂Sn[Co(CO)₄]₂	CDCl₃	8·88	45·7	43·6	169
Me₂Sn[Mo(CO)₃C₅H₅]₂	CDCl₃	9·15	37·4	35·8	168
ClMe₂SnMo(CO)₃C₅H₅	CCl₄	9·02	47·7	45·6	168

(d) Compounds with 1 Sn—C bond

	CCl₄	9·06	63	59	249
MeSnCl₃	CCl₄	8·31-8·32	98-100	95·5-95·6	138, 232
MeSnBr₃	CCl₄	8·05-8·15	88·6-89	84·3-85	138, 232
MeSnI₃	CCl₄	7·52-7·68	73-73·4	70	138, 232
MeSn[N(Me)SiMe₃]₃	CCl₄	9·53	73·5	70·1	193
MeSn(SMe)₃	CHCl₃, CCl₄	9·03-9·06	66·5-66·6	63·5-64·5	2, 234
ClMeSnOx₂	CHBr₃	8·98	106·6	101·7	110
ClMeSn(acac)₂	CHCl₃	9·12	120·9	111·5	110
MeSn[Co(CO)₄]₃	CDCl₃	8·98		33·0ᵃ	169
ClMeSn[Co(CO)₄]₂	CDCl₃	8·48		42·2ᵃ	169

ᵃ Mean of $J(^{119}\mathrm{Sn}-\mathrm{C}-\mathrm{H})$ and $J(^{117}\mathrm{Sn}-\mathrm{C}-\mathrm{H})$.

ᶜ acacH = acetylacetone.

ᵇ OxH = 8-hydroxyquinoline.

ᵃ picH = picolinic acid.

This method has been used, for example, to study the association of organotin carboxylates[207, 208]. Inspection of Table VII shows that the lowest J values are obtained for compounds in which tin is bonded to a transition metal and this implies that, in these compounds, there is less than 25% s-character in the tin–carbon bonds. Graham[168, 169] has shown that it would be reasonable to expect a concentration of s-character in the transition metal–tin bonds with a consequent reduction in s-electron density in the Sn—C bonds. Unfortunately the relationship between the tin–proton coupling constants and the hybridisation at tin does not always hold and Tobias has suggested that it is of most use when comparing closely related compounds where the structure of one is known[229]. The generalisation has been made that the relationship holds for compounds in which the hybridisation at tin is sp^n ($n = 1, 2, 3$) but not for sp^3d^m ($m = 1, 2$) hybridisation[233]. A striking illustration of the value of coupling constants in structure determinations is given by dimethyltin dioxinate which has been quoted as an example of a compound where the relationship breaks down since the $J(^{119}Sn—C—H)$ values are very similar to those in Me_2SnCl_2 yet it was known that the oxine groups are chelating[119, 142]. However, this conflicting evidence was reconciled when a subsequent X-ray crystal structure determination showed that the geometry of the hexacoordinate dioxinate was so distorted that the Me—Sn—Me angle was close to tetrahedral (Chapter 12). The indirect tin–proton coupling constants in trimethyltin chloride adducts, $Me_3SnCl.L$, are linearly related to the enthalpies of formation and $J(^{117, 119}Sn—C—H)$ values are used as a measure of the interaction between a ligand and Me_3SnCl^{12}.

If the alkyl group R in Me_3SnR is varied systematically it is found that the $J(^{117, 119}Sn—C—H)$ values for the methyl protons vary according to the inductive effect of R^{14}. Similarly the coupling constants for the methyl protons in $Me_3SnC_6H_4X$ depend upon the Taft σ^* values of the X substituents[14]. Empirical methods of calculating coupling constants have been shown to be applicable to the indirect $J(^{117, 119}Sn—C—H)$ constants[210, 243].

The observation of two proton resonances apparently due to the tin-bound methyl groups in $[Me_2Sn.4Me_2SO]^{2\oplus}$ is of interest particularly as the intensities are unequal. The latter point indicates that this splitting cannot be accounted for by postulating a highly distorted octahedron with environmentally distinct methyl groups but the observations could be explained in terms of ion pairs of differing geometries[130]. The values of the $J(^{117,119}Sn$—C—H) coupling constants for $Me_2SnCl_2 . 2Me_2SO$ show substantial variations according to whether the measurements are made in dimethyl sulphoxide or chloroform, these variations are thought to be caused by differing configurations of the complex in the two solvents[120].

The $J(^{119}Sn$—$^{13}C)$ coupling constants have been determined for several methyltin compounds and there is an approximately linear relationship between $J(^{119}Sn$—$^{13}C)$ and $J(^{119}Sn$—C—H)[141, 248]. The factors influencing $J(^{13}C$—H) coupling constants have been investigated for methyl derivatives of a number of elements including tin[104, 140, 248].

The chemical shifts of Sn—Me and N—Me protons in N,N-dimethyldithiocarbamate (=DMTC) derivatives of the type $XYSn(DMTC)_2$ (X = Me, Y = Me or halogen) have been measured in benzene and in chloroform and the different results obtained in the two solvents accounted for in terms of specific solvent-solute interactions[98]. By pmr measurements on compound IV it was possible to distinguish protons from three types of methyl groups (labelled a, b and c in the Figure). Assignment of the two types of Sn—Me peaks was made by using variable temperature measurements. At 85° the c peak disappears and is incorporated into the b peak due to the dissociation $IV \to V$[36].

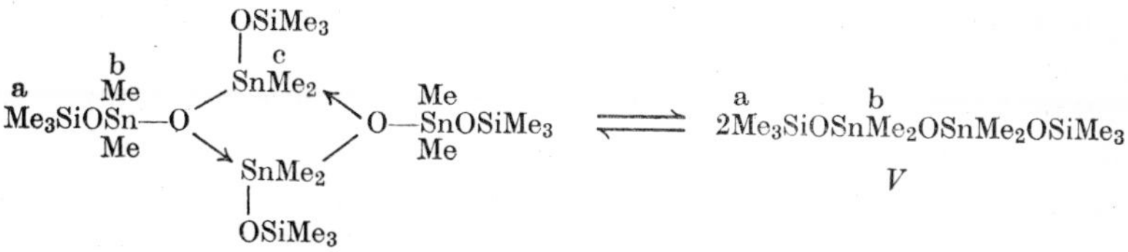

The temperature dependancy of the line width of the pmr signal in tetramethyltin has been investigated[212].

Proton magnetic resonance measurements have been carried out predominantly on methyltin compounds with substantially less work reported for aryl- and higher alkyl-tin derivatives. The protons of a phenyl group attached to tin show chemical shifts of $\tau \sim 2.7$ ppm[128] and o-bis(trimethylstannyl)benzene showed a complex multiplet at $\tau = 2.47–3.1$ ppm due to the protons of the aromatic ring[60]. Although separated by two carbon atoms, spin–spin coupling between the

o-protons and the ^{119}Sn and ^{117}Sn nuclei can be observed in the spectra of phenyltin compounds. Values for the coupling constants in phenyltin chlorides are given in Table VIII. Coupling between tin nuclei and the *meta* protons has also been observed in phenyltin compounds[128].

TABLE VIII

Tin–proton coupling constants for the o-protons in phenyltin chlorides[238]

Compound	Temp.[a]	J(^{119}Sn—C—C—H) (cps)	J(^{117}Sn—C—C—H) (cps)
Ph₃SnCl	100°	61·7	59
Ph₂SnCl₂	45°	81·8	78·6
PhSnCl₃	22°	122	117·5

[a] Temperatures chosen so that measurements could be carried out on the liquid compounds.

Some results of proton magnetic resonance measurements on ethyltin compounds[138, 160, 237] are given in Table IX. It can be seen that the longer range coupling constants, J(Sn—C—C—H), are always greater than the J(Sn—C—H) values and that the chemical shifts of the methylene protons are more sensitive than those of the methyl protons to changes in the extent of halogenation at the tin atom.

Results of pmr measurements on other types of organotin compounds are given in Table X. Included in this Table is a selection of the comprehensive list of pmr parameters for olefinic tin compounds obtained by Leusink and his coworkers[134]. It is clear that this is a valuable method for assigning configurations to alkenyltin compounds[50a, 199, 236] for example the tin–proton coupling constants in the unit SnC=CH are approximately doubled when the relative position of tin and hydrogen changes from *cis* to *trans*[134].

Proton magnetic resonance measurements on cyclopentadienyltin compounds are of interest in that, even down to −60°, there is only a single peak due to the C₅H₅ protons[67]. This is thought to be caused by a valence bond tautomerism of the type:

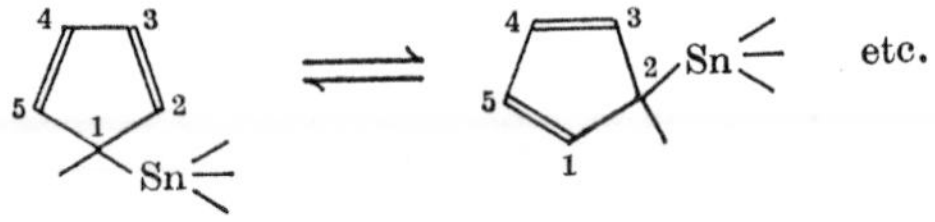

TABLE IX

Proton magnetic resonance measurements on ethyltin compounds

Compound	Solvent	$J(^{119}Sn-C-C-H)^{a}$ (cps)	$J(^{117}Sn-C-C-H)^{a}$ (cps)	$J(^{119}Sn-C-H)^{b}$ (cps)	$J(^{117}Sn-C-H)^{b}$ (cps)	τHa	τHb	Reference
Et$_4$Sn	none	71·2	68·1	32·2	30·8	8·846	9·314	237
Et$_3$SnF	MeOH	95·8	91·5	60·7	58·0	8·70	8·87	138
Et$_3$SnCl	none	92	88	40·6	38·7	8·547	8·513	237
Et$_3$SnBr	CHCl$_3$	94·2	90·0	34·5	33·0	8·71	8·71	138
Et$_3$SnI	CHCl$_3$	101·5	97·6	33·5	31·9	8·76	8·68	138
Et$_2$SnF$_2$	MeOH	—		—	—	8·67	8·50	138
Et$_2$SnCl$_2$	none	130·8	124·8	51·7	49·5	8·304	7·938	237
Et$_2$SnBr$_2$	CHCl$_3$	136·0	130·0	54·0	52·0	8·58	8·18	138
Et$_2$SnI$_2$	CHCl$_3$	137·0	131·0	44·0	42·0	8·70	8·08	138
EtSnCl$_3$	none	242	231·2	83·35	81·5	8·04	7·05	237
EtSnBr$_3$	CHCl$_3$	224	214	75·6	71·8	8·58	7·61	138
EtSnI$_3$	CHCl$_3$	216	206	58·0	55·5	8·89	7·42	138

TABLE X

Proton magnetic resonance measurements on miscellaneous organotin compounds. The results refer to the hydrogen atoms which are underlined

Compound	Solvent	$J(^{119}\mathrm{Sn}—(\mathrm{C})_n—\mathrm{H})$ (cps)	$J(^{117}\mathrm{Sn}—(\mathrm{C})_n—\mathrm{H})$ (cps)	τ	Reference
(MeCH₂CH₂)₂SnCl₂	*	48·4	47	8·09	240
MeCH₂CH₂SnCl₃	*	85	81	7·55	240
MeCH₂CH₂CH₂SnCl₃	*	88	85	7·70	240
Ph₃SnCH₂CH₂CN	*	51·5	49·5	8·39	240
Ph₃SnCH₂CH₂Ph	*	55·5	53·5	8·25	240
Ph₃SnCH₂CH₂COOMe	*		56†	8·45	240
Ph₃SnCH₂CH₂OPh	*	63·5	61·5	8·13	240
Ph₃SnCH₂CH₂OAc	*	56	54	8·21	240
(ClCH₂)₃SnCl	CHCl₃/CCl₄	18·3†		6·31	241
(ClCH₂)₂SnCl₂	CHCl₃/CCl₄	18·7†		6·17	241
ClCH₂SnCl₃	CHCl₃/CCl₄	19·4†		5·92	241
(BrCH₂)₃SnBr	CHCl₃/CCl₄	16·4†		6·67	241
(BrCH₂)₂SnBr₂	CHCl₃/CCl₄	16·4†		6·47	241
(PhCH₂)₄Sn	None	58	55·9	8·069	239
(PhCH₂)₃SnCl	None	64·8	62·0	7·385	239
(PhCH₂)₃SnCl	CHCl₃	66·9	64·3	7·340	239
(PhCH₂)₂SnCl₂	CHCl₃	79·3	76·3	6·800	239
(PhCH₂)₂SnI₂	CHCl₃	64·9	62·2	6·543	239
(PhCH₂)₂Sn(OAc)₂	CHCl₃	90·1	86·1	6·943	239

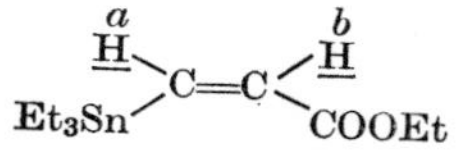

Compound	Solvent				Ref.
H_a, Et₃Sn / C=C / H_b, COOEt	CCl₄	64·0a / 114·0b	61·0a / 109·0b	2·90a / 3·29b	134
Et₃Sn, H_a / C=C / H_b, COOEt	CCl₄	64·0a / 56·5b	61·0a / 54·0b	2·40a / 3·76b	134
H_b, H_a / C=C / SnMe₃, COOEt $_a^b$	CCl₄	128·0a / 63·0b	123·0a / 60·5b	3·14a / 4·09b	134
Me₃SnCH₂CH₂COOEt	CCl₄	53·0a / 69·5b	50·5a / 66·5b	9·06a / 7·50b	134
Me, H / C=C / SnBu₃, COOEt	CCl₄	112·5	108·0	2·59	134
H, Me / C=C / SnBu₃, COOEt	CCl₄	64·5	61·5	8·00	134
EtOOC, H / C=C / SnMe₃, COOEt	CCl₄	97·0	92·5	3·19	134

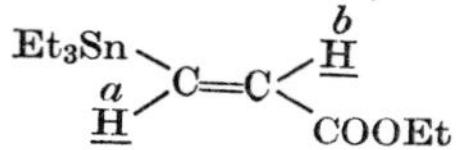

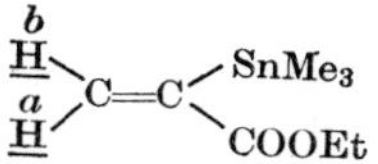

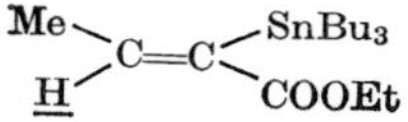

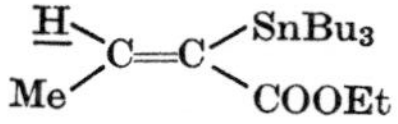

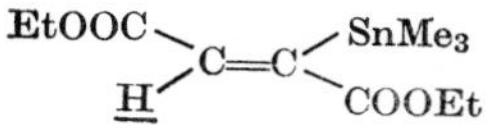

TABLE X—cont.

Compound	Solvent	$J(^{119}\text{Sn}-(\text{C})_n-\text{H})$ (cps)	$J(^{117}\text{Sn}-(\text{C})_n-\text{H})$ (cps)	τ	Reference
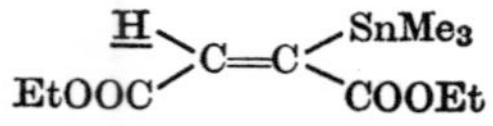 $\underline{\text{H}}$ $>$C=C$<$ SnMe$_3$ / EtOOC COOEt	CCl$_4$	60·5	58·0	4·03	134
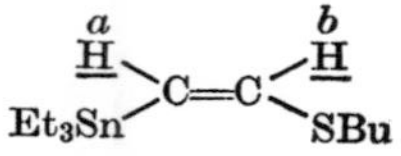 b $\underline{\text{H}}$ $>$C=C$<$ / a $\underline{\text{H}}$ Et$_3$Sn SBu	CCl$_4$	62·5a / 127·5b	59·5a / 122·0b	4·00a / 2·93b	134
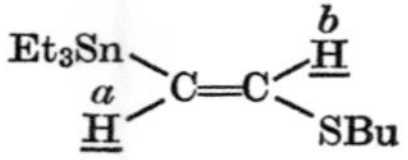 Et$_3$Sn b $\underline{\text{H}}$ $>$C=C$<$ / a $\underline{\text{H}}$ SBu	CCl$_4$	63·0a / 61·0b	60·0a / 58·0b	4·09a / 3·58b	134
C$_5$H$_5$SnMe$_3$	CS$_2$	22·7	21·7	4·13	67
(C$_5$H$_5$)$_4$Sn	CS$_2$	26·9	25·9	4·15	67
Ph$_3$SnC≡CH	CCl$_4$	34·2	32·7	7·68	206
Ph$_3$SnC≡CC≡CH	CCl$_4$		5·2†	8·06	206

* J values measured on neat liquids, τ values measured on solutions in carbon tetrachloride.

† Coupling with the ^{119}Sn and ^{117}Sn nuclei not resolved.

This idea has subsequently received support from theoretical calculations[200]. Further support was sought from pmr measurements on the methylcyclopentadienyl compound *VI*, the results purported to show that the presence of the methyl group inhibited the valence bond tautomerism[68]. These results were reinterpreted and shown to be consistent with a tautomeric molecule in which migration of the tin atom is not inhibited by the presence of the methyl group[49].

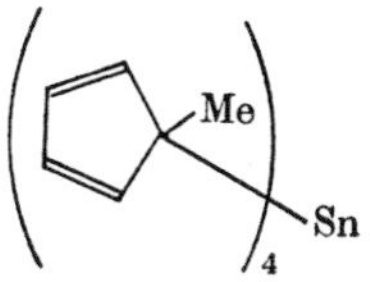

VI

Coupling constants have been measured in compounds where the proton and tin atom are separated by carbon and nitrogen and, for example, the $J(^{119}Sn$—N—C—H) values for $[Me_3SiN(Me)]_nSnMe_{4-n}$ vary between 45·5 and 66·8 cps depending upon the value of n[193]. The $J(^{119, 117}Sn$—N—C—H) constants have been shown to be a valuable tool for investigating exchange processes, which involve Sn—N bond fission, in the stannylamines $R_nSn(NR'R'')_{4-n}$[184]. Similarly, coupling constants have been reported for compounds in which the proton and the tin nucleus are separated by carbon and sulphur, thus for Me_3SnSMe, $J(^{119}Sn$—S—C—H) = 37·5 cps and $J(^{117}Sn$—S—C—H) = 36·0 cps[2]. Long range spin–spin coupling between protons of the acetylacetonate group and tin has been measured in compounds of the type $XYSn(acac)_2$ (X = halogen or alkyl, Y = alkyl)[112, 115]. It has been suggested that this type of complex is too labile for stereochemical assignments to be made on the basis of nmr studies[13].

The technique of proton magnetic resonance is particularly well suited to the study of organotin hydrides and a considerable amount of work on this topic is summarised in recent papers[55, 129, 144]. For the alkyltin hydrides, R_nSnH_{4-n}, as neat liquids, τ values for the hydride protons in the range 4·70 to 5·86 have been obtained. Lower values are observed for phenyltin hydrides as indicated by the following figures which were obtained on the neat liquids, Ph_3SnH $\tau = 3·17$, Ph_2SnH_2 $\tau = 3·91$, $PhSnH_3$ $\tau = 5·07$[6] (somewhat different values are obtained from solution measurements[187]). The changes in chemical shift when the hydrogens of SnH_4 are progressively replaced by organic groups have been discussed in terms of the effective shielding values of alkyl[114] and phenyl[187] groups. The direct tin–proton coupling constants for the organostannanes are in the ranges $J(^{119}Sn—H)$ 1506–1936 cps and $J(^{117}Sn—H)$ 1437–1851 cps[55, 129, 144]. For a given R group in a series of alkyltin hydrides R_nSnH_{4-n} ($n = 0, 1, 2, 3$) there is a linear relationship between $J(^{119}Sn—H)$ and the infrared stretching frequency $\nu(Sn—H)$ but this relationship does not hold for the phenylstannanes Ph_nSnH_{4-n}[144]. There has been some interest in pmr studies of alkyltin halohydrides R_2SnXH (X = F, Cl, Br, I), chemical shifts for the tin–bound proton are in the range $\tau = 2.44–3.92$ and the ranges for the coupling constants are $J(^{119}Sn—H) = 1736–2228$ cps and $J(^{117}Sn—H) = 1663–1877$ cps[111, 190, 191]. Using the double resonance technique it was shown that there is coupling between a proton attached to tin and a methyl proton in $(t\text{-}Bu)_2SnH_2$ with $J(H—Sn—C—C—H) = 0.7$ cps[148].

A few reports have been made of ^{19}F nmr studies of compounds containing fluorocarbon groups bound to tin[29, 147] and there is particular interest in the perfluorophenyltin compounds where the degree of resolution possible in the ^{19}F nmr spectrum depends upon the number of C_6F_5 groups attached to tin[62]. In compounds such as $(C_6F_5)_2SnMe_2$ the methyl protons couple only with the fluorine nuclei which are ortho to the tin atom[18]. There is a straight line relationship between the values of the o-p fluorine coupling constants, $J(2,4)$, and the chemical shifts of the p-fluorine atoms in a series of compounds C_6F_5NAB; the relative positions of $C_6F_5NH_2$ and $C_6F_5NHSnMe_3$ on this line are consistent with the absence of π-character in the Sn—N bond of the latter compound[97].

A measure of the basicities of Me_3SnNEt_2, Me_3SnSR, $(Me_3Sn)_2O$ and $(Me_3Sn)_2S$ has been obtained by measuring the proton chemical shift of chloroform dissolved in these compounds. Large steric effects were observed and the results are meaningful only when compounds in an isostructural series are compared[1].

The spin–spin coupling constant $J(^{119}Sn—^{15}N)$ for Bu_3SnNCS has been reported[103]. From measurements on $(PhMe_2CCH_2)_3SnF$ it was shown that the tin fluorine coupling constant is negative[141a].

13.3. MÖSSBAUER SPECTROSCOPY

This subject has been discussed in a number of excellent review articles[52, 56, 65, 76, 76a, 83] and only a brief and somewhat simplified account of the technique and its application to the study of organotin compounds will be given here.

Mössbauer spectroscopy is γ-ray resonance spectroscopy. The radiation source is the radioactive ^{119m}Sn isotope which has a half-life of 250 days and which, in decaying from the nuclear first excited state to the ground state emits a γ-ray of energy 23·8 Kev. In principle, an absorber containing tin atoms in the same chemical environment as the source will, as a consequence of the presence of 8·62% (natural abundance) of the ^{119}Sn isotope, absorb this 23·8 Kev radiation and hence become raised from the ground state to the first excited state. In practice this resonance absorption of γ-rays may not be observed because an emitting nucleus undergoes recoil and the energy of the γ-quantum emitted is diminished by the amount of the nuclear recoil energy. Similarly an absorber nucleus suffers recoil and the net energy available is insufficient to bring about excitation so that no absorption can occur. Mössbauer showed that, if these nuclei occupied lattice positions in a solid, then a significant proportion of emissions and absorptions could occur without recoil and hence the resonance absorption of γ-rays was observed. For the study of tin compounds sources are available which give an acceptable fraction of recoil-less emissions at room temperature, but it is necessary to cool the absorber (usually to liquid nitrogen temperature) in order to increase the proportion of recoil-less absorptions.

When the environments of the tin atoms in the source and absorber are different the nuclear energy levels no longer coincide and absorption can only occur when the energy of the source is modulated. This is done by using the Doppler effect and moving the source relative to the absorber. A range of velocities is scanned until absorption occurs and the magnitude of the velocity applied, in mm. sec^{-1}, at the point of maximum absorption is the isomer shift $= \delta$ (also known as the chemical shift and the chemical isomer shift). This is a measure of the difference in excitation energies between the source and absorber. Isomer shifts arise because the charge radii of the ground and first excited states are not the same so that the excitation energy is sensitive to changes in the nuclear electron density. Hence direct information is obtained only about the density of s-electrons, other electrons being insufficiently penetrating. By making certain assumptions[254] it can be shown that the isomer shift is given by $\delta = $ const. $\Delta R/R \, (|\Psi_s(O)|_a^2 - |\Psi_s(O)|_s^2)$ where $\Delta R/R$ is the fractional change in the nuclear charge radius on excitation and $|\Psi_s(O)|_a^2$ and $|\Psi_s(O)|_s^2$ are, respectively, the total s-electron densities at the nuclei of the absorber and source. After some controversy it is now accepted that

17

$\Delta R/R$ is positive[11, 133, 165] so that the isomer shift increases with increasing electron density at the tin nucleus.

The first excited state of ^{119}Sn has nuclear spin $I = 3/2$ and deviation of the nuclear charge distribution from spherical symmetry can lead to a splitting of the energy levels. Hence, because the tin atom in Me_4Sn has Td symmetry only a single band is observed in the Mössbauer spectrum but the lower symmetry of Me_3SnCl gives a two-band spectrum and the separation between the bands, in mm sec^{-1}, is known as the quadrupole splitting $= \Delta E_Q$. It can be shown that $\Delta E_Q = $ const. Qq where Q is the nuclear quadruple moment and q is the electric field gradient, the latter has been found to be of positive sign for Me_2SnCl_2 and Me_2SnMoO_4[81a].

It would be inappropriate to give here details of the equipment used (these can be obtained from the review articles cited at the beginning of this section) though it is pertinent to comment briefly on some source materials. Most of the early work was done with the ^{119m}Sn isotope incorporated into a stannic oxide matrix giving a source which has a large recoil-less fraction at room temperature but which suffers from the disadvantage of having a broad line-width due to a small amount of quadrupole splitting[91, 156, 220]. Several measurements have been carried out with Mg_2Sn sources which show a narrow line width[135] but a weak effect. Some compromise between these two extremes was obtained by the use of palladium–tin alloy sources[92]. The best general-purpose source material to date is, however, barium stannate which combines a narrow line-width with a high proportion of recoil-less γ-emissions at room temperature[173, 189]. According to the source employed different zero positions may be used for isomer shift measurements, factors for converting other scales to the SnO_2 scale are given in the footnotes to Table XI.

Since this technique is relatively new and data are scattered widely in the literature the opportunity has been taken of making a fairly extensive survey of Mössbauer parameters in Table XI. The errors in the δ and ΔE_Q values given in the Table are of the order of ± 0.1 mm sec^{-1} but, even allowing for this, there is occasionally poor agreement when the Mössbauer parameters have been determined by different groups.

It can be seen from Table XI that, contrary to expectations, the presence of different substituents at the tin atom does not necessarily lead to quadrupole splitting. As examples we may note that the hydrides R_nSnH_{4-n} ($n = 1,2,3$), $Ph_3SnSnPh_3$ and Ph_3SnLi all have

$\Delta E_Q = 0$. The generalisation has been made[73, 90] that quadrupole interactions will only be observed when there is an imbalance in the presence of subsidiary π-interactions. The source of π-electron density may differ in the various bonds (e.g. halogen, oxygen, sulphur, aromatic heterocyclic nitrogen) without observable quadrupole effects but if it is present in some bonds and absent from others then a quadrupole splitting will be observed. Although this rule gives correct predictions in the majority of cases an increasing number of exceptions have been found[3, 73, 166, 172, 219]. It has been recently suggested[166] that electronegativity considerations rather than π-bonding determine whether or not a quadrupole splitting will

TABLE XI

Mössbauer parameters for organotin compounds

Compound	δ^a (mm sec^{-1})	ΔE_Q (mm sec^{-1})	Reference
(a) Compounds with 4 Sn—C bonds			
Me$_4$Sn	1·22	0	90
Me$_4$Sn	1·21	0	221
Me$_4$Sn	1·59	0	37
Me$_4$Sn	1·29	0	166
Me$_3$SnCH=CH$_2$	1·30	0	4
Me$_3$SnCF$_3$	1·31	1·38	166
Me$_3$SnPh	1·16	0	27a
Me$_3$SnC$_6$F$_5$	1·27	1·31	166
Me$_3$SnC$_6$H$_4$F-p	1·23	0	27a
Me$_3$SnC$_6$F$_4$H-p	1·24	1·08	27a
Me$_3$SnC$_6$H$_4$CF$_3$-o	1·21	0·66	27a
1,2-(Me$_3$Sn)$_2$C$_6$F$_4$	1·26	0·85	27a
1,4-(Me$_3$Sn)$_2$C$_6$F$_4$	1·20	1·20	27a
Me$_3$SnC$_6$Cl$_5$	1·32	1·09	166
1,4-(Me$_3$Sn)$_2$C$_6$Cl$_4$	1·26	1·10	27a
Me$_2$Sn(C$_6$F$_4$Br-o)$_2$	1·25	1·41	27a
Me$_2$Sn(C$_6$F$_5$)$_2$	1·23	1·56	27a
Me$_3$SnC≡CPh	1·23	1·17	166
Et$_3$SnC≡CH	1·44	1·42	170
Et$_3$SnC≡CMe	1·37	1·22	170
Et$_3$SnC≡CEt	1·35	1·05	170
Et$_3$SnC≡CPh	1·38	1·48	170
Et$_3$SnC≡CCl	1·39	1·75	170
Et$_3$SnC≡CP(O)(OEt)$_2$	1·42	2·40	170
Bu$_3$SnC≡CH	1·40	1·42	170

TABLE XI—cont.

Compound	δ^a (mm sec^{-1})	ΔE_Q (mm sec^{-1})	Reference
$(C_6H_{11})_4Sn$	1·52	0	94
$(neo)_4Sn^b$	1·34	0	94
HC—CSnPr$_3$ with B$_{10}$H$_{10}$	1·45	1·65	3
Pr$_3$SnC—CSnPr$_3$ with B$_{10}$H$_{10}$	1·45	1·50	3
PhC—CSnPr$_3$ with B$_{10}$H$_{10}$	1·35	1·50	3
(PhC—C—B$_{10}$H$_{10}$)$_2$—SnBu$_2$	1·20	1·70	3
Bu$_2$Sn—SnBu$_2$ bridged by two B$_{10}$H$_{10}$ carborane cages	1·20	1·60	3
Ph$_4$Sn	1·21	0	94
Ph$_4$Sn	1·15	0	73
Ph$_4$Sn	1·22	0	219
Ph$_4$Sn	1·27	0	166
Ph$_3$SnC$_6$F$_5$	1·25	0·98	219
Ph$_3$SnC$_6$F$_5$	1·30	0·90	166
Ph$_3$SnCH=CH$_2$	1·28	0	166
Ph$_3$SnC$_6$Cl$_5$	1·27	0·84	166
Ph$_2$Sn(C$_6$F$_5$)$_2$	1·22	1·11	219
PhSn(C$_6$F$_5$)$_3$	1·16	0·92	219
PhSn(CH=CH$_2$)$_3$	1·25	0	166
$(C_6F_5)_4Sn$	1·04	0	219
$(C_6F_5)_3SnC_6H_4Me$-p	1·18	1·02	219
$(C_6F_5)_2Sn(C_6H_4Me$-$p)_2$	1·22	1·18	219
$(C_6F_5)_3SnMe$	1·19	1·14	219
$(C_6F_5)_2SnMe_2$	1·25	1·48	219
$(p$-$FC_6H_4)_4Sn$	1·33	0	94
$(p$-$CF_3C_6H_4)_4Sn$	1·29	0	94
$(m$-$CF_3C_6H_4)_4Sn$	1·28	0	94

TABLE XI—*cont.*

Compound	δ^a (mm sec^{-1})	ΔE_Q (mm sec^{-1})	Reference
HC—CSnPh$_3$ \\ B$_{10}$H$_{10}$	1·05	0·95	3
Ph$_3$SnC—CSnPh$_3$ \\ B$_{10}$H$_{10}$	0·95	0·70	3
PhC—CSnPh$_3$ \\ B$_{10}$H$_{10}$	1·30	1·20	3
(b) Compounds with 3 Sn—C bonds			
Me$_3$SnH	1·24	0	90
Me$_3$SnH	1·15	0	37
Me$_3$SnF	1·18	3·47	93
Me$_3$SnF	1·28	3·86	37
Me$_3$SnCl	1·41	3·41	219
Me$_3$SnCl	1·43	3·41	37
Me$_3$SnCl	1·44	3·01	50
Me$_3$SnBr	1·38	3·28	219
Me$_3$SnBr	1·30	2·98	161
Me$_3$SnBr	1·49	3·25	37
Me$_3$SnBr	1·45	3·40	17a
Me$_3$SnI	1·48	3·05	37
Me$_3$SnNCS	1·40	3·77	70a
Me$_3$SnCN	1·39	3·12	70a
Me$_3$SnNa	1·38	0	37
Me$_3$SnSnMe$_3$	1·46	0	166
Me$_3$SnOH	1·07	2·71	93
Me$_3$SnOH	1·19	2·91	37
Me$_3$SnN$_3$	1·24	3·23	93
Me$_3$SnOCOH	1·35	3·52	93
Me$_3$SnOAc	1·34	3·50	93
Me$_3$SnNO$_3$	1·44	4·14	37
1-(trimethylstannyl)imidazole	1·16	2·76	93
1-(trimethylstannyl)benzimidazole	1·20	2·87	95
1-(trimethylstannyl)-1,2,3-benztriazole	1·31	3·18	95
1-(trimethylstannyl)-1,2,4-triazole	1·27	2·96	93
Me$_3$SnOCOC≡CCOOSnMe$_3$	1·40	3·80	93
Et$_3$SnBr	1·38	2·82	161
Et$_3$SnBr	1·62	3·45	17a
Et$_3$SnNCS	1·57	3·80	70a

TABLE XI—*cont.*

Compound	δ^a (mm sec^{-1})	ΔE_Q (mm sec^{-1})	Reference
Et$_3$SnCN	1·41	3·19	70a
Pr$_3$SnBr	1·46	2·92	17a
Bu$_3$SnH	1·41	0	90
Bu$_3$SnCl	1·58	3·40	90
Bu$_3$SnCl	1·36	2·78	161
Bu$_3$SnBr	1·60	3·30	17a
Bu$_3$SnNCS	1·60	3·69	70a
Bu$_3$SnCN	1·37	3·27	70a
Bu$_3$SnOAc	1·38	3·67	93
Bu$_3$SnOCOC≡CCOOSnBu$_3$	1·25	3·62	93
(neo)$_3$SnF^b	1·33	2·79	94
(neo)$_3$SnCl	1·41	2·63	94
(neo)$_3$SnBr	1·42	2·64	94
(neo)$_3$SnI	1·41	2·40	94
(neo)$_3$SnN$_3$	1·33	2·48	94
(neo)$_3$SnOH	1·13	1·08	94
Ph$_3$SnH	1·39	0	90
Ph$_3$SnH	1·45	0	4
Ph$_3$SnF	1·25	3·53	166
Ph$_3$SnCl	1·37	2·45	94
Ph$_3$SnCl	1·35	2·5	80
Ph$_3$SnCl	1·34	2·46	221
Ph$_3$SnCl	1·31	2·56	166
Ph$_3$SnBr	1·37	2·48	166
Ph$_3$SnBr	1·40	2·40	17a
Ph$_3$SnI	1·20	2·25	166
Ph$_3$SnI	1·31	2·05	17a
Ph$_3$SnNCS	1·35	3·50	70a
Ph$_3$SnOH	1·35	2·7	80
Ph$_3$SnOAc	1·20	3·17	93
(Ph$_3$Sn)$_2$O	1·08	2·15	95
(Ph$_3$Sn)$_2$S	1·22	1·17	95
Ph$_3$SnN$_3$	1·40	3·19	95
1-(triphenylstannyl)-1,2,3-benztriazole	1·31	2·98	95
1-(triphenylstannyl)benzimidazole	1·19	2·59	95
1-(triphenylstannyl)-1,2,4-triazole	1·29	2·76	95
Ph$_3$SnLi	1·40	0	4
Ph$_3$SnSnPh$_3$	1·41	0	94
Ph$_3$SnSnPh$_3$	1·38	0	221
Ph$_3$SnSnPh$_3$	1·30	0	73

TABLE XI—*cont.*

Compound	δ^a (mm sec^{-1})	ΔE_Q (mm sec^{-1})	Reference
$(Ph_3Sn)_4Sn$	1·33	0	73
$(Ph_3Sn)_4Sn$	1·56	0	221
$(Ph_3Sn)_4Ge$	1·13	0	73
$(Ph_3Sn)_4Pb$	1·39	0	73
$(Ph_3Sn)_2SnPh_2$	1·06	0	73
$Ph_3SnFe(CO)_2C_5H_5$	1·50	0	89
$(p\text{-}ClC_6H_4)_3SnCl$	1·37	2·49	94
$(p\text{-}FC_6H_4)_3SnH$	1·37	0	94
$[(m\text{-}CF_3C_6H_4)_3Sn]_2$	1·40	0	94
$[(p\text{-}ClC_6H_4)_3Sn]_2$	1·44	0	94
$[(p\text{-}FC_6H_4)_3Sn]_2$	1·33	0	95
$(C_6F_5)_3SnBr$	1·26	1·60	17a
$(p\text{-}ClC_6H_4)_3SnCl$	1·37	2·49	95
$(p\text{-}FC_6H_4)_3SnI$	1·23	1·92	95
$Me_3SnCl.pyridine$	1·33	3·44	221
$Me_3SnCl.pyridine$	1·45	3·44	37
$Me_3SnCl\cdot pyridine$	1·42	3·35	95
$Et_4N^{\oplus}[Me_3SnCl_2]^{\ominus}$	1·24	3·23	50
$[Me_3Sn(H_2O)_2]^{\oplus}BPh_4^{\ominus}$	1·42	4·10	50
$Me_3SnBr.pyridine$	1·30	3·18	161
$Me_3SnBr.quinoline$	1·34	3·20	161
$Me_3SnBr.isoquinoline$	1·34	3·14	161
$Me_3SnBr.acridine$	1·34	3·26	161
$Me_3SnBr.1\text{-azaph}^c$	1·32	3·38	161
$Me_3SnBr.4\text{-azaph}^d$	1·43	2·86	161
$Me_3SnBr.9\text{-azaph}^e$	1·30	3·06	161
$Et_3SnBr.pyridine$	1·43	3·06	161
$Et_3SnBr.quinoline$	1·11	2·60	161
$Et_3SnBr.isoquinoline$	1·33	2·66	161
$Et_3SnBr.acridine$	1·30	3·00	161
$Et_3SnBr.1\text{-azaph}$	1·39	2·92	161
$Et_3SnBr.4\text{-azaph}$	1·26	2·48	161
$Et_3SnBr.9\text{-azaph}$	1·29	2·52	161
$Bu_3SnCl.pyridine$	1·36	2·84	161
$Bu_3SnCl.quinoline$	1·25	2·76	161
$Bu_3SnCl.isoquinoline$	1·21	2·82	161
$Bu_3SnCl.acridine$	1·29	2·76	161
$Bu_3SnCl.1\text{-azaph}$	1·32	2·82	161
$Bu_3SnCl.4\text{-azaph}$	1·36	2·80	161
$Bu_3SnCl.9\text{-azaph}$	1·33	2·82	161
Ph_3SnOx^f	1·07	1·75	177

TABLE XI—*cont.*

Compound	δ^a (mm sec^{-1})	ΔE_Q (mm sec^{-1})	Reference
(c) Compounds with 2 Sn—C bonds			
Me$_2$SnH$_2$	1·23	0	90
Me$_2$SnCl$_2$	1·52	3·62	219
Me$_2$SnCl$_2$	1·54	3·33	50
Me$_2$SnCl$_2$	1·61	3·55	37
Me$_2$SnBr$_2$	1·59	3·41	219
Me$_2$Sn(NCS)$_2$	1·48	3·87	70a
Me$_2$Sn(OMe)$_2$	0·99	2·31	95
Me$_2$SnO	0·92	1·82	80
Me$_2$SnO	0·92	1·92	95
Me$_2$Sn(OCOH)$_2$	1·45	4·72	93
Me$_2$Sn(OCOH)$_2$	1·14	4·47	221
Me$_2$Sn(OCOPh)$_2$	1·40	3·96	93
Me$_2$(OAc)SnOSn(OAc)Me$_2$	1·38	3·57	93
Me$_2$Sn(OH)NO$_3$	1·28	3·52	37
Me$_2$Sn(NO$_3$)$_2$	1·62	4·12	185a
Me$_2$SnOx$_2$	0·88	1·98	177
Me$_2$SnOx$_2$	0·85	1·93	85
Me$_2$SnOx$_2$	0·77	1·98	93
Me$_2$SnOx$_2$	0·84	1·81	37
Me$_2$SnClOx	1·26	3·12	177
Me$_2$Sn(EDT)^g	1·35	2·33	59
Me$_2$Sn(TDT)^h	1·36	2·62	59
Me$_2$Sn(acac)$_2$^{j}	1·18	3·93	93
Et$_2$SnI$_2$	1·77	3·21	17a
Et$_2$Sn(NCS)$_2$	1·56	3·96	70a
Et$_2$SnOx	0·99	2·02	177
Et$_2$SnClOx	1·34	3·13	177
Et$_2$SnBrOx	1·39	3·08	177
Et$_2$SnIOx	1·43	2·85	177
Et$_2$Sn(NCS)Ox	1·31	3·07	177
Pr$_2$SnClOx	1·31	2·78	177
Pr$_2$SnO	1·10	2·10	80
Pr$_2$SnOx$_2$	0·98	2·08	177
Pr$_2$SnOx$_2$	1·02	2·20	64
Bu$_2$Sn(NCS)$_2$	1·56	3·96	70a
Bu$_2$Sn(NCS)$_2$	1·54	3·90	158
Bu$_2$SnH$_2$	1·42	0	90
Bu$_2$SnH$_2$	1·45	0	4
Bu$_2$SnHCl	1·56	3·34	90
Bu$_2$SnO	0·98	2·06	80
Bu$_2$SnO	1·15	2·08	73

TABLE XI—*cont.*

Compound	δ^a (mm sec^{-1})	ΔE_Q (mm sec^{-1})	Reference
Bu$_2$Sn(OAc)$_2$	1·34	3·50	80
Bu$_2$(OAc)SnOSn(OAc)Bu$_2$	1·30	3·24	80
Bu$_2$Sn(OCOC$_2$H$_5$)$_2$	1·49	3·70	125
Bu$_2$Sn(OCOC$_4$H$_9$)$_2$	1·47	3·23	125
Bu$_2$Sn(OCOC$_6$H$_{13}$)$_2$	1·46	3·64	125
Bu$_2$Sn(OCOC$_7$H$_{15}$)$_2$	1·35	3·45	80
Bu$_2$Sn(OCOC$_{11}$H$_{23}$)$_2$	1·34	3·35	80
Bu$_2$Sn(OCOC$_{11}$H$_{23}$)$_2$	1·48	3·41	125
Bu$_2$Sn(OCOC$_{17}$H$_{35}$)$_2$	1·36	3·56	80
Bu$_2$Sn(OCOPh)$_2$	1·62	3·44	93
Bu$_2$SnOx$_2$	1·02	2·04	177
Bu$_2$SnOx$_2$	0·93	2·05	158
Bu$_2$SnClOx	1·40	3·21	177
Bu$_2$Sn(NCS)Ox	1·33	3·25	158
(C$_5$H$_{11}$)$_2$SnCl$_2$	1·59	3·40	95
(C$_8$H$_{17}$)$_2$SnOx$_2$	1·13	1·86	177
(C$_8$H$_{17}$)$_2$SnClOx	1·56	3·36	177
Ph$_2$SnH$_2$	1·38	0	90
Ph$_2$SnCl$_2$	1·37	2·76	158
Ph$_2$SnCl$_2$	1·34	2·89	219
Ph$_2$SnCl$_2$	1·37	2·76	158
Ph$_2$SnCl$_2$	1·38	2·75	50
Ph$_2$SnCl$_2$	1·31	2·66	95
Ph$_2$SnO	0·88	1·73	80
Ph$_2$Sn(NCS)$_2$	1·45	3·96	158
Ph$_2$SnOx$_2$	0·78	1·64	177
Ph$_2$SnOx$_2$	0·83	1·78	64
Ph$_2$SnOx$_2$	0·72	1·63	158
Ph$_2$SnOx$_2$	0·64	1·53	93
Ph$_2$SnClOx	1·12	2·40	177
Ph$_2$SnClOx	1·08	2·39	158
Ph$_2$Sn(NCS)Ox	0·98	2·48	158
Ph$_2$Sn[O$_2$P(OEt)$_2$]$_2$	1·04	4·22	64
Ph$_2$Sn(EDT)	1·36	1·69	59
Ph$_2$Sn(TDT)	1·33	1·93	59
Ph$_2$Sn(acac)$_2$	0·74	2·14	64
(p-ClC$_6$H$_4$)$_2$SnO	0·88	1·73	80
(p-BrC$_6$H$_4$)$_2$SnO	0·92	1·83	80
(p-IC$_6$H$_4$)$_2$SnO	0·84	1·73	80
[Me$_2$SnCl$_2$]$_2$.terpy^k	1·38	3·31	50
[Me$_2$SnCl.terpy]$^{\oplus}$BPh$_4^{\ominus}$	1·46	3·58	50
[Me$_2$SnCl.terpy]$^{\oplus}$I$^{\ominus}$	1·38	3·56	50

TABLE XI—*cont.*

Compound	δ^a (mm sec^{-1})	ΔE_Q (mm sec^{-1})	Reference
Me$_2$SnCl.terpy]$^{\oplus}$ClO$_4^{\ominus}$	1·38	3·50	50
Et$_4$N$^{\oplus}$[Me$_2$SnCl$_3$]$^{\ominus}$	1·40	3·30	50
[PyH]$^{\oplus}$[Me$_2$SnCl$_4$]$^{2\ominus m}$	1·59	4·32	64
Me$_2$SnCl$_2$.2pyridine	1·37	3·83	37
Me$_2$SnCl$_2$.bipy^n	1·45	4·02	37
Me$_2$SnCl$_2$.bipy	1·55	4·09	93
Me$_2$SnCl$_2$.bipy	1·39	4·08	221
Me$_2$SnCl$_2$.phen^p	1·32	4·03	93
Me$_2$Sn(EDT).pyridine	1·25	2·25	59
Me$_2$Sn(TDT).pyridine	1·26	2·28	59
Me$_2$Sn(TDT).phen	1·00	2·05	59
Et$_2$SnCl$_2$.dipyam^q	1·68	3·78	143
Et$_2$SnBr$_2$.dipyam	1·72	3·64	143
Bu$_2$SnCl$_2$.phen	1·59	4·07	157
Bu$_2$SnBr$_2$.phen	1·63	3·94	157
Bu$_2$SnI$_2$.phen	1·69	3·75	157
Bu$_2$Sn(NCS)$_2$.phen	1·42	4·18	158
Bu$_2$SnCl$_2$.bipy	1·56	3·83	157
Bu$_2$SnBr$_2$.bipy	1·62	3·95	157
Bu$_2$SnI$_2$.bipy	1·70	3·82	157
Bu$_2$Sn(NCS)$_2$.bipy	1·43	4·04	158
Ph$_2$SnCl$_2$.2pyridine	1·32	3·39	143
Ph$_2$SnBr$_2$.2pyridine	1·34	3·49	143
Ph$_2$SnCl$_2$.bipy	1·26	3·51	143
Ph$_2$SnCl$_2$.bipy	1·22	3·39	158
Ph$_2$SnBr$_2$.bipy	1·33	3·52	143
Ph$_2$SnI$_2$.bipy	1·41	3·35	143
Ph$_2$Sn(NCS)$_2$.bipy	0·82	2·13	158
Ph$_2$SnCl$_2$.phen	1·21	3·37	158
Ph$_2$Sn(NCS)$_2$.phen	0·81	2·34	158
Ph$_2$SnCl$_2$.dipyam	1·23	3·58	143
Ph$_2$SnBr$_2$.dipyam	1·34	3·45	143
Ph$_2$SnCl$_2$.tripyam^r	1·29	3·59	143
Ph$_2$SnBr$_2$.tripyam	1·38	3·62	143
Ph$_2$Sn(EDT).pyridine	1·22	1·76	59
Ph$_2$Sn(EDT).phen	1·00	1·76	59
Ph$_2$Sn(TDT).pyridine	1·24	1·76	59
[PyH]$_2^{\oplus}$[Ph$_2$SnCl$_4$]$^{2\ominus}$	1·44	3·80	64
[Ph$_2$SnCl$_2$]$_2$.terpy	1·17	2·88	50
[Ph$_2$SnCl.terpy]$^{\oplus}$BPh$_4^{\ominus}$	1·20	3·24	50
[Ph$_2$SnCl.terpy]$^{\oplus}$ClO$_4^{\ominus}$	1·24	3·01	50
Et$_4$N$^{\oplus}$[Ph$_2$SnCl$_3$]$^{\ominus}$	1·25	2·62	50

TABLE XI—*cont.*

Compound	δ^a (mm sec^{-1})	ΔE_Q (mm sec^{-1})	Reference
(d) Compounds with 1 Sn—C bond			
MeSnH$_3$	1·24	0	90
MeSnBr$_3$	1·41	1·91	219
i-PrSnH$_3$	1·46	0	90
BuSnH$_3$	1·44	0	90
BuSnCl$_3$	1·31	1·83	50
BuSn(NCS)$_3$	1·43	1·46	70a
BuSnClOx$_2$	0·84	1·67	177
BuSnOx$_3$	0·69	1·82	177
PhSnH$_3$	1·40	0	90
PhSnCl$_3$	1·27	1·80	219
[BuSnCl$_3$]$_3$. 2terpy	1·07	1·94	50
[BuSnCl$_2$. terpy]$^\oplus$BPh$_4^\oplus$	1·09	1·76	50
[BuSnCl$_2$. terpy]$^\oplus$ClO$_4^\ominus$	0·92	1·74	50
[Et$_4$N]$_2^\oplus$[BuSnCl$_5$]$^{2\ominus}$	1·07	1·86	50

^{a} All chemical shifts are given relative to SnO$_2$. Where the literature values refer to other absorbers they have been corrected to the SnO$_2$ scale by the addition of the following factors; α (grey) tin $+ 2\cdot10^{88a}$, β (white) tin $+ 2\cdot70^{88a}$, Mg$_2$Sn $+ 1\cdot82^{88a}$, Pd/Sn $+ 1\cdot52^{172}$, for the BaSnO$_3$ standard no correction was applied^{189}.

^{b} neo = PhC(Me)$_2$CH$_2$—.

^{c} 1-azaph = 1-azaphenanthrene.

^{d} 4-azaph = 4-azaphenanthrene.

^{e} 9-azaph = 9-azaphenanthrene.

^{f} OxH = 8-hydroxyquinoline.

^{g} EDTH$_2$ = ethane-1,2-dithiol.

^{h} TDTH$_2$ = toluene-3,4-dithiol.

^{j} acacH = acetylacetone.

^{k} terpy = 2,2′,6′,2″-terpyridyl.

^{m} [PyH]$^\oplus$ = pyridinium.

^{n} bipy = 2,2′-bipyridyl.

^{p} phen = 1,10-phenanthroline.

^{q} dipyam = 2,2′-dipyridylamine.

^{r} tripyam = 2,2′,2″-tripyridylamine.

occur, though a satisfactory explanation of all the existing results has yet to be found. The extent of the splitting is strongly influenced by the tendency of compounds containing tin bonded to atoms carrying lone-pairs to associate, thus decreasing the symmetry at the tin atom.

Although both the anion and the cation of [Me$_2$SnCl . terpyridyl]$^\oplus$ [Me$_2$SnCl$_3$]$^\ominus$ would be expected to show a doublet only a two-line spectrum was obtained. This was shown to be caused by a fortuitous overlap of the two doublets[50]. Occasionally asymmetry is observed in a quadrupole doublet and, although this could be due to impurities in the sample[205] it could also be caused by a non-random alignment of crystals in the sample since anisotropy of the Mössbauer effect in single crystals has been demonstrated[78, 218]. The extent of the quadrupole splitting has been found useful in assigning configurations to hexacoordinate complexes of the type R_2SnX_4 (R is alky or aryl, X_4 is made up from Cl, O or N). In compounds with *cis* R groups $\Delta E_Q = \sim 2$ mm sec^{-1}, and for the corresponding *trans* compounds $\Delta E_Q = \sim 4$ mm sec^{-1} [64].

The Mössbauer spectra of tin compounds are usually measured with the samples at liquid nitrogen temperature (bubbling of this coolant can cause vibrations leading to errors[217]) though certain compounds show a measurable effect at room temperature[9, 59]. Herber and his coworkers[93, 221] have shown that polymeric tin compounds are most likely to show an effect at ambient temperatures and this has now become a useful method of detecting the presence of polymeric structures. Although originally based on empirical observations, some theoretical justification for correlating room-temperature Mössbauer effects with polymeric structures has recently been obtained[216a].

Despite some criticism[161] the magnitude of the ratio ($=\rho$) of the quadrupole splitting to the isomer shift has been found useful in indicating the coordination number at tin[95]. If ρ is > 2.1 then tin probably has a coordination number of 5 or 6, though it now appears unlikely that the converse is true.

Although Mössbauer measurements are confined to the solid state the study of frozen solutions is possible. This technique has been used to obtain a measure of the solvating power of various solvents, as indicated by the magnitude of the isomer shift, towards dibutyltin dichloride[81]. Information concerning the coordination number of tin can often be obtained by comparing the Mössbauer parameters shown by a compound in the solid state and in a (frozen) donor solvent such as pyridine[163].

Mössbauer spectroscopy has been used to study a number of other problems including the adsorption of tetramethyltin on alumina[106, 108, 109], the structure of di(9-phenanthryl)tin[8], and the nature of transition metal–tin bonds[107]. Compounds containing Sn—Fe bonds, such as $[C_5H_5Fe(CO)_2]_nSnPh_{4-n}$ ($n = 1, 2, 3, 4$), have been examined by both ^{57}Fe and ^{119}Sn γ-ray resonance spectroscopy[77, 88a, 89]. Reports have also appeared on the effect on the Mössbauer spectra of applying an external magnetic field to diphenyltin dichloride[21] and on the development of a relationship between the Mössbauer chemical shifts and the spin–spin coupling constants obtained from nmr spectroscopy[153]. A number of approaches to the theoretical calculation of Mössbauer isomer shifts have been examined[79, 84].

13.4. MISCELLANEOUS PHYSICAL MEASUREMENTS

Some ultraviolet absorption bands which have been recorded for organotin compounds are given in Table XII. It is now generally accepted that, in organodistannanes, there is intense absorption associated with the Sn—Sn bond which is not dependent upon the presence of aromatic groups joined to tin[38, 53, 87] although no observation of λ_{max} in hexabutyldistannane was possible[53]. Similar absorption bands are observed with compounds in which tin is joined to other Group IVb metals. As the length of the metal atom chain increases the absorption shifts to higher wavelengths until, with five or more metal atoms, an edge of the band reaches the visible region. This is the explanation of the yellow colour of compounds such as $(Et_2Sn)_6$ and $(Et_2Sn)_9$[38]. Assignment of absorption bands to metal–metal bonds has been aided by measuring the spectra at different temperatures[54]. The intense absorptions recorded for some simple butyltin compounds are remarkable (see Table XII) and possible origins of these bands have been discussed[43]. Detailed analyses have been made of the ultraviolet absorption spectra of the phenyltin chlorides[86, 151], of some vinyltin compounds[171], of compounds of the type $Me(SnMe_2)_nMe$[203] and also the corresponding ethyl compounds[53a]. Comparison of the spectra of benzene and anisole with those of phenyl- and p-methoxyphenyl–tin compounds has been used to obtain an estimate of the π-character in aryl–tin bonds[159].

Some dipole moments are given in Table XIII and this technique has also been used[99, 100] to obtain evidence for the presence of $d\pi$–$p\pi$ bonding in phenyltin compounds (see Chapter 1). The polarity of the Sn—Cl bond has been estimated from the dipole moments of a series of organotin chlorides[137]. Some group moments have been calculated, for example, Bu_3SnO— $1\cdot38D$[44], Bu_3SnS— $1\cdot91D$[44], Me_3Sn— $0\cdot69D$[100]. The dipole moments of the four compounds Ph_2SnCh_2 (ChH = 8-hydroxyquinoline, 1,3-diphenylpropane-1,3-dione, pentane-2,4-dione, 1-phenylbutane-1,3-dione) were all in the

TABLE XII

Ultraviolet absorption spectra of some organotin compounds

Compound	λ_{max} (nm)	log ϵ	Reference
Bu_4Sn	202	3·99	43
p-$Me_3SnC_6H_4Ph$	254·5	4·40	45
	202·5	4·70	
Ph_4Sn	~210	4·70	86
	~260	3·00	
$(Bu_3Sn)_2O$	206·5	4·19	43
$(Bu_3Sn)_2S$	208	4·32	43
	246·5	3·67	
$(PhCH_2)_3SnCl$	247·5	4·35	87
$Ph_3SnSnPh_3$	247·5	4·53	87
$(PhCH_2)_3SnSn(CH_2Ph)_3$	246·5	4·70	87
$(p$-$MeC_6H_4)_3SnSn(C_6H_4Me$-$p)_3$	247·5	4·63	87
$Et_3SnGePh_3$	242	4·30	38
$Ph_3SnGeBu_3$	<240	—	38
$Bu_3GeSnPh_2GeBu_3$	<240	—	38
$Ph_3GeSnEt_2GePh_3$	252	4·49	38
$Ph_3GeSnEt_2SnPh_3$	258	4·52	38
$Ph_3SnGePh_2SnPh_3$	254	4·50	38
$EtSn(GePh_3)_3$	262	5·09	38
	269	5·09	
$(Ph_3Sn)_4Sn$	277	4·90	53
$(Ph_3Sn)_4Ge$	276	4·86	53
$(Ph_3Sn)_4Pb$	298	4·77	53
$Ph_3Ge(SnEt_2)_3GePh_3$	293sh	4·36	38
$Ph_3GeSnEt_2GePh_2SnEt_2GePh_3$	298	4·47	38
$Et_{12}Sn_5$	310	4·33	38
$Ph_3GeSnEt_2SnPh_2SnEt_2GePh_3$	303	4·75	38
$Ph_3GeSnEt_2(SnPh_2)_2SnEt_2GePh_3$	302sh	4·41	38
$Et_{14}Sn_6$	325	4·44	38

range 4–5D[162]. These values indicate *trans* phenyl groups, an assignment which was supported by Mössbauer measurements[64].

The [35]Cl nuclear quadrupole resonance spectra of a number of organotin chlorides have been investigated[82, 197, 214, 222, 252]. A correlation has been established between [81]Br nqr frequencies and Mössbauer quadrupole splittings of some organotin bromides[17a]. The electron spin resonance spectrum has been reported of tetraethyltin and triethyltin iodide after irradiation with γ-rays[61].

TABLE XIII

Dipole moments of some organotin compounds

Compound	μ (Debye)	Reference
Me_3SnPh	0·51	100
Me_3SnCH_2Ph	0·91	99, 100
$Me_3SnC_6H_4F\text{-}p$	1·98	100
$Me_3SnC_6H_4Cl\text{-}p$	2·16	100
$Me_3SnC_6H_4Br\text{-}p$	2·15	100
$Me_3SnC_6H_4Me\text{-}p$	0·48	100
$Me_3SnC_6H_4OMe\text{-}p$	1·46	100
$Me_3SnC_6H_4NMe_2\text{-}p$	1·83	100
$Ph_3SnC_6H_4Cl\text{-}p$	1·89	100
$R_3SnC\equiv CH(R = Et,Pr,Bu)$	1·25–1·27	155
$Et_3SnC\equiv CMe$	0·86	155
$Et_3SnC\equiv CEt$	0·61	155
$Et_3SnC\equiv CCl$	1·84	155
$Et_3SnC\equiv CBr$	1·77	155
$Et_3SnC\equiv CP(O)(OEt)_2$	4·59	155
Me_3SnCl	3·50	100, 232
Me_3SnBr	3·45	232
Me_3SnI	3·37	232
$(Bu_3Sn)_2O$	1·60	44
$(Bu_3Sn)_2S$	2·45	44
Me_2SnCl_2	4·10	232
Me_2SnBr_2	3·86	232
Me_2SnI_2	3·76	232
$MeSnCl_3$	3·64	232
$MeSnBr_3$	3·24	232
$MeSnI_3$	3·24	232

A good deal of molar refractivity data for liquid organotin compounds have been assembled by Sayre[192] who has shown that the Eisenlohr-Denbigh system of bond refractivities[242] gives better correlations than other methods. Hydrides are not included in this compilation but values for the Sn—H bond refractivity have been reported elsewhere[28, 174]. Optical exaltation has been noted for a number of organotin compounds[132]. The parachors of some organotin compounds have been measured and compared with calculated values, best results were obtained when atomic- rather than bond-constants were used[182]. Surface tension values have been reported for a number of butyltin compounds in the temperature range 20–125°[213].

Interpretation of the mass spectra of organotin compounds is complicated by the ten isotopes of tin but the use of this technique is steadily increasing[15, 24, 25, 26, 27, 51, 74, 75, 88, 121, 126, 181].

References

1. Abel, E. W., Armitage, D. A. and Tyfield, S. P., *J. Chem. Soc.* (*A*), 1967, 554.
2. Abel, E. W. and Brady, D. B., *J. Organometallic Chem.*, 1968, **11**, 145.
3. Aleksandrov, A. Yu., Bregadze, V. I., Gol'danskii, V. I., Sakharkin, L. I., Okhlobystin, O. Yu. and Khrapov, V. V., *Dokl. Akad. Nauk S.S.S.R.*, 1965, **165**, 593.
4. Aleksandrov, A. Yu, Okhlobystin, O. Yu., Polak, L. S. and Shpinel, V. S., *Dokl. Akad. Nauk S.S.S.R.*, 1964, **157**, 934.
5. Alleston, D. L., Davies, A. G., Hancock, M. and White, R. F. M., *J. Chem. Soc.*, 1965, 5469.
6. Amberger, E., Fritz, H. P., Kreiter, C. G. and Kula, M. R., *Chem. Ber.*, 1963, **96**, 3270.
7. Atwell, W. H. and Weyenberg, D. R., *J. Org. Chem.*, 1967, **32**, 885.
8. Baranovskii, V. I., Dzevitskii, B. E., Krizhanskii, L. M. and Rogozev, B. I., *Zh. Strukt. Khim.*, 1966, **7**, 808 (*Chem. Abs.*, 1967, **66**, 37244w).
9. Bearden, A. J., Marsh, H. S. and Zuckerman, J. J., *Inorg. Chem.*, 1966, **5**, 1260.
10. Beattie, I. R. and McQuillan, G. P., *J. Chem. Soc.*, 1963, 1519.
11. Bocquet, J. P., Chu, Y. Y., Kistner, O. C., Perlman, M. L. and Emery, G. T., *Phys. Rev. Let.*, 1966, **17**, 809.
12. Bolles, T. F. and Drago, R. S., *J. Amer. Chem. Soc.*, 1966, **88**, 5730.
13. Bonati, F., *Organometallic Chem. Rev.*, 1966, **1**, 379.
14. Boué, S., Gielen, M. and Nasielski, J., *Bull. Soc. Chim. Belges*, 1967, **76**, 559.
15. Boué, S., Gielen, M. and Nasielski, J., *Bull. Soc. Chim. Belges*, 1968, **77**, 43.
16. Brown, D. H., Mohammed, A. and Sharp, D. W. A., *Spectrochim. Acta*, 1965, **21**, 1013.
17. Brown, T. L. and Stark, K., *J. Phys. Chem.*, 1965, **69**, 2679.
17a. Bryuchova, E. V., Semin, G. K., Gol'danskii, V. I. and Khrapov, V. V., *Chem. Comm.*, 1968, 491.
18. Burdon, J., *Tetrahedron*, 1965, **21**, 1101.
19. Burke, J. J. and Lauterbur, P. C., *J. Amer. Chem. Soc.*, 1961, **83**, 326.
20. Butcher, F. K., Gerrard, W., Mooney, E. F., Rees, R. G., Willis, H. A., Anderson, A. and Gebbie, H. A., *J. Organometallic Chem.*, 1964, **1**, 431.
21. Bykov, G. A., Ryasnyl, G. K. and Shpinel, V. S., *Fiz. Tverd. Tela*, 1965, **7**, 1657 (*Chem. Abs.*, 1965, **63**, 7794e).
22. Carey, N. A. D. and Clark, H. C., *Chem. Comm.*, 1967, 292.
23. Cattanach, C. J. and Mooney, E. F., *Spectrochim. Acta, Pt. A.*, 1968, **24**, 407.
24. Chambers, D. B. and Glockling, F., *J. Chem. Soc.* (*A*), 1968, 725.

25. Chambers, D. B., Glockling, F. and Light, J. R. C., *Quart. Rev.*, 1968, **22**, 317.

26. Chambers, D. B., Glockling, F., Light, J. R. C. and Weston, M., *Chem. Comm.*, 1966, 281.

27. Chambers, D. B., Glockling, F. and Weston, M., *J. Chem. Soc. (A)*, 1967, 1759.

27a. Chivers, T. and Sams, J. R., *Chem. Comm.*, 1969, 249.

28. Christopher, P. M. and Fitzgerald, J. M., *Austral. J. Chem.*, 1965, **18**, 1705.

29. Clark, H. C., Cyr, N. and Tsai, J. H., *Canad. J. Chem.*, 1967, **45**, 1073.

30. Clark, H. C. and Goel, R. G., *J. Organometallic Chem.*, 1967, **7**, 263.

31. Clark, H. C. and Tsai, J. H., *Inorg. Chem.*, 1966, **5**, 1407.

32. Clark, J. P., Langford, V. M. and Wilkins, C. J., *J. Chem. Soc. (A)*, 1967, 792.

33. Clark, J. P. and Wilkins, C. J., *J. Chem. Soc. (A)*, 1966, 871.

34. Clark, R. J. H., Davies, A. G. and Puddephatt, R. J., *J. Chem. Soc. (C)*, 1968, 1828.

35. Clark, R. J. H. and Williams, C. S., *Spectrochim. Acta*, 1965, **21**, 1861.

36. Considine, W. J., Baum, G. A. and Jones, R. C., *J. Organometallic Chem.*, 1965, **3**, 308.

37. Cordey-Hayes, M., Peacock, R. D. and Vucelic, M., *J. Inorg. Nucl. Chem.*, 1967, **29**, 1177.

38. Creemers, H. M. J. C. and Noltes, J. G., *J. Organometallic Chem.*, 1967, **7**, 237.

39. Cummins, R. A., *Austral. J. Chem.*, 1963, **16**, 985.

40. Cummins, R. A., *Austral. J. Chem.*, 1965, **18**, 98.

41. Cummins, R. A., *Austral. J. Chem.*, 1965, **18**, 985.

42. Cummins, R. A. and Evans, J. V., *Spectrochim. Acta*, 1965, **21**, 1016.

43. Cumper, C. W. N., Melnikoff, A. and Vogel, A. I., *J. Chem. Soc. (A)*, 1966, 242.

44. Cumper, C. W. N., Melnikoff, A. and Vogel, A. I., *J. Chem. Soc. (A)*, 1966, 246.

45. Curtis, M. D., Lee, R. K. and Allred, A. L., *J. Amer. Chem. Soc.*, 1967, **89**, 5150.

46. Davies, A. G., *Chemistry in Britain*, 1968, **4**, 403.

46a. Davies, A. G., Harrison, P. G., Kennedy, J. D., Mitchell, T. N., Puddephatt, R. J. and McFarlane, W., *J. Chem. Soc. (C)*, 1969, 1136.

47. Davies, A. G., Harrison, P. G. and Silk, T. A., *Chem. and Ind.*, 1968, 949.

48. Davies, A. G. and Mitchell, T. N., *J. Organometallic Chem.*, 1966, **6**, 568.

49. Davison, A. and Rakita, P. E., *J. Amer. Chem. Soc.*, 1968, **90**, 4479.

50. Debye, N. W. G., Rosenberg, E. and Zuckerman, J. J., *J. Amer. Chem. Soc.*, 1968, **90**, 3234.

50a. Delmas, M., Maire, J. C. and Santamaria, J., *J. Organometallic Chem.*, 1969, **16**, 405.

51. De Ridder, J. J. and Dijkstra, G., *Rec. Trav. Chim.*, 1967, **86**, 737.

52. Devoe, J. R. and Spijkerman, J. J., *Analytical Chem.*, 1966, **38**, 382R; 1968, **40**, 472R.

53. Drenth, W., Janssen, M. J., van der Kerk, G. J. M. and Vliegenthart, J. A., *J. Organometallic Chem.*, 1964, **2**, 265.

53a. Drenth, W., Noltes, J. G., Bulten, E. J. and Creemers, H. M. J. C., *J. Organometallic Chem.*, 1969, **17**, 173.

54. Drenth, W., Willemsens, L. C. and van der Kerk, G. J. M., *J. Organometallic Chem.*, 1964, **2**, 279.

55. Dufermont, J. and Maire, J. C., *J. Organometallic Chem.*, 1967, **7**, 415.

56. Duncan, J. F. and Golding, R. M., *Quart. Rev.*, 1965, **19**, 36.

57. Egorov, Yu. P., *Teor. i Eksperim. Khim. Akad. Nauk Ukr. S.S.R.*, 1965, **1**, 30 (*Chem. Abs.*, 1965, **63**, 7773).

58. Engelhardt, G., Reich, P. and Schumann, H., *Z. Naturforsch*, 1967, **22B**, 352.

59. Epstein, L. M. and Straub, D. K., *Inorg. Chem.*, 1965, **4**, 1551.

60. Evnin, A. B. and Seyferth, D., *J. Amer. Chem. Soc.*, 1967, **89**, 952.

61. Fenrick, H. W. and Willard, J. E., *J. Amer. Chem. Soc.*, 1966, **88**, 412.

62. Fenton, D. E., Massey, A. G., Jolley, K. W. and Sutcliff, L. H., *Chem. Comm.*, 1967, 1097.

63. Finch, A., Poller, R. C. and Steele, D., *Trans. Farad. Soc.*, 1965, **61**, 2628.

64. Fitzsimmons, B. W., Seeley, N. J. and Smith, A. W., *Chem. Comm.*, 1968, 390; *J. Chem. Soc. (A)*, 1969, 143.

65. Fluck, E., The Mössbauer Effect and Its Application in Chemistry, *Advances in Inorganic Chemistry and Radiochemistry*, Academic Press, New York and London, No. 6, 1964, 433.

66. Friebe, E. and Kelker, H., *Z. Anal. Chem.*, 1963, **192**, 267.

67. Fritz, H. P. and Kreiter, C. G., *J. Organometallic Chem.*, 1964, **1**, 323.

68. Fritz, H. P. and Kreiter, C. G., *J. Organometallic Chem.*, 1965, **4**, 313.

69. Gager, H. M., Lewis, J. and Ware, M. J., *Chem. Comm.*, 1966, 616.

70. Galasso, V., de Alti, G. and Bigotto, A., *Z. Phys. Chem.*, 1968, **57**, 132.

70a. Gassenheimer, B. and Herber, R. H., *Inorg. Chem.*, 1969, **8**, 1120.

71. Gastilovich, E. A., Shigorin, D. N. and Komarov, N. V., *Tr. Komis. po. Spektroskopii, Akad. Nauk S.S.S.R.*, 1964, **3**, 70 (*Chem. Abs.*, 1966, **65**, 593f).

72. Geissler, H. and Kriegsmann, H., *J. Organometallic Chem.*, 1968, **11**, 85.

73. Gibb, T. C. and Greenwood, N. N., *J. Chem. Soc. (A)*, 1966, 43.

74. Gielen, M. and Mayenc, G., *J. Organometallic Chem.*, 1968, **12**, 363.

75. Gielen, M. and Nasielski, J., *Bull. Soc. Chim. Belges*, 1968, **77**, 5.

76. Gol'danskii, V. I., *The Mössbauer Effect and its Application in Chemistry*, New York Consultants Bureau, 1964.

76a. Gol'danskii, V. I., Khrapov, V. V. and Stukan, R. A., *Organometallic Chem. Revs.*, A, 1969, **4**, 225.

77. Gol'danskii, V. I., Borshagovskii, B. V., Makarov, E. F., Stukan, R. A., Anisimov, K. N., Kolobova, N. E. and Skripkin, V. V., *Teor. Eksp. Khim.*, 1967, **3**, 478 (*Chem. Abs.*, 1968, **68**, 100418).

78. Gol'danskii, V. I., Makarov, E. F. and Khrapov, V. V., *Physics Letters*, 1963, **3**, 344.

79. Gol'danskii, V. I., Makarov, E. F. and Stukan, R. A., *J. Chem. Phys.*, 1967, **47**, 4048.

80. Gol'danskii, V. I., Makarov, E. F., Stukan, R. A., Trukhtanov, V. A. and Khrapov, V. V., *Dokl. Akad. Nauk S.S.S.R.*, 1963, **151**, 357.

81. Gol'danskii, V. I., Okhlobystin, O. Yu., Rochev, V. Ya., and Khrapov, V. V., *J. Organometallic Chem.*, 1965, **4**, 160.

81a. Goodman, B. A. and Greenwood, N. N., *Chem. Comm.*, 1969, 1105.

82. Green, P. J. and Graybeal, J. D., *J. Amer. Chem. Soc.*, 1967, **89**, 4305.
83. Greenwood, N. N., *Chemistry in Britain*, 1967, **3**, 56.
84. Greenwood, N. N., Perkins, P. G. and Wall, D. H., *Symp. Faraday. Soc.*, 1967, **1**, 51 (*Chem. Abs.*, 1968, **69**, 101526).
85. Greenwood, N. N. and Ruddick, J. N. R., *J. Chem. Soc.* (*A*), 1967, 1679.
86. Griffiths, V. S. and Derwish, G. A. W., *J. Mol. Spec.*, 1959, **3**, 165.
87. Hague, D. N. and Prince, R. H., *J. Chem. Soc.*, 1965, 4690.
88. Heldt, E., Hoeppner, K., and Krebs, K. H., *Z. Anorg. Allgem. Chem.*, 1966, **347**, 95.
88a. Herber, R. H., *Tech. Rept. Ser.*, Intern. At. Energy Agency, 1966, No. 50, 121.
89. Herber, R. H. and Goscinny, Y., *Inorg. Chem.*, 1968, **7**, 1293.
90. Herber, R. H. and Parisi, G. I., *Inorg. Chem.*, 1966, **5**, 769.
91. Herber, R. H. and Spijkerman, J. J., *J. Chem. Phys.*, 1965, **42**, 4312.
92. Herber, R. H. and Spijkerman, J. J., *J. Chem. Phys.*, 1965, **43**, 4057.
93. Herber, R. H. and Stoeckler, H. A., *Tech. Rept. Ser.*, Intern. At. Energy Agency, 1966, No. 50, 110.
94. Herber, R. H. and Stoeckler, H. A., *Trans. N.Y. Acad. Sci.*, 1964, **26**, 929.
95. Herber, R. H., Stoeckler, H. A. and Reichle, W. T., *J. Chem. Phys.*, 1965, **42**, 2447.
96. Hester, R. E. and Jones, K., *Chem. Comm.*, 1966, 317.
97. Hogben, M. G., Oliver, A. J. and Graham, W. A. G., *Chem. Comm.*, 1967, 1183.
97a. Holmes, J. R. and Kaesz, H. D., *J. Amer. Chem. Soc.*, 1961, **83**, 3903.
98. Honda, M., Kawasaki, Y. and Tanaka, T., *Tetrahedron Letters*, 1967, 3313.
99. Huang, H. H. and Hui, K. M., *J. Organometallic Chem.*, 1964, **2**, 288.
100. Huang, H. H. and Hui, K. M., *J. Organometallic Chem.*, 1966, **66**, 504.
101. Hunter, B. K. and Reeves, L. W., *Canad. J. Chem.*, 1968, **46**, 1399.
102. Jakobsen, R. J. and Bentley, F. F., *Applied Spectroscopy*, 1964, **18**, 88.
103. Jenks, G. J., Aust. Commonwealth Dep. Supply, Def. Stand Lab., Tech. Note 1966, No. 87 (*Chem. Abs.*, 1968, **69**, 6989).
104. Jouve, P., *Compt. Rend.*, 1966, **262B**, 815.
105. Kaesz, H. D., *J. Amer. Chem. Soc.*, 1961, **83**, 1514.
106. Karasev, A. N., Kolbanovskii, Yu. A., Polak, L. S. and Shlikhter, E. B., *Kinet. Katal.*, 1967, **8**, 232 (*Chem. Abs.*, 1967, **67**, 6014n).
107. Karasev, A. N., Kolobova, N. E., Polak, L. S., Shpinel, V. S. and Anisimov, K. N., *Teor. i Eksperim. Khim.*, Akad. Nauk Ukr. S.S.R., 1966, **2**, 126 (*Chem. Abs.*, 1966, **65**, 6334d).
108. Karasev, A. N., Polak, L. S., Shlikhter, E. B. and Shpinel, V. S., *Kinetika i Kataliz*, 1965, **6**, 710 (*Chem. Abs.*, 1965, **63**, 15603c).
109. Karasev, A. N., Polak, L. S., Shlikhter, E. B. and Shpinel, V. S., *Zh. Fiz. Khim.*, 1965, **39**, 3117 (*Chem. Abs.*, 1966, **64**, 8961d).
110. Kawakami, K. and Okawara, R., *J. Organometallic Chem.*, 1966, **6**, 249.
111. Kawakami, K., Saito, T. and Okawara, R., *J. Organometallic Chem.*, 1967, **8**, 377.
112. Kawasaki, Y., *J. Inorg. Nucl. Chem.*, 1967, **29**, 840.
113. Kawasaki, Y., Hori, M. and Uenaka, K., *Bull. Chem. Soc. Japan*, 1967, **40**, 2463.

114. Kawasaki, Y., Kawakami, K. and Tanaka, T., *Bull. Chem. Soc. Japan*, 1965, **38**, 1102.
115. Kawasaki, Y., Tanaka, T. and Okawara, R., *Bull. Chem. Soc. Japan*, 1967, **40**, 1562.
116. Kawasaki, Y., Tanaka, T. and Okawara, R., *J. Organometallic Chem.*, 1966, **6**, 95.
117. Kawasaki, Y., Tanaka, T. and Okawara, R., *Spectrochim. Acta*, 1966, **22**, 1571.
118. Kimmel, H. and Dillard, C. R., *Spectrochim. Acta*, Pt. A, 1968, **24**, 909.
119. Kitching, W., *J. Organometallic Chem.*, 1966, **6**, 586.
120. Kitching, W., *Tetrahedron Letters*, 1966, **31**, 3689.
121. Kostyanovskii, R. G., *Izv. Akad. Nauk S.S.S.R., Ser. Khim.*, 1967, 2784.
122. Kriegsmann, H. and Geissler, H., *Z. Anorg. Allgem. Chem.*, 1963, **323**, 170.
123. Kriegsmann, H., Hoffmann, H. and Pischtschan, S., *Z. Anorg. Allgem. Chem.*, 1962, **315**, 283.
124. Kriegsmann, H. and Pauly, S., *Z. Anorg. Allgem. Chem.*, 1964, **330**, 275.
125. Krizhanskii, L. M., Okhlobystin, O. Yu., Popov, A. V. and Rogozev, B. I., *Dokl. Akad. Nauk S.S.S.R.*, 1965, **160**, 1121.
126. Kuehlein, K. and Neumann, W. P., *J. Organometallic Chem.*, 1968, **14**, 317.
127. Kuivila, H. G., *Advances in Organometallic Chemistry*, Academic Press, New York and London, 1964, **1**, 47.
128. Kula, M. R., Amberger, E. and Mayer, K. K., *Chem. Ber.*, 1965, **98**, 634.
129. Kula, M. R., Amberger, E. and Rupprecht, H., *Chem. Ber.*, 1965, **98**, 629.
130. Kumar Das, V. G. and Kitching, O. W., *J. Organometallic Chem.*, 1967, **10**, 59.
131. Kushlefsky, B., Simmons, I. and Ross, A., *Inorg. Chem.*, 1963, **2**, 187.
132. Lapkin, I. I. and Dumler, V. A., *Uch. Zap., Permsk. Gos. Univ.*, 1964, **111**, 190 (*Chem. Abs.*, 1966, **64**, 11068g).
133. Lees, J. K. and Flinn, P. A., *J. Chem. Phys.*, 1968, **48**, 882.
134. Leusink, A. J., Budding, H. A. and Marsman, J. W., *J. Organometallic Chem.*, 1967, **9**, 285.
135. I-Fa Li, Ho Tsu K'o Hsueh, 1965, **4**, 3 (*Chem. Abs.*, 1966, **65**, 16276d).
136. Lohmann, D. H., *J. Organometallic Chem.*, 1965, **4**, 382.
137. Lorberth, J. and Noeth, H., *Chem. Ber.*, 1965, **98**, 969.
138. Lorberth, J. and Vakrenkamp, H., *J. Organometallic Chem.*, 1968, **11**, 111.
139. Luijten, J. G. A., Janssen, M. J. and van der Kerk, G. J. M., *Rec. Trav. Chim.*, 1962, **81**, 202.
140. Lunazzi, L. and Taddei, F., *Boll. Sci. Fac. Chim. Ind. Bologna*, 1965, **23**, 359 (*Chem. Abs.*, 1966, **64**, 18736c).
141. McFarlane, W. M., *J. Chem. Soc. (A)*, 1967, 528.
141a. McFarlane, W. M. and Wood, R. J., *Chem. Comm.*, 1969, 262.
142. McGrady, M. M. and Tobias, R. S., *J. Amer. Chem. Soc.*, 1965, **87**, 1909.
143. McWhinnie, W. R., Poller, R. C., Ruddick, J. N. R. and Thevarasa, M., *J. Chem. Soc. (A)*, 1969, 2327.
144. Maddox, M. L., Flitcroft, N. and Kaesz, H. D., *J. Organometallic Chem.*, 1965, **4**, 50.
145. Maddox, M. L., Stafford, S. L. and Kaesz, H. D., Applications of NMR to the Study of Organometallic Compounds, *Advances in Organometallic Chemistry*, Academic Press, London and New York, 1965, **3**, 1.

146. Maeda, Y. and Okawara, R., *J. Organometallic Chem.*, 1967, **10**, 247.
147. Maire, J. C., *J. Organometallic Chem.*, 1967, **9**, 271.
148. Maire, J. C., and Dufermont, J., *J. Organometallic Chem.*, 1967, **10**, 369.
149. Maire, J. C. and Ouaki, R., *Helv. Chim. Acta*, 1968, **51**, 1151.
150. Marchand, A., Mendelsohn, J. and Valade, J., *Compt. Rend.*, 1964, **259**, 1737.
151. Marrot, J., Maire, J. C. and Cassan, J., *Compt. Rend.*, 1965, **260**, 3931.
152. Matsubayashi, G., Kawasaki, Y., Tanaka, T. and Okawara, R., *Bull. Chem. Soc. Japan*, 1967, **40**, 1566.
153. May, L. and Spijkerman, J. J., *J. Chem. Phys.*, 1967, **46**, 3272.
154. Mendelsohn, J., Marchand, A. and Valade, J., *J. Organometallic Chem.*, 1966, **6**, 25.
155. Mingaleva, K. S., Ionin, B. I., Zavgorodnii, V. S., Sharanina, L. G. and Petrov, A. A., *Zh. Obshch. Khim.*, 1968, **38**, 606 (*Chem. Abs.*, 1968, **69**, 76480).
156. Mitrofanov, K. P., Plotnikov, M. V. and Shpinel, V. S., *Zh. Eksperim. i Theo. Fiz.*, 1965, **48**, 791 (*Chem. Abs.*, 1965, **63**, 2581c).
157. Mullins, M. A. and Curran, C., *Inorg. Chem.*, 1967, **6**, 2017.
158. Mullins, M. A. and Curran, C., *Inorg. Chem.*, 1968, **7**, 2584.
159. Musker, W. K. and Savitsky, G. B., *J. Phys. Chem.*, 1967, **71**, 431.
160. Narasimhan, P. T. and Rogers, M. T., *J. Chem. Phys.*, 1961, **34**, 1049.
161. Nasielski, J., Sprecher, N., Devooght, J. and Lejeuene, S., *J. Organometallic Chem.*, 1967, **8**, 97.
162. Nelson, W. H. and Martin, D. F., *J. Inorg. Nucl. Chem.*, 1965, **27**, 89.
163. Nesmeyanov, A. N., Gol'danskii, V. I., Khrapov, V. V., Rochev, V. Ya., Kravtsov, D. N. and Rokhlina, El. M., *Izv. Akad. Nauk S.S.S.R.*, *Ser. Khim.*, 1968, 793 (*Chem. Abs.*, 1968, **69**, 76380).
164. Okawara, R. and Ohara, M., *J. Organometallic Chem.*, 1964, **1**, 360.
165. Okawara, R. and Yasuda, K., *J. Organometallic Chem.*, 1964, **1**, 356.
166. Parish, R. V. and Platt, R. H., *Chem. Comm.*, 1968, 1118.
167. Pascal, P. *Nouveau Traité de Chimie Minerale*, Masson et Cie, Paris, 1963, **8**, 285.
168. Patil, H. R. H. and Graham, W. A. G., *Inorg. Chem.*, 1966, **5**, 1401.
169. Patmore, D. J. and Graham, W. A. G., *Inorg. Chem.*, 1967, **6**, 981.
170. Petrov, A. A., Rogozev, B. I., Krizhanskii, L. M., and Zavgorodnii, V. S., *Zh. Obshch. Khim.*, 1968, **38**, 1196 (*Chem. Abs.*, 1968, **69**, 56134).
171. Petukhov, V. A., Mironov, V. F. and Kravchenko, A. L., *Izv. Akad. Nauk S.S.S.R.*, *Ser. Khim.*, 1966, 156.
172. Philip, J., Mullins, M. A. and Curran, C., *Inorg. Chem.*, 1968, **7**, 1895.
173. Plotnikova, M. V., Mitrofanov, K. P. and Shpinel, V. S., *Zh. Eksperim. i Teor. Fiz.*, *Pis'ma v Ridaktsiyu*, 1966, **3**, 323 (*Chem. Abs.*, 1966, **65**, 3122h).
174. Pohl, J. J., *Allg. Prakt. Chem.*, 1968, **19**, 84.
175. Poller, R. C., *J. Inorg. Nucl. Chem.*, 1962, **24**, 593.
176. Poller, R. C., *Spectrochim Acta*, 1966, **22**, 935.
177. Poller, R. C. and Ruddick, J. N. R., *J. Chem. Soc. (A)* 1969, 2273.
178. Poller, R. C. and Spillman, J. A., *J. Chem. Soc. (A)*, 1966, 1024.
179. Poller, R. C. and Toley, D. L. B., *J. Chem. Soc. (A)*, 1967, 1578.
180. Poller, R. C. and Toley, D. L. B., *J. Organometallic Chem.*, 1968, **14**, 453.

268   THE CHEMISTRY OF ORGANOTIN COMPOUNDS

181. Preston, P. N., Rice, P. J. and Weir, N. A., *Int. J. Mass Spectrom. Ion Phys.*, 1968, **1**, 303 (*Chem. Abs.*, 1968, **69**, 91220).
182. Putnam, R. C., *Canad. J. Chem.*, 1966, **44**, 1343.
183. Randall, E. W., Ellmer, J. J. and Zuckerman, J. J., *Inorg. Nucl. Chem. Letters*, 1966, **1**, 109.
184. Randall, E. W., Yoder, C. H. and Zuckerman, J. J., *J. Amer. Chem. Soc.*, 1967, **89**, 3438.
185. Ruby, S. L., Kalvius, G. M., Beard, G. B. and Snyder, R. E., *Phys. Rev.*, 1967, **159**, 239.
185a. Ruddick, J. N. R., M.Sc Thesis, University of Newcastle-upon-Tyne, 1966.
186. Ruidisch, I. and Schmidt, M., *J. Organometallic Chem.*, 1963, **1**, 160.
187. Ryan, M. T. and Lehn, W. L., *J. Organometallic Chem.*, 1965, **4**, 455.
188. Sacher, R. E., Lemmon, D. H. and Miller, F. A., *Spectrochim Acta*, 1967, **23A**, 1169.
189. Sano, H. and Herber, R. H., *J. Inorg. Nucl. Chem.*, 1968, **30**, 409.
190. Sawyer, A. K. and Brown, J. E., *J. Organometallic Chem.*, 1966, **5**, 438.
191. Sawyer, A. K., Brown, J. E. and Hanson, E. L., *J. Organometallic Chem.*, 1965, **3**, 464.
192. Sayre, R., *J. Chem. and Eng. Data*, 1961, **6**, 560.
193. Scherer, O. J. and Hornig, P., *J. Organometallic Chem.*, 1967, **8**, 465.
194. Schmidbauer, H. and Hussek, H., *J. Organometallic Chem.*, 1964, **1**, 244.
195. Schumann, H., Jutzi, P., Roth, A., Schwabe, P. and Schauer, E., *J. Organometallic Chem.*, 1967, **10**, 71.
196. Schumann, H. and Schmidt, M., *J. Organometallic Chem.*, 1965, **3**, 485.
197. Semin, G. K., Babushkina, T. A., Prokof'ev, A. K. and Kostyanovskii, R. G., *Izv. Akad. Nauk S.S.S.R., Ser. Khim.*, 1968, 1401 (*Chem. Abs.*, 1968, **69**, 63419).
198. Seyferth, D. and Evnin, A. B., *J. Amer. Chem., Soc.*, 1967, **89**, 1468.
199. Seyferth, D. and Vaughan, L. G., *J. Organometallic Chem.*, 1963, **1**, 138.
200. Shchembelov, G. A. and Ustynyuk, Yu. A., *Dokl. Akad. Nauk S.S.S.R.*, 1967, **173**, 847.
201. Shergina, N. I., Golovanova, N. I., Komarov, N. V. and Misyunas, V. K., *Primen. Mol. Spektrosk. Khim., Sb. Dokl. Sib. Soveshch, 3rd Krasnoyarsk, U.S.S.R.*, 1964, 93 (*Chem. Abs.*, 1968, **69**, 14227).
202. Shergina, N. I., Golovanova, N. I., Mirskov, R. G. and Vlasov, V. M., *Izv. Akad. Nauk S.S.S.R., Ser. Khim.*, 1967, 1378 (*Chem. Abs.*, 1968, **68**, 38836).
203. Shorygin, P. P., Petukhov, V. A., Nefedov, O. M., Kolesnikov, S. P. and Shiryaev, V. I., *Teor. i Eksperim. Khim., Akad. Nauk Ukr. S.S.R.*, 1966, **2**, 190 (*Chem. Abs.*, 1966, **65**, 14660g).
204. Shostakovskii, M. F., Sherlina, N. I., Golovanova, N. I., Komarov, N. V., Brodskaya, E. I. and Misyunas, V. K., *Zh. Obshch. Khim.*, 1965, **35**, 1768.
205. Shpinel, V. S., Aleksandrov, A. Yu., Ryasnyi, G. K. and Okhlobystin, O. Yu., *Zh. Eksperim. i. Teor. Fiz.*, 1965, **48**, 69 (*Chem. Abs.*, 1965, **62**, 14066d).
206. Simonnin, M. P., *J. Organometallic Chem.*, 1966, **5**, 155.
207. Simons, P. B. and Graham, W. A. G., *J. Organometallic Chem.*, 1967, **8**, 479.
208. Simons, P. B. and Graham, W. A. G., *J. Organometallic Chem.*, 1967, **10**, 457.
209. Sisido, K. and Kozima, S., *J. Org. Chem.*, 1964, **29**, 907.

210. Sisido, K., Miyanisi, T., Nabika, K. and Kozima, S., *J. Organometallic Chem.*, 1968, **11**, 281.
211. Smith, A. L., *Spectrochim. Acta, Part A*, 1968, **24**, 695.
212. Smith, G. W., Liquids, Struct., Properties Solid Interactions, *Proc. Symp. Warren, Mich.*, 1963, 219 (Pub. 1965) (*Chem. Abs.*, 1966, **64**, 16864c).
213. Spichale, W., Kapitza, H. and Utschick, H., *Z. Chem.*, 1967, **7**, 442.
214. Srivastava, T. S., *J. Organometallic Chem.*, 1967, **10**, 373.
215. Srivastava, T. S., *J. Organometallic Chem.*, 1969, **16**, P53.
216. Steingross, W. and Zeil, W., *J. Organometallic Chem.*, 1966, **6**, 109; 464.
216a. Stoeckler, H. A. and Sano, H., *Chem. Comm.*, 1969, 954.
217. Stoeckler, H. A. and Sano, H., *Nucl. Instr. Methods*, 1966, **44**, 103 (*Chem. Abs.*, 1966, **65**, 16312f).
218. Stoeckler, H. A. and Sano, H., *Phys. Lett. A*, 1967, **25**, 550.
219. Stoeckler, H. A. and Sano, H., *Trans. Farad. Soc.*, 1968, **64**, 577.
220. Stoeckler, H. A., Sano, H. and Herber, R. H., *J. Chem. Phys.*, 1966, **45**, 1182.
221. Stoeckler, H. A., Sano, H. and Herber, R. H., *J. Chem. Phys.*, 1967, **47**, 1567.
222. Swiger, E. D. and Graybeal, J. D., *J. Amer. Chem., Soc.*, 1965, **87**, 1464.
223. Taimsalu, P. and Wood, J. L., *Spectrochim. Acta*, 1964, **20**, 1043.
224. Taimsalu, P. and Wood, J. L., *Trans. Farad. Soc.*, 1963, **59**, 1754.
225. Tanaka, T., *Inorganica Chim. Acta*, 1967, **1**, 217.
226. Tanaka, T., Komura, M., Kawasaki, Y. and Okawara, R., *J. Organometallic Chem.*, 1964, **1**, 484.
227. Tanaka, T., Matsumura, Y., Okawara, R. and Musya, Y., *Bull. Chem. Soc. Japan*, 1968, **41**, 1497.
228. Thayer, J. S. and Strommen, D. P., *J. Organometallic Chem.*, 1966, **5**, 383.
229. Tobias, R. S., *Organometallic Chem., Rev.*, 1966, **1**, 93.
230. Tobias, R. S. and Freidline, C. E., *Inorg. Chem.*, 1965, **4**, 215.
231. van den Berghe, E. V. and van der Kelen, G. P., *Bull. Soc. Chim. Belges*, 1965, **74**, 479.
232. van den Berghe, E. V. and van der Kelen, G. P., *J. Organometallic Chem.*, 1966, **6**, 515.
233. van den Berghe, E. V. and van der Kelen, G. P., *J. Organometallic Chem.*, 1968, **11**, 479.
234. van den Berghe, E. V., van de Vondel, D. F. and van der Kelen, G. P., *Inorganica Chim. Acta*, 1967, **1**, 97.
235. van der Kelen, G. P., *Nature*, 1962, **193**, 1069.
236. Vaughan, L. G. and Seyferth, D., *J. Organometallic Chem*, 1966, **5**, 295.
237. Verdonck, L. and van der Kelen, G. P., *Ber. Bunsenges Physik. Chem.* 1965, **69**, 478.
238. Verdonck, L. and van der Kelen, G. P., *Bull. Soc. Chim. Belges*, 1965, **74**, 361.
239. Verdonck, L. and van der Kelen, G. P., *J. Organometallic Chem.*, 1966, **5**, 532.
240. Verdonck, L. and van der Kelen, G. P., *J. Organometallic Chem.*, 1968, **11**, 491.

241. Verdonck, L., van der Kelen, G. P. and Eeckhaut, Z., *J. Organometallic Chem.*, 1968, **11**, 487.
242. Vickery, B. C. and Denbigh, K. G., *Trans. Farad. Soc.*, 1949, **45**, 61.
243. Vladimiroff, T. and Malinowski, E. R., *J. Chem. Phys.*, 1965, **42**, 440.
244. Vyshinskii, N. N., Kozlova, T. V. and Rudnevskii, N. K., *Tr. Komis. po Spektroskopii Akad. Nauk S.S.S.R.*, 1964, 451 (*Chem. Abs.*, 1965, **63**, 13027d).
245. Vyshinskii, N. N. and Rudnevskii, N. K., *Optics and Spectroscopy (USSR)*, 1961, **10**, 421.
246. Vyshinskii, N. N. and Rudnevskii, N. K., *Spectroskopiya, Metody i Primenenie, Akad. Nauk S.S.S.R. Sibirsk. Otd.*, 1964, 115 (*Chem. Abs.*, 1965, **62**, 3533e).
247. Wada, M. and Okawara, R., *J. Organometallic Chem.*, 1965, **4**, 487.
248. Weigert, F. J., Winokur, M. and Roberts, J. D., *J. Amer. Chem. Soc.*, 1968, **90**, 1566.
249. Wieber, M. and Schmidt, M., *J. Organometallic Chem.*, 1964, **2**, 129.
250. Yasuda, K., Matsumoto, H. and Okawara, R., *J. Organometallic Chem.*, 1966, **6**, 528.
251. Yasuda, K. and Okawara, R., *J. Organometallic Chem.*, 1965, **3**, 76.
252. Zeil, W. and Haas, B., *Z. Naturforsch.*, 1967, **22A**, 2011.
253. Zemlyanskii, N. N., Gol'dshtein, I. P., Gur'yanova, E. N., Panov, E. M., Slovokhotova, N. A. and Kocheshkov, K. A., *Dokl. Akad. Nauk S.S.S.R.*, 1964, **156**, 131.
254. Zuckerman, J. J., *J. Inorg. Nucl. Chem.*, 1967, **29**, 2191.

14

BIOLOGICAL EFFECTS
AND CONSEQUENT APPLICATIONS

Many organotin compounds are toxic to various organisms and they are used to a limited but significant extent as biocidal agents in agriculture and technology[7, 84].

The toxicology of tin compounds was reviewed in 1959[12] and 1964[7] and this topic is included in a more recent article[10] dealing with the toxicology of organometallic compounds. A striking feature of this interesting subject is that, unlike, for example, lead, mercury or arsenic, toxicity is only manifest in certain organotin compounds, inorganic tin being completely non-toxic[12, 102a, 107, 125].

Detailed figures for mammalian toxicity are given below but, when handling normal quantities of organotin compounds, the experimental chemist need only observe normal good laboratory practice such as washing hands after possible contact, use of fume cupboards and avoiding the inhalation of dusts or the vapours of more volatile compounds. Particular care should be taken with the vesicant lower alkyltin halides and industrial workers handling large amounts of materials may need special protection[106].

14.1. MAMMALIAN TOXICITY

The toxicity of organotin compounds varies considerably according to the number of organic groups attached to tin, the highest toxicity being shown by compounds having three tin–carbon bonds. Acute toxicity values for some of these compounds are given in Table I. Toxicity is also strongly dependent upon the nature of the organic groups attached to tin; triethyltin compounds are the most toxic and trioctyltin acetate is essentially non-toxic when given by mouth to rats[11]. There is wide variation in the LD_{50} figures reported for triphenyltin acetate though it is clear that this compound shows a toxicity which is intermediate in level and is species dependent. There have been extensive studies of the toxicity of this compound because of its widespread use as an agricultural fungicide (see below). In chronic toxicity studies the lowest dietary level of triphenyltin acetate causing growth retardation in the rat was 25 ppm compared with 5 ppm of triethyltin hydroxide[123].

By using triphenyltin acetate labelled with ^{113}Sn it was shown that about 80% of the compound administered to guinea-pigs was excreted within 10 days.[47] The ^{113}Sn isotope was detected in the brain of rats

TABLE I

Acute toxicities of some trialkyl(aryl)tin compounds

Compound	Species	Administration route[a]	LD$_{50}$ (mg/Kg body wt)	Reference
Me$_3$SnOAc	Rat	P.O.	9·1	11
Et$_3$SnOAc	Rat	P.O.	4·0	11
Pr$_3$SnOAc	Rat	P.O.	118·3	11
i-Pr$_3$SnOAc	Rat	P.O.	44·1	11
Bu$_3$SnOAc	Rat	P.O.	380·2	11
(Bu$_3$Sn)$_2$O	Rat	P.O.	~200	13
(Bu$_3$Sn)$_2$O	Mouse	P.O.	122	7
Bu$_3$SnOH	Rat	P.O.	38	70
(C$_6$H$_{13}$)$_3$SnOAc	Rat	P.O.	1000	11
(C$_8$H$_{17}$)$_3$SnOAc	Rat	P.O.	>1000	11
Ph$_3$SnOAc	Rat	P.O.	125–150	7
Ph$_3$SnOAc	Rat	P.O.	429–491	115
Ph$_3$SnOAc	Rat	I.P.	8·5–11·9	115
Ph$_3$SnOAc	Guinea-pig	P.O.	23·5–41·2	115
Ph$_3$SnOAc	Guinea-pig	I.P.	3·7	115
Ph$_3$SnOAc	Mouse	P.O.	200–300	121
Ph$_3$SnOAc	Mouse	P.O.	81·3	115
Ph$_3$SnOAc	Mouse	P.O.	9·1	11
Ph$_3$SnOAc	Mouse	I.P.	15	121
Ph$_3$SnOAc	Mouse	I.P.	7·9	115
(Ph$_3$Sn)$_2$S	Mouse	P.O.	⩾680	121
(Ph$_3$Sn)$_2$S	Mouse	P.O.	⩾1470	121

[a] P.O. = oral; I.P. = interperitoneal.

and guinea-pigs after oral or interperitoneal administration of Ph$_3$^{113}SnOAc and it disappeared slowly with a half-life of several days[47]. Cattle fed with green fodder containing residues of triphenyltin acetate rapidly excrete most of the compound, small amounts which are absorbed are broken down and the products excreted within a few weeks[48].

Typical signs of acute toxicity were evident 4 hours after oral administration of tributyltin -chloride, -acetate, -benzoate or -oleate at 500 mg/Kg to mice though, in the case of tributyltin laurate at this dose level, the signs were delayed for a further 4 hours[92]. In poisoning by trialkyltin compounds inhibition of oxidative phosphorylation occurs; the deactivation of adenosinetriphosphatases *in vitro* has been demonstrated[63, 83] as has the slowing down of ^{32}P incorporation into the phospholipids of experimental animals. It was shown that $Et_3{}^{113}SnCl$ does not become uniformly distributed in the rat, guinea-pig and hamster, the highest concentrations being found in the blood of the rat and the liver of all three species[103]. Studies of the optic- and peripheral-nerve response to intoxication by triethyltin compounds in the rabbit have been reported[105], the cerebral oedema induced can be reduced by injection of glycerol[80] or dexamethasone[120]. The effect of triethyltin compounds upon plasma platelet aggregation[89] and also upon the oxidation of glucose and pyruvate in brain cortex slices[23] has been reported.

The use of trialkyltin compounds as rodent repellants has been proposed[146]. An unexpected application is the suggestion that small quantities of trialkyltin compounds should be added to animal feeds to improve food utilisation and weight gain[134].

Dialkyltin compounds may be somewhat less toxic though there are discrepancies in the published figures and acute oral LD_{50} values for dibutyltin compounds fed to rats vary from 100–1000 mg/Kg[7]. Dioctyltin compounds are considerably less toxic (LD_{50} values for compounds administered orally to rats are, dioctyltin dilaurate > 6000 mg/Kg[68], 'dioctyltin mercaptide' 750–1450 mg/Kg[52]) and this is particularly important since it is the dialkyltin compounds which are used to stabilise poly(vinyl chloride).

The potential risks in using PVC films for food packaging, where additives in the plastic may be partly extracted by the food, has been realised for some time[54]. Similar considerations apply to the use of PVC in medicine[35] and pharmacy[42, 60]. The use of limited quantities of certain dioctyltin stabilisers in PVC for food packaging does seem to be acceptable[29] and, in the United States, the Federal Food and Drug Administration now permits the use of a maximum of 0·55% dioctyltin *bis*(2-ethylhexyl maleate) for this purpose[2]. Analytical procedures for screening PVC for the presence of the more toxic dibutyltin stabilisers have been developed and a sensitive direct method for the general screening of plastics, based on tissue implants, has been described[87].

The metabolism of diethyltin compounds by the rat has been

studied and it was shown that dealkylation to monoethyltin compounds occurs[15]. The mode of toxic action of dibutyltin diacetate in rats and mice was thought to involve combination with dithiol groupings in enzymes necessary for the production of mitochondrial adenosinetriphosphate[19].

There is little information about the toxicity of monoalkyltin compounds though this is probably low[10, 12]; it seems that ethyltin trichloride is not metabolised by the rat[15].

Only isolated accounts[74] have been reported of toxicity studies on tetraalkyl(aryl)tin compounds since the subject was reviewed by Barnes and Stoner[10, 12]. It appears that tetraalkyltin compounds are dealkylated *in vivo* so that, although toxic symptoms may be delayed, the final effects are the same as those produced by trialkyltin compounds.

14.2. FUNGICIDAL ACTIVITY

Fungicidal activity is strikingly dependent upon the extent of alkylation of the tin atom, being at a maximum in compounds with 3 Sn—C bonds. This is shown by the following figures which give the minimum concentrations, in parts per million, causing complete inhibition of the growth of *Penicillium italicum*: Et_3SnCl, 2; Et_4Sn, > 1000; Et_2SnCl_2, 100; $EtSnCl_3$, > 1000[122]. (However patents exist covering the use of butyltin-triacetate[154] and -triformate[138] and some dibutylacetoxytin mercaptides, $Bu_2Sn(OAc)SR$[190], as fungicides.) Activity against fungi is also influenced by the nature of the organic groups as illustrated by the following minimum concentrations, in ppm, causing complete inhibition of the growth of *Penicillium funiculosum*: Pr_3SnCl, 1; Bu_3SnCl, 4; Ph_3SnCl, 8; Et_3SnCl, 16; $(C_5H_{11})_3SnCl$, 250; $(C_8H_{17})_3SnCl$, > 125; Me_3SnCl, > 500[131]. Much experimental work has been carried out with the tributyl compounds, Bu_3SnX, and fungicidal activity shown to be largely independent of the group X[34]. There are some exceptions to this generalisation, however, and it does not hold for other R_3SnX compounds[24]. High antifungal activity has been claimed for tributyltin fluoroborate and some related fluoro-compounds[57, 155], triethyltin oxinate[28], hexalkyldistannanes[172] and for the addition product of tributylstannyl N-dodecyl-3-aminobutanoic acid with tolylene diisocyanate[165]. The fungicidal activity of a group of Ph_3SnX compounds were shown to be highest when X = NCO or NCS[113]; triphenylstannyl titanates and some related compounds are claimed to be efficient fungicides[191].

The influence of tributyltin naphthenate on phosphorus metabolism during the germination of spores of *Rhizopus nigricans* has been investigated[58].

14.2.1. *Fungicides in agriculture and horticulture*

Trialkyltin compounds have not found any appreciable application as agricultural fungicides, partly because of their high phytotoxicity[86] and to some extent their mammalian toxicity, but also because these compounds are often less effective in the field than in laboratory trials[7]. On the other hand the triphenyltin compounds, with their much reduced phytotoxicity, are now used to a considerable extent for the control of *Phytophthora infestans* on potatoes and *Cercospora beticola* on sugar beet. The compounds most widely used are triphenyltin acetate and triphenyltin hydroxide, the potency and range of effectiveness of these compounds is at least comparable with that of the copper-containing fungicides[30, 84, 94, 114]. It is claimed that organotin fungicides are degraded on weathering to give, eventually, inorganic tin so that, unlike the copper and mercury fungicides, they do not leave toxic residues in the soil[111]. The Federal Food and Drug Administration in the U.S.A. has recently agreed to establish a tolerance of 0·05 ppm for residues of triphenyltin hydroxide on or in potatoes[3].

There has been a considerable amount of work directed at obtaining organotin fungicides with decreased phytotoxicity but promising compounds are usually found to have other disadvantages. This is exemplified by triphenyltin -sulphide and -disulphide which, although much less phytotoxic than triphenyltin acetate, are only active against a narrow range of fungi[121]. Among compounds which have been examined recently, *N*-triphenylstannylsulphamates,[163] triphenylstannyl-phosphinates[76] and -benzoates[148] appear to have a favourable ratio of fungicidal to phytocidal activity; this is also the case for adducts of triphenyltin chloride with phosphine oxides[152], pyridine *N*-oxides[144] and sulphoxides[151].

Experimentation on *Phytophthora* and *Cercospora* control continues[78, 94] and includes the evaluation of triphenyltin compounds for the protection of stored sugar beets[129]. Other applications include the use of triphenyltin compounds against *Piricularia oryzae* causing rice blast disease[117] and against *Dothistroma* needle blight of pines[50]. The control of scab (*Fusicladium effusum*) and other fungus diseases of pecans[27] and also of fungi in peanuts[59] by triphenyltin hydroxide has been reported. Although Ph_3SnOH was effective in controlling fungus diseases of cucumbers it caused foliage damage[61]; similarly triphenyltin acetate was too phytotoxic for the treatment of banana leaf spot (*Mycosphaerella musicola*)[75]. The use of tributyltin compounds has been proposed for fungicidal seed dressings[187] and triphenyltin hydroxide for the post-harvest fungicidal treatment of gladioli corms[76]. Tributyltin trichloroacetate was one of the more promising of 55 compounds evaluated for the control of pea root rot[46] and N-tributylstannyl-p-toluenesulphonamide was beneficial in the treatment of winter wheat infected with *Tilletia tritici*[170]. Triphenyltin acetate offered no advantage over copper-based fungicides in the treatment of Arabica coffee for leaf rust (*Hemileia vastatrix*)[51].

The toxicology of organotin fungicides was treated in a recent review[67].

14.2.2. *Other fungicidal applications*

Tributyltin compounds are almost ideal wood preservatives in that they are highly active against both fungi and insects which attack wood. Further, they do not increase inflammability or impart undesirable colours or odours and they are very resistant to leaching[40, 84, 100]. Compounds of the type Bu_3SnX are used for wood preservation, the effectiveness of these compounds being largely independent of the group X. The activity of $(Bu_3Sn)_2O$ is 30–40 times that of pentachlorophenol towards wood-destroying fungi[100]. Tributyltin compounds at concentrations of 100–300 g/ton, can be used to inhibit fungus growth in wood pulp[93]. Recent patents cover the use of trialkyltin -sulphamates[150, 157], -borates[147] and adducts of tertiary amines with trialkyltin compounds[175] against wood-destroying fungi. [As noted above the same compounds are active against insects which attack wood. Thus, wood is protected against the common furniture beetle (*Anobium punctatum*) when impregnated with $(Bu_3Sn)_2O$ at 1·39–2·93 Kg/m^3 and against the powder post beetle (*Lyctus brunneus*) by surface treatment with the same compound at the rate of 0·75 g/m^2 [8]]. New methods have been described for the comparative evaluation of wood preservatives[99] and for the estimation of tributyltin terephthalate in wood[56].

In temperate climates moulds do not usually appear on the hard gloss surfaces left by oil-based paints but films of emulsion paints readily support the growth of moulds and fungi. For this reason tributyltin oxide is added to emulsion paints at levels of 0·05–0·1%

for interior use to as much as 1% where the paint will be exposed to particularly warm and humid conditions[131]. An informative discussion on the use of organotin additives in paints was published in 1966[131] and aspects of the subject have been covered in more recent papers[39, 90]. Coating compositions with fungicidal properties are prepared by introducing large quantities (20–40%) of tributyltin oxide into polyurethan and epoxide resins[166, 167].

Fabrics which have been impregnated with tributyltin oxide and then treated with a resin have improved properties including increased resistance to fungal attack[97].

14.3. BACTERIOSTATIC ACTIVITY

Trialkyl- and triaryl-tin compounds show pronounced bacteriostatic properties towards gram-positive bacteria. Concentrations of compounds, in ppm, causing complete inhibition of the growth of *Staphylococcus aureus* are: Ph_3SnCl, 1; $(C_5H_{11})_3SnCl$, 1; Bu_3SnCl, 2; Pr_3SnCl, 8; Et_3SnCl, 16; Me_3SnCl, > 500[131]. Trialkyl(aryl)tin compounds have only very low activity against gram-negative bacteria[45].

Applications in this area are few since the activity of the organotin compounds is mainly limited to bacteriostasis and the compounds involved have high toxicities. (There was a tragic experience in France following the marketing, in 1948, of a proprietary preparation, for the treatment of furuncles, which contained di- and tri-ethyltin iodide and which lead to the deaths of 100 people[12].)

Compounds which have been tested recently for antibacterial effects include trialkyltin chlorides[136], oxides[136], hydroxides[145], O-oxime derivatives[160, 161, 185], sulphamates[158], carboxylates[159] and adducts of trialkyl(aryl)tin chlorides with long chain amines[192] and with phosphine oxides[174]. Compounds of the type R_3Sn—HgR' have higher bacterial activity per unit weight of mercury than simple organomercury compounds[82, 156]. The antibacterial effects of mixtures of organotin compounds with phenols[128] and quaternary ammonium salts[71] have been reported.

Antibacterial activity is not confined to compounds with three Sn—C bonds and dialkyltin compounds inhibit the growth of several species of gram-positive and gram-negative bacteria. The mode of action of diethyltin dichloride on *Escherichia coli* has been studied and the effects on metabolism are similar to those of arsenite[62]. The antibacterial effects of tetraalkyltin compounds are less than those of trialkyltin derivatives[9]. An interesting recent development has been the demonstration that certain polymeric tin compounds are active against bacteria[69, 72, 104, 132, 139].

An important antimicrobial use of organotin compounds is for slime control in paper manufacture[64, 84, 110, 194]. Large quantities of water are used in the manufacture of paper and this water carries cellulosic fibres, phosphates, sizes, starches, etc. and also certain bacteria and fungi. The metabolism of these microorganisms produces slime which can cause production difficulties in paper mills.

Several other applications of organotin compounds, based upon their antimicrobial properties, have been suggested including the use of tributyltin carboxylates and sulphonates in concentrates for the washing of textiles[164] and tributyltin benzoate as a disinfectant for the treatment of fabrics[108], wood[108], skins[133] and furs[133]. The incorporation of suitable organotin compounds confers antimicrobial properties upon synthetic polymers such as PVC[173], polyacrylonitrile[188], nylon[162], polyurethan foams[142] and several others[140]. Trialkyltin compounds may be included as antibacterial agents in water-repellant formulations for the treatment of masonry[137].

14.4. INSECTICIDAL ACTIVITY

This topic was comprehensively reviewed in 1964 by Ascher and Nissim[7]. Tables II and III show that insecticidal activity is largely confined to compounds with three Sn—C bonds. Variations in the nature of the fourth group attached to tin do not produce striking effects except that the fluorides, R_3SnF, appear to be inactive, possibly because the polymeric nature of these compounds makes absorption difficult. Although hexabutyldistannane has a low toxicity to house flies the corresponding hexamethyl compound, when applied at 1 ppm to mosquito larvae, caused 100% mortality within 48 hours[193]. Essentially similar results to those given in Tables II and III were obtained when 24 organotin compounds were tested against the larvae of the mosquito *Culex pipiens pipiens*; the R_3SnX types were the most effective and the larvicidal effect was independent of the group X[43].

The mode of absorption by the insect depends upon the compound, thus triphenyltin compounds function as stomach poisons rather than contact insecticides[36] whereas tributyltin oxide acts as a fumigant[37]. For trialkyl(aryl)tin compounds there is a good correlation between

TABLE II

Insecticidal activity of organotin compounds. Twenty-four hour LD_{50} *values of organotin compounds applied topically to sensitive house flies* [14]

Compound	LD_{50} (Moles × 10^{-10}/fly)	Compound	LD_{50} (Moles × 10^{-10}/fly)
Me_4Sn	>560	$Bu_3SnSnBu_3$	>210
Bu_4Sn	290	Ph_3SnCl	3·9
Ph_4Sn	>290	$(CH_2=CH)_2BuSnCl$	12·0
Me_3SnBr	4·5	$(CH_2=CH)_2PhSnF$	>500
Et_3SnOH	11·0	Me_2SnCl_2	120
Et_3SnOAc	11·0	Bu_2SnCl_2	340
Et_3SnSPh	5·0	$Bu_2Sn[OCOCHEt(CH_2)_3Me]_2$	130
Bu_3SnCl	4·6	Ph_2SnCl_2	290
$Bu_3SnOSnBu_3$	11·0	$BuSn[OCOCHEt(CH_2)_3Me]_3$	>160
$Bu_3SnOCO(CH_2)_6CH_3$	5·6	$PhSnCl_3$	>390

TABLE III

Insecticidal activity of organotin compounds. Minimum concentrations, in g/m^2, *to cause* 100% *mortality* [70]

Compound	Housefly	Bed bug	Roach	Mosquito	Flea
Me_3SnOH	0·1	—	1·0	0·5	1·0
Me_3SnOAc	0·01	0·1	0·5	0·01	0·1
$Me_3SnOCOCH_2Cl$	0·1	0·5	>1·0	0·05	—
$Et_3SnOCOC(Me)=CH_2$	0·1	1·0	0·5	0·05	0·5
$Pr_3SnOCOC(Me)=CH_2$	2·0	1·5	1·0	0·7	—
Bu_3SnOAc	0·5	>2·0	2·0	—	—
$Bu_3SnOSnBu_3$	0·2	0·7	0·8	0·5	—
Ph_3SnOAc	0·5	0·5	0·5	—	—
$Me_2Sn[OCOC(Me)=CH_2]_2$	>1·5	—	—	—	—
Et_2SnO	>1·0	—	—	—	—
Bu_2SnO	>2·0	Inactive	Inactive	Inactive	Inactive

toxicity to house flies and ability to inhibit *in vitro* adenosinetriphosphatase activity[95].

In other studies trimethyltin acetate was found to have the highest insecticidal activity in a series of trialkyltin compounds[69] and Me_3SnCl was shown to be toxic to house flies which are resistant to

19

DDT and organophosphorus insecticides[38]. However, house flies do develop resistance to organotin insecticides; the gene which is responsible for this resistance (and which also intensifies resistance to other insecticides) has been located[55, 96].

Promising results were obtained with Me_3SnOAc, Me_3SnOH, $(Bu_3Sn)_2O$ and $Et_2(C_8H_{17})SnOAc$ against the lepidopterans *Heliothis zea and H. virescens*[44, 127] and with trialkyltin borates against the mosquito *Aedes aegypti*[153]. The insecticidal activity of a series of organotin phosphorus compounds increased with increasing solubility in organic solvents[76]. Although dibutyltin dilaurate had a reduced insecticidal effect compared with tributyltin chloride against the rice stem borer *Chilo suppressalis* its action was sustained and its phytotoxicity lower[73]. Phytotoxicity was a problem in attempts to use triphenyltin hydroxide for the control of the tobacco leaf beetle *Epitrix hirtipennis*[118] and infestation of lima beans by the seed corn maggot[77].

Some triphenyltin compounds of relatively low insecticidal activity may still be used to protect crops from insect attack in that they have the property of deterring insects from feeding; these compounds have been termed antifeedants[6]. The antifeedant activity of triphenyltin-acetate and -hydroxide toward larvae of the noctuid moth *Prodenia litura* was assessed by measuring the degree of protection afforded to sugar beet leaves and also by the degree of starvation exhibited by the larvae[6]. The same triphenyltin compounds have a moderate antifeedant effect on house fly larvae and the striped maize borer *Chilo agamemnon*[5]. Although triphenyltin acetate gave good protection to potato leaves from attack by larvae of the potato tuber moth *Gnorimoschema operculella*, attack on the tubers was undeterred[81]. During 24 hours, triphenyltin benzoate caused a 2% decrease in the weight of the larvae of the Colorado potato beetle *Leptinotarsa decemlineata* while the controls increased in weight by 69%[18]. Other triphenyltin compounds have similar antifeedant effects[16].

Another recent development in the use of organotin compounds for insect control has been the demonstration of chemosterilant activity. Several triphenyltin derivatives act as reproduction inhibitors to the house fly and the most active compounds, for example Ph_3SnOH, $Ph_3SnCH_2CH=CH_2$ and $(Ph_3Sn)_2S$, sterilise adult flies at well below lethal concentrations[65]. Triphenyltin-acetate and -chloride, however, were considered to be too toxic to house flies at the sterilant threshold[66]. The more active compounds such as triphenyltin hydroxide sterilise female house flies at lower concentrations than are required for males[65, 168]. Sterility induced in male house flies occurs by a slow poisoning of the sperm[4]. Triphenyltin compounds also suppress reproduction in the German cockroach *Blatella germanica* [65], the confused flour beetle *Tribolium confusum* [65], the Colorado potato beetle *Leptinotarsa decemlineata*[16, 17] and the beetle *Callosobruchus chinensis* [85].

Triaryltin compounds are effective in the control of certain arachnids particularly the citrus rust mite *Phyllocoptruta oleivora*[7, 98, 184] and also *Aculus pelekassi* on citrus fruits[98]. Trialkyltin compounds are active against sensitive and resistant strains of *Tetranychus althaeae* on beans and citrus fruits[149], *Metatetranychus ulmi* on cherry trees[149] and are effective toxicants for scorpions[181].

14.5. MISCELLANEOUS BIOLOGICAL APPLICATIONS

The use of compounds such as dibutyltin dilaurate and dibutyltin maleate for the eradication of tape worms in poultry is well established[7]. The anthelmintic activity and toxicity to chickens of a number of dialkyl(aryl)tin compounds have been examined recently. In general, these compounds are effective against cestode parasites, particularly species of *Raillietina* and *Choanotaenia* but are much less active against nematodes[41]. The effects of dibutyltin compounds alone, or combined with other compounds which are present in veterinary anthelmintic formulations such as phenothiazine, on chicken fertility and egg quality have been examined[32, 33, 91, 126]. These compounds may be administered as capsules or as emulsified formulations which can be added to drinking water[141]. Apart from poultry, organotin compounds have not found wide application as veterinary anthelmintics because their therapeutic index for animals is unsatisfactory although their use in mice[177] and cats[119] has been reported.

Protective coatings applied to surfaces which are in contact with sea water often contain organotin compounds which are active against marine borers and also reduce fouling by marine organisms. Comparative figures for the use of triphenyl- and tributyl-tin compounds have been published[130] and Bu_3SnX compounds are effective against teredine borers and *Martesia* but less so against the isopod *Limnoria*[49]. There is interest in antifouling compositions containing tributyltin compounds for corrosion-resistant steel hulls[169, 176, 180] and the leaching rate of $(Bu_3Sn)_2O$ from antifouling paints applied to bakelite surfaces has been studied[22]. Special analytical techniques which have been developed to study the leaching of organotin compounds from antifouling compositions include colorimetric procedures[21] and a sensitive biological method based on the ability of these compounds to inhibit the growth of the unicellular green alga *Chlamydomonas*[102]. It has been suggested that there may be a physiological hazard in the prolonged handling of antifouling paints containing simple organotin compounds and that this can be obviated by the use of non-toxic polymeric organotin derivatives[143]. However, patents covering antifouling compositions usually specify monomeric tributyltin compounds[135, 178, 179, 183]. Orthodox treatments to prevent fouling of ships' bottoms usually meet with limited success since, eventually, the toxic agent becomes inactive due to a coating of dead marine organisms. A much more effective system of antifouling treatment has recently been developed in which an envelope of air bubbles carrying traces of tributyltin oxide is continuously circulated to the underwater surfaces of the ship[7, 171].

Certain organotin compounds are effective in controlling the snails which transmit the parasite responsible for the disease bilharziasis. Concentrations of trialkyltin compounds of about 1 ppm are lethal to the snail *Australorbis glabratus*[20, 26, 31]; reports that triphenyltin compounds are much less effective[31, 101] do not appear to be confirmed[20, 109]. The mortality of the snails *Helisoma tenue* and *H. subscrenatum* was 100% in water containing 0·06 ppm of $Bu_3SnSC_6Cl_5$[186] but, in general, the high toxicity of organotin compounds makes it unlikely that they will find widespread use in natural waters containing fish[25, 26]. A number of triaryltin compounds have been shown to have high activity against the snail *Biomphalaria glabrata* with a low phytotoxicity to swamp rice[53]. Trimethyltin compounds have been used for plankton control in water reservoirs[116].

An unusual application for trivinyltin hydroxide is its use to generate continuously small quantities of ethylene which can be employed in the control of ripening of stored fruit and the sprouting of potatoes[182].

References

1. Aldridge, W. N., *Proc. Roy. Soc. Med.*, 1965, **58**, 599 (*Chem. Abs.*, 1965, **63**, 10566b).
2. Anon., U.S. Federal Register, Washington D.C., 1964, **29**, 18055 (Dec. 19, 1964) (*Chem. Abs.*, 1965, **62**, 7028a).

3. Anon., U.S. Federal Register, Washington D.C., 1968, **33**, 1 (April 20, 1968) (*Tin and Its Uses*, 1968, No. 78, 1).
4. Ascher, K. R. S., Meisner, J. and Nissim, S., *World Review of Pest Control*, 1968, **7**, 84.
5. Ascher, K. R. S., Moscowitz, J. and Nissim, S., *Tin and Its Uses*, 1967, **73**, 8 (*Chem. Abs.*, 1967, **67**, 63281n).
6. Ascher, K. R. S. and Nissim, S., *Internat. Pest Control*, 1965, **7**, 21.
7. Ascher, K. R. S. and Nissim, S., *World Review Pest Control*, 1964, **3**, 188.
8. Baker, J. M. and Taylor, J. M., *Ann. Appl. Biol.*, 1967, **60**, 181 (*Chem. Abs.*, 1968, **68**, 28714).
9. Balandin, A. A. and Gindin, L. G., *Biofizika*, 1965, **10**, 986 (*Chem. Abs.*, 1966, **64**, 7299e).
10. Barnes, J. M., *Organometallic Chem. Rev.*, 1968, **3**, 137.
11. Barnes, J. M. and Stoner, H. B., *Brit. Journal of Industrial Medicine*, 1958, **15**, 15.
12. Barnes, J. M. and Stoner, H. B., *Pharmacol. Rev.*, 1959, **11**, 211.
13. Bennett, R. F. and Zedler, R. J., *J. Oil Colour Chem. Assoc.*, 1966, **49**, 928.
14. Blum, M. S. and Pratt, J. J., *J. Econ. Entomol.*, 1960, **53**, 445.
15. Bridges, J. W., Davies, D. S. and Williams, R. T., *Biochem. J.*, 1966, **98**, 14P; 1967, **105**, 1261.
16. Byrdy, S., *Rocz. Nauk Roln.*, *Ser. A*, 1968, **93**, 789 (*Chem. Abs.*, 1968, **69**, 51184).
17. Byrdy, S., Ejmocki, Z. and Eckstein, Z., *Bull. Acad. Polon. Sci.*, *Ser. Sci. Chim.*, 1965, **13**, 683 (*Chem. Abs.*, 1966, **64**, 13324h).
18. Byrdy, S., Ejmocki, Z. and Eckstein, Z., *Meded. Rijksfac. Landbouwwetensch. Gent.*, 1966, **31**, 876 (*Chem. Abs.*, 1968, **69**, 66441).
19. Calley, D., Guess, W. L. and Autian, J., *J. Pharm. Sci.*, 1967, **56**, 1267.
20. Camey, T. and Paulini, E., *Rev. Brasil Malariol. Doencas Trop.*, 1964, **16**, 487 (*Chem. Abs.*, 1966, **64**, 20554b).
21. Chromy, L. and Uhacz, K., *J. Oil Colour Chem. Assoc.*, 1968, **51**, 494.
22. Cooksley, M. V. and Parnham, D. N., *Surface Coatings*, 1966, **2** (8), 280 (*Chem. Abs.*, 1967, **66**, 76969f).
23. Cremer, J. E., *Biochem. J.*, 1967, **104**, 212.
24. Czerwinska, E., Eckstein, Z., Ejmocki, Z. and Kowalik, R., *Bull. Acad. Pol. Sci.*, *Ser. Sci. Chim.*, 1967, **15**, 335 (*Chem. Abs.*, 1968, **68**, 11909).
25. Deschiens, R., Brottes, H. and Mvogo, L., *Bull. Soc. Pathol. Exot.*, 1966, **59**, 231 (*Chem. Abs.*, 1967, **66**, 94208m).
26. Deschiens, R., Floch, H. and Floch, T., *Bull. Soc. Pathol. Exot.*, 1964, **57**, 454 (*Chem. Abs.*, 1965, **62**, 16903f).
27. Diener, U. L. and Garrett, F. E., *Plant Dis. Rep.*, 1967, **51** (3), 185 (*Chem. Abs.*, 1967, **66**, 104276y).
28. Foldesi, I. and Straner, Gy., *Ann. Univ. Sci. Budapest, Rolando Eotvos Nominatae, Sect. Chem.*, 1965, **7**, 89 (*Chem. Abs.*, 1966, **65**, 5437g).
29. Franzen, V., *Kunstoffe*, 1965, **55**, 327 (*Chem. Abs.*, 1965, **63**, 11792e).
30. Fraselle, J., *Rev. Agr.* (*Brussels*), 1967, **20**, 1013 (*Chem. Abs.*, 1968, **68**, 28709).
31. Frick, L. P. and de Jimenez, W. Q., *Bull. World Health Organ.*, 1964, **31**, 429 (*Chem. Abs.*, 1965, **62**, 11053e).

32. Fry, J. L. and Wilson, H. R., *Poultry Sci.*, 1967, **46**, 319 (*Chem. Abs.*, 1967, **67**, 92763q).
33. Fry, J. L. and Wilson, H. R., *Sunshine State Agr. Res. Rept.*, 1965, **10**, 18 (*Chem. Abs.*, 1965, **62**, 16675a).
34. Fuse, G. and Nishimoto, K., *Mokuzai Kenkyu*, 1964, **32**, 15 (*Chem. Abs.*, 1966, **65**, 9646g).
35. Galley, D. J., Guess, W. L. and Autian, J., *J. Pharm. Sci.*, 1967, **56**, 240 (*Chem. Abs.*, 1967, **66**, 63872u).
36. Gardiner, B. G. and Poller, R. C., *Bull. Ent. Res.*, 1964, **55**, 17.
37. Gardiner, B. G. and Poller, R. C., unpublished observations.
38. Georghiou, G. P., Metcalf, R. L. and von Zboray, E. P., *Bull. World Health Organ.*, 1965, **33**, 479 (*Chem. Abs.*, 1966, **64**, 5690g).
39. Giesen, M., *Congr.*, 1966, **8**, 185 (*Chem. Abs.*, 1966, **65**, 17625h).
40. Goya, Y., *Zairyo*, 1965, **14**, 675 (*Chem. Abs.*, 1967, **67**, 72709u).
41. Graber, M. and Gras, G., *Rev. Elevage Med. Vet. Pays Trop.*, 1965, **18**, 405 (*Chem. Abs.*, 1966, **65**, 4498a).
42. Gras, G. and Castel, J., *Trav. Soc. Pharm. Montpellier*, 1966, **25**, 178 (*Chem. Abs.*, 1966, **65**, 3672h).
43. Gras, G. and Rioux, I. A., *Arch. Inst. Pasteur Tunis*, 1965, **42**, 9 (*Chem. Abs.*, 1966, **65**, 6223a).
44. Graves, J. B., Bradley, J. R. and Bagent, J. L., *J. Econ. Entomol.*, 1965, **58**, 583 (*Chem. Abs.*, 1965, **63**, 2339c).
45. Gruen, L., *Gesundheitsw. Disinfek.*, 1966, **58**, 81 (*Chem. Abs.*, 1967, **66**, 610n).
46. Haglund, W. A., *Plant Disease Rept.*, 1965, **49**, 793 (*Chem. Abs.*, 1965, **63**, 18959b).
47. Heath, D. F., *Radioisotop. Detection Pestic. Residues, Proc. Panel, Vienna*, 1965, 18 (*Chem. Abs.*, 1967, **67**, 1828w).
48. Herok, J. and Gotte, H., *International J. Appl. Rad. and Isotopes*, 1963, **14**, 461.
49. Hochman, H. and Roe, T., *U.S. Dept. Com., Office Tech. Serv.*, AD401,215, 1963 (*Chem. Abs.*, 1965, **62**, 5827d).
50. Hocking, D., *Ann. Appl. Biol.*, 1967, **59**, 363 (*Chem. Abs.*, 1967, **67**, 72722t).
51. Hocking, D. and Freeman, G. H., *Trop. Agr.* (*London*), 1968, **45**, 141 (*Chem. Abs.*, 1968, **69**, 26251).
52. Homrowski, S., *Rocz. Panstw. Zakl. Hig.*, 1968, **19**, 329 (*Chem. Abs.*, 1968, **69**, 85100).
53. Hopf, H. S., Duncan, J., Beesley, J. S. S., Webley, D. J. and Sturrock, R. F., *Bull. World Health Organ.*, 1967, **36**, 955.
54. Houska, M. and Majlath, L., *Ernaehrungsforschung*, 1966, **11**, 427 (*Chem. Abs.*, 1967, **66**, 74486x).
55. Hoyer, R. F. and Plapp, F. W., *J. Econ. Entomol.*, 1968, **61**, 1269 (*Chem. Abs.*, 1968, **69**, 95403).
56. Ishii, Y., Kawamura, H. and Yagi, S., *Bunseki Kagaku*, 1968, **17**, 3 (*Chem. Abs.*, 1968, **68**, 96897).
57. Iwamoto, H. and Kikuchi, M., *Hakko Kyokaishi*, 1964, **22**, 218 (*Chem. Abs.*, 1965, **63**, 18706g).

58. Iwamoto, H. and Kikuchi, M., *Kogyo Gijutsuin Hakko Kenkyusho Kenkyu Hokoku*, 1963, **24**, 1 (*Chem. Abs.*, 1965, **62**, 15363f).

59. Jackson, C. R., *Plant Disease Rept.*, 1965, **49**, 928 (*Chem. Abs.*, 1966, **64**, 4195d).

60. Jaminet, Fr., *Tribune CEDED-EAU* (*Centre Belge Etude Doc. Eaux*), 1965, **18**, 580 (*Chem. Abs.*, 1966, **64**, 17364a).

61. Jones, J. P. and Everett, P. H., *Plant Disease Rept.*, 1965, **49**, 29; 1966, **50**, 340.

62. Kahana, L. and Sijpesteijn, A. K., *Antonie van Leeuwenhoek, J. Microbiol. Serol.*, 1967, **33**, 427 (*Chem. Abs.*, 1967, **67**, 106167h).

63. Kameyama, T. and Sekine, T., *J. Biochem.* (*Tokyo*), 1965, **58**, 420 (*Chem. Abs.*, 1966, **64**, 2603g).

64. Kamitani, I. and Hasegawa, N., *Kami-pa Gikyoshi*, 1966, **20**, 55 (*Chem. Abs.*, 1966, **64**, 9941h).

65. Kenaga, E. E., *J. Econ. Entomol.*, 1965, **58**, 4 (*Chem. Abs.*, 1965, **62**, 9715d).

66. Kissam, J. B. and Hays, S. B., *J. Econ. Entomol.*, 1966, **59**, 748 (*Chem. Abs.*, 1966, **65**, 2939h).

67. Klimmer, O. R., *Pflanzenschutzberichte*, 1968, **37**, 57 (*Chem. Abs.*, 1968, **69**, 18227).

68. Klimmer, O. R. and Nebel, I. U., *Arzneimittel-Forsch.*, 1960, **10**, 44.

69. Kochkin, D. A. and Azerbaev, I. N., *Vestn. Akad. Nauk Kaz. S.S.R.*, 1966, **22**, 53 (*Chem. Abs.*, 1967, **67**, 64467y).

70. Kochkin, D. A., Vashkov, V. I. and Dremova, V. P., *J. Gen. Chem. U.S.S.R.*, 1964, **34**, 325.

71. Kochkin, D. A., Vashkov, V. I., Kiryutkin, G. V. and Savel'eva, A. R., *Zh. Obshch. Khim.*, 1964, **34**, 4027.

72. Kochkin, D. A., Zubov, P. I., Rzaev, Z. M., Azerbaev, I. N., Vashkov, V. I., Koveleva, N. I. and Shcheglova, G. V., *Vestn. Akad. Nauk Kaz. S.S.R.*, 1966, **22**, 45 (*Chem. Abs.*, 1967, **66**, 76330r).

73. Koike, H., *Botyu-Kagaku*, 1961, **26**, 51 (*Chem. Abs.*, 1965, **63**, 12255a).

74. Kolla, V. E. and Zalesov, V. S., *Uch. Zap. Permskii Gos. Univ.*, 1964, **111**, 196 (*Chem. Abs.*, 1966, **64**, 1247b).

75. Kranz, J., *Phytopathol. Z.*, 1965, **52**, 59; 335.

76. Kubo, H., *Agricultural and Biological Chemistry*, 1965, **29**, 43.

77. McEwen, F. L. and Davis, A. C., *J. Econ. Entomol.*, 1965, **58**, 369 (*Chem. Abs.*, 1965, **62**, 15367e).

78. McIntosh, A. H. and Eveling, D. W., *Ann. Appl. Biol.*, 1965, **55**, 397 (*Chem. Abs.*, 1965, **63**, 13963b).

79. Magie, R. O., *Proc. Florida State Hort. Soc.*, 1964, **77**, 549 (*Chem. Abs.*, 1965, **63**, 13965f).

80. Mandell, S., Taylor, J. M., Kotsilimbas, D. G. and Scheinberg, L. C., *J. Neurosurg.*, 1966, **24**, 984 (*Chem. Abs.*, 1966, **65**, 9573d).

81. Meisner, J. and Ascher, K. R. S., *Z. Pflanzenkrankh. Pflantzenschutz*, 1965, **72**, 458 (*Chem. Abs.*, 1965, **63**, 18964d).

82. Miller, V. L. and Jerstad, A. C., *J. Med. Chem.*, 1966, **9**, 208 (*Chem. Abs.*, 1966, **64**, 11588e).

83. Minakami, S., Saito, T., Kume, S. and Yoshikawa, H., *J. Biochem.* (*Tokyo*), 1965, **57**, 221 (*Chem. Abs.*, 1965, **63**, 3489g).

84. Mufti, A. S. and Poller, R. C., *Science and Industry* (*Karachi*), 1966, **4**, 157.
85. Nagasawa, S., Shinohara, H. and Shiba, M., *Bochu-Kagaku*, 1965, **30**, 91 (*Chem. Abs.*, 1965, **66**, 18960c).
86. Nakamura, K. and Fukunishi, S., *Takamine Kenkyusho Nempo*, 1961, **13**, 245 (*Chem. Abs.*, 1965, **63**, 4880g).
87. Nimni, M., J. *Pharm. Sci.*, 1964, **53**, 1262 (*Chem. Abs.*, 1965, **62**, 990d).
88. Nishimoto, K. and Fuse, G., *Wood Research*, 1965, **34**, 118.
89. O'Brien, J. R., J. *Chim. Pathol.*, 1964, **17**, 275; *Nature*, 1965, **207**, 306.
90. Pauli, O., *Deut. Farben-Z.*, 1967, **21** (2), 64 (*Chem. Abs.*, 1967, **66**, 96258b).
91. Peardon, D. L., Haberman, W. O., Marr, J. E., Garland, F. W. and Wilcke, H. L., *Poultry Sci.*, 1964, **44**, 413 (*Chem. Abs.*, 1965, **62**, 16674a).
92. Pelikan, Z. and Cerny, E., *Arch. Toxicol.*, 1968, **23**, 283 (*Chem. Abs.*, 1968, **69**, 104751).
93. Pelissier, J. P., *Papeterie*, 1964, **86**, 1537 (*Chem. Abs.*, 1967, **66**, 56754d).
94. Picco, D., *Notiz. Mal. Piante*, 1965, **72/73**, 3 (*Chem. Abs.*, 1966, **65**, 7929e).
95. Pieper, G. R. and Casida, J. E., J. *Econ. Entomol.*, 1965, **58**, 392 (*Chem. Abs.*, 1965, **63**, 2334e).
96. Plapp, F. W. and Hoyer, R. F., J. *Econ. Entomol.*, 1968, **61**, 1298 (*Chem. Abs.*, 1968, **69**, 95405).
97. Powlowska, Z., *Prace Central. Inst. Ochrony. Pracy*, 1966, **16**, 79 (*Chem. Abs.*, 1966, **65**, 4019b).
98. Reed, D. K., Crittenden, C. R. and Lyon, D. J., J. *Econ. Entomol.*, 1967, **60**, 668 (*Chem. Abs.*, 1967, **67**, 20977k).
99. Richardson, B. A., *Int. Pest. Control.*, 1968, **10**, 14 (*Chem. Abs.*, 1968, **68**, 94834).
100. Richardson, B. A., *Tin and Its Uses*, 1964, **64**, 5; *Wood*, 1964, 57.
101. Ritchie, L. S., Berrios-Duran, L. A., Frick, L. P. and Fox, I., *Bull. World Health Organ.*, 1964, **31**, 147 (*Chem. Abs.*, 1965, **62**, 11093d).
102. Rivett, P., J. *Appl. Chem.*, 1965, **15**, 469.
102a. Roe, F. J. C., Boyland, E. and Millican, K., *Food Cosmet. Toxicol.*, 1965, **3**, 277 (*Chem. Abs.*, 1966, **64**, 16518c).
103. Rose, M. S. and Aldridge, W. N., *Biochem. J.*, 1968, **106**, 821 (*Chem. Abs.*, 1968, **68**, 85779).
104. Rzaev, Z. M., Kochkin, D. A. and Zubov, P. I., *Dokl. Akad. Nauk S.S.S.R.*, 1967, **172**, 364.
105. Scheinberg, L. C., Taylor, J. M., Herzog, I. and Mandell, S., J. *Neuropath. Exper. Neurol.*, 1966, **25**, 202 (*Chem. Abs.*, 1967, **66**, 9594n).
106. Schmutzler, G., *Plaste Kaut.*, 1966, **13**, 462 (*Chem. Abs.*, 1967, **67**, 46950n).
107. Schroeder, H. A. and Balassa, J. L., J. *Nutr.*, 1967, **92**, 245 (*Chem. Abs.*, 1967, **67**, 42153r).
108. Schulz-Utermoehl, H. and Weissenstein, H., *Gesundheitsw. Desinfekt.*, 1965, **57**, 41 (*Chem. Abs.*, 1965, **63**, 6793d).
109. Seiffer, E. A. and Schoof, H. F., *Public Health Rept.*, 1967, **82**, 833 (*Chem. Abs.*, 1968, **68**, 2250).
110. Shimono, K., Katsuyama, A. and Katsuji, M., *Shizuoka'ken Seishi Kogyo Shikensho Shiken Hokoku*, 1967, **19**, 1 (*Chem. Abs.*, 1968, **69**, 53016).
111. Sijpesteijn, A. K., *Meded. Landb. Hoogesch. Gent.*, 1959, **24**, 850.

112. Sone, N., J. Biochem. (Tokyo), 1964, 56, 151.
113. Srivastava, T. N. and Tandon, S. K., Naturwissenschaften, 1968, 55, 391.
114. Stallknecht, G. F. and Calpouzos, L., Phytopathology, 1968, 58, 788 (Chem. Abs., 1968, 69, 34910).
115. Stoner, H. B., Brit. J. Ind. Med., 1966, 23, 222 (Chem. Abs., 1966, 65, 11220h).
116. Stroganov, N. S., Kochkin, D. A., Khobot'sv, V. G. and Kolosova, L. V., Dokl. Akad. Nauk S.S.S.R., 1966, 170, 1189.
117. Tamura, H., Nogyo Gijutsu Kenkyusho Hokoku, Byori Konchu, 1965, 18, 135; 19, 47 (Chem. Abs., 1965, 63, 18958c,h).
118. Tappan, W. B., J. Econ. Entomol., 1965, 58, 730 (Chem. Abs., 1965, 63, 7600c).
119. Tareeva, A. I. and Borodina, G. M., Farmakol. Toksikol., 1967, 30, 207 (Chem. Abs., 1967, 67, 20351b).
120. Taylor, J. M., Levy, W. A., McCoy, G. and Scheinberg, L., Nature. 1964. 204, 4961.
121. van der Kerk, G. J. M., Fungicides in Agriculture and Horticulture, S.C.I. Monograph, No. 15, 1961, 67.
122. van der Kerk, G. J. M. and Luijten, J. G. A., J. App. Chem., 1954, 4, 314.
123. Verscuuren, H. G., Kroes, R., Vink, H. H. and van Esch, G. J., Food Cosmet. Toxicol., 1966, 4, 35 (Chem. Abs., 1966, 65, 12764h).
124. Visintin, B., Pepe, A. and Giuseppe, S. A., Ann. Ist. Super. Sanita, 1965, 1, 767 (Chem. Abs., 1966, 65, 12343f).
125. Walters, M. and Roe, F. J. C., Food Cosmet. Toxicol., 1965, 3, 271 (Chem. Abs., 1966, 64, 16518c).
126. Wilson, H. R., Fry, J. L. and Jones, J. E., Poultry Sci., 1967, 46, 304 (Chem. Abs., 1967, 66, 92762p).
127. Wolfenbarger, D. A., Guerra, A. A. and Lowry, W. L., J. Econ. Entomol., 1968, 61, 78 (Chem. Abs., 1968, 68, 58702).
128. Yoshikawa, K., Kurose, K. and Teramoto, S., Kogyo Kagaku Zasshi, 1964, 67, 1418 (Chem. Abs., 1965, 62, 4359h).
129. Zahradnicek, J., Schmidt, L. and Havranek, A., Listz Cukrov, 1967, 83, 193 (Chem. Abs., 1968, 68, 14258f).
130. Zedler, R. J., Trav. Centre Rech. Etud. Oceanogr. (Paris), 1965, 6, 401 (Chem. Abs., 1967, 66, 105934y).
131. Zedler, R. J. and Beiter, C. B., Soap and Chemical Specialities, 1962, 75.
132. Belgium Patent 637,090, Dec. 31, 1963 (Chem. Abs., 1965, 62, 6687d).
133. Belgium Patent 657,068, April 1, 1965 (Chem. Abs., 1966, 64, 20024g).
134. Belgium Patent 666,631, Jan. 10, 1966 (Chem. Abs., 1966, 65, 6206d).
135. British Patent 1,001,369, Aug. 18, 1965 (Chem. Abs., 1965, 63, 13569f).
136. British Patent 1,022,025, March 9, 1966 (Chem. Abs., 1966, 64, 20554d).
137. British Patent 1,026,692, April 20, 1966 (Chem. Abs., 1966, 65, 919h).
138. British Patent 1,081,969, Sept. 6, 1967 (Chem. Abs., 1968, 68, 2256).
139. British Patent 1,089,428, Nov. 1, 1967 (Chem. Abs., 1968, 68, 13889).
140. British Patent 1,107,929, March 27, 1968 (Chem. Abs., 1968, 69, 3461).
141. Czech Patent 113,485, Jan. 15, 1965 (Chem. Abs., 1965, 63, 16140h).
142. French Patent 1,354,803, March 13, 1964 (Chem. Abs., 1965, 62, 2676f).
143. French Patent 1,388,533, Feb. 5, 1965 (Chem. Abs., 1965, 63, 13568g).

144. French Patent 1,389,821, Feb. 19, 1965 (*Chem. Abs.*, 1965, **63**, 1816e).
145. French Patent 1,396,634, April 23, 1965 (*Chem. Abs.*, 1966, **64**, 8678).
146. French Patent 1,400,314, May 21, 1965 (*Chem. Abs.*, 1966, **64**, 4203g).
147. French Patent 1,482,847, June 2, 1967 (*Chem. Abs.*, 1968, **68**, 49780).
148. German Patent 1,155,630, Oct. 10, 1963 (*Chem. Abs.*, 1965, **62**, 1032a).
149. German Patent 1,181,977, Nov. 19, 1964 (*Chem. Abs.*, 1965, **62**, 4555e).
150. German Patent 1,214,684, April 21, 1966 (*Chem. Abs.*, 1966, **65**, 3908g).
151. German Patent 1,215,709, May 5, 1966 (*Chem. Abs.*, 1966, **65**, 5489f).
152. German Patent 1,216,300, May 12, 1966 (*Chem. Abs.*, 1966, **65**, 5490b).
153. German Patent 1,246,732, Aug. 10, 1967 (*Chem. Abs.*, 1967, **67**, 90938x).
154. German (East) Patent 56,133, May 30, 1967 (*Chem. Abs.*, 1967, **67**, 81484u).
155. Japanese Patent 5,810, March 24, 1965 (*Chem. Abs.*, 1965, **63**, 1815g).
156. Japanese Patent 12,788, June 22, 1965 (*Chem. Abs.*, 1966, **64**, 6694d).
157. Japanese Patent 17,141, Sept. 29, 1966 (*Chem. Abs.*, 1967, **66**, 11049e).
158. Japanese Patent 1,977, Jan. 30, 1967 (*Chem. Abs.*, 1967, **66**, 115803u).
159. Japanese Patent 13,867, Aug. 5, 1967 (*Chem. Abs.*, 1968, **68**, 39817).
160. Japanese Patent 24,573, Nov. 25, 1967 (*Chem. Abs.*, 1968, **69**, 44022).
161. Japanese Patent 26,296, Dec. 13, 1967 (*Chem. Abs.*, 1968, **69**, 52304).
162. Japanese Patent 01,505, Jan. 19, 1968 (*Chem. Abs.*, 1968, **69**, 20337).
163. Netherlands Patent 103,105, Nov. 15, 1962 (*Chem. Abs.*, 1965, **63**, 1175e).
164. Netherlands Patent 301,027, Sept. 27, 1965 (*Chem. Abs.*, 1966, **64**, 5139c).
165. Netherlands Patent 6,405,136, Nov. 11, 1964 (*Chem. Abs.*, 1965, **62**, 11988a).
166. Netherlands Patent 6,408,317, Jan. 25, 1965 (*Chem. Abs.*, 1965, **65**, 800d).
167. Netherlands Patent 6,408,318, Jan. 25, 1965 (*Chem. Abs.*, 1965, **65**, 800d).
168. Netherlands Patent 6,412,559, April 29, 1965 (*Chem. Abs.*, 1965, **63**, 15483h).
169. Netherlands Patent 6,510,567, Feb. 14, 1966 (*Chem. Abs.*, 1966, **65**, 13962c).
170. Netherlands Patent 6,511,285, March 2, 1966 (*Chem. Abs.*, 1966, **65**, 2944c).
171. Netherlands Patent 6,511,311, March 1, 1967 (*Chem. Abs.*, 1967, **67**, 53032b).
172. Netherlands Patent 6,515,201, May 31, 1966 (*Chem. Abs.*, 1966, **65**, 15428e).
173. Netherlands Patent 6,604,853, Nov. 14, 1966 (*Chem. Abs.*, 1967, **66**, 95825x).
174. Netherlands Patent 6,612,312, March 6, 1967 (*Chem. Abs.*, 1967, **67**, 73687x).
175. Netherlands Patent 6,614,955, April 24, 1967 (*Chem. Abs.*, 1967, **67**, 81485v).
176. U.S. Patent 3,211,680, Oct. 12, 1965 (*Chem. Abs.*, 1965, **63**, 18476g).
177. U.S. Patent 3,226,292, Dec. 28, 1965 (*Chem. Abs.*, 1966, **64**, 6415d).
178. U.S. Patent 3,227,563, Jan. 4, 1966 (*Chem. Abs.*, 1966, **64**, 8499f).
179. U.S. Patent 3,234,032, Feb. 8, 1966 (*Chem. Abs.*, 1966, **64**, 12966b).
180. U.S. Patent 3,236,793, Feb. 22, 1966 (*Chem. Abs.*, 1966, **64**, 16135e).
181. U.S. Patent 3,264,177, Aug. 2, 1966 (*Chem. Abs.*, 1966, **65**, 14364h).
182. U.S. Patent 3,265,756, Aug. 9, 1966 (*Chem. Abs.*, 1966, **65**, 15223d).
183. U.S. Patent 3,268,347, Aug. 23, 1966 (*Chem. Abs.*, 1966, **65**, 18861h).
184. U.S. Patent 3,268,395, Aug. 23, 1966 (*Chem. Abs.*, 1966, **65**, 14363b).
185. U.S. Patent 3,282,672, Nov. 1, 1966 (*Chem. Abs.*, 1967, **66**, 28891c).
186. U.S. Patent 3,284,296, Nov. 8, 1966 (*Chem. Abs.*, 1967, **66**, 45790m).
187. U.S. Patent 3,297,523, Jan. 10, 1967 (*Chem. Abs.*, 1967, **66**, 54562r).

188. U.S. Patent 3,306,955, Feb. 28, 1967 (*Chem. Abs.*, 1967, **66**, 96178a).
189. U.S. Patent 3,312,725, April 4, 1967 (*Chem. Abs.*, 1967, **67**, 10686a).
190. U.S. Patent 3,328,441, June 27, 1967 (*Chem. Abs.*, 1968, **68**, 49768).
191. U.S. Patent 3,334,119, Aug. 1, 1967 (*Chem. Abs.*, 1968, **69**, 77511).
192. U.S. Patent 3,346,607, Oct. 10, 1967 (*Chem. Abs.*, 1968, **68**, 69130).
193. U.S. Patent 3,400,202, Sept. 3, 1968 (*Chem. Abs.*, 1968, **69**, 105341).
194. U.S.S.R. Patent 220,389, June 28, 1968 (*Chem. Abs.*, 1968, **69**, 88142).

15

APPLICATIONS BASED UPON
CHEMICAL PROPERTIES

This topic was reviewed in 1965[53] and, while established applications of organotin compounds such as in polymer stabilisation and catalysis continue to grow, some new uses are beginning to emerge. The majority of publications in this field are in the patent literature, a medium where immoderate claims are commonplace, so that assessment of the significance of some of the newer applications is not easy. Most of the space in this chapter is devoted to established uses but an attempt is made to mention, however fleetingly, all patents in the years 1965–1968 covering non-biological applications of organotin compounds.

15.1. THE STABILISATION OF POLYMERS

Before dealing with current developments in this field it is appropriate to give a brief general account of the stabilisation of poly(vinyl chloride). PVC begins to flow at about 180° and processing of the polymer is carried out in this temperature region. In the absence of stabilisers rapid dehydrochlorination would occur at these temperatures leading to severe darkening and loss of mechanical properties. In addition to this thermal reaction, dehydrochlorination can also be induced by exposure to light.

The nature of the dehydrochlorination reaction is uncertain and both free-radical and ionic mechanisms have been postulated. Much early work has been summarised by Geddes[21] who concludes that the reaction occurs by a radical process, a view which is supported by a more recent publication[4a]. It is known that elimination of hydrogen chloride does not occur randomly along the polymer chain. A double bond activates adjacent atoms so that loss of the next HCl molecule gives a conjugated double bond and, eventually, a conjugated polyene chain is formed leading to discolouration and finally blackening of the polymer. A reasonable mechanism for such a process is that suggested by Winkler.[64]

$$S \cdot + -CHClCH_2CHClCH_2CHClCH_2CHClCH_2- \longrightarrow$$
$$SH + -CHCl\dot{C}HCHCHClCH_2CHClCH_2CHClCH_2- \longrightarrow$$
$$Cl \cdot + -CHClCH{=}CHCH_2CHClCH_2CHClCH_2- \longrightarrow$$
$$HCl + -CHClCH{=}CH\dot{C}HCHCHClCH_2CHClCH_2- \longrightarrow$$
$$Cl \cdot + -CHClCH{=}CHCH{=}CHCH_2CHClCH_2- \longrightarrow \quad etc$$

However, it is known that Lewis acids catalyse the dehydrochlorination and a powerful case has been made for regarding the degradation as an ionic process[31]. It should be remembered that the term poly-(vinyl chloride) covers a wide range of materials, made by different methods, showing wide variations in molecular weight and containing different types and amounts of structural irregularities. These materials undoubtedly lose hydrogen chloride by more than one mechanism. Attention is focused here on dehydrochlorination but several other types of reaction contribute to polymer degradation including oxidation, cross-linking and chain-scission processes.

While several classes of compound are used to stabilise PVC[10], the organotin stabilisers (for which the first patent was issued in 1940[209]) are the most effective[55]. With so much uncertainty attached to the dehydrochlorination reaction it is not surprising that the mechanism of stabilisation is also somewhat obscure. Stabilisers are selected by empirical methods and their efficacy depends upon the number of alkyl groups attached to tin as follows, $R_2SnX_2 > RSnX_3 > R_3SnX \gg R_4Sn$, the tetraalkyltin compounds being almost totally ineffective. The alkyl groups are usually butyl or, if toxicity considerations are important, octyl. The most common types of compound used are dialkyltin esters, particularly maleates, and dialkyltin mercaptides or similar compounds containing two Sn—S bonds. By studying the incorporation of radioactively labelled fragments of organotin compounds into PVC a number of reactions between polymer and stabiliser were postulated[19]. Some of these reactions have now been shown to be chemically feasible; thus organotin maleates undergo Diels-Alder reactions with model dienes[39]. Also, at 180°, ester and mercaptide groups attached to tin are exchanged with chlorine atoms at allylic and tertiary sites in model compounds but no exchange occurs with secondary chloro compounds[4]. (There is other evidence that dehydrochlorination is initiated at points where structural irregularities occur in the polymer chain.)

It is clear that organotin compounds are very effective stabilisers because they are multi-functional and can deal, in various ways, with a number of undesirable situations in a polymer that is degrading or about to degrade. The behaviour of an organotin stabiliser can be summarised as follows.

(1) The stabiliser, R_2SnX_2, is compatible with the polymer.

(2) It can exchange X groups with reactive chlorine atoms in the polymer and thereby inhibit the initiation of the dehydrochlorination reaction.

(3) If the X groups are maleate residues the stabiliser could interrupt the conjugation in a polyene chain by undergoing a Diels-Alder reaction with a diene unit.

(4) The stabiliser acts as an acceptor for hydrogen chloride by the reaction below; the organotin product being compatible with the polymer.

$$R_2SnX_2 + 2HCl \longrightarrow R_2SnCl_2 + 2XH$$

(5) The above reaction releases, precisely at the point where active dehydrochlorination is occurring, XH molecules which may further assist in the inhibition of undesirable reactions. Thus if the stabiliser is a mercaptide, $R_2Sn(SR')_2$, then $R'SH$ is released which is an effective transfer agent for the termination of radical chain reactions. Alternatively if XH is maleic acid, or an alkyl hydrogen maleate, this would undergo a Diels-Alder reaction more readily than the stabiliser itself.

The reason why compounds with two alkyl groups attached to tin are the most effective stabilisers is not clear. It is probable that this represents a balance between the amount of hydrogen chloride that can be absorbed and the deleterious effects of the organotin chlorides which are thus formed. The catalytic effect on dehydrochlorination of Lewis acids has already been mentioned, organotin chlorides promote this reaction[31, 38] with the most likely sequence for catalytic activity being $R_3SnCl < R_2SnCl_2 < RSnCl_3$.

PVC degrades more slowly in a nitrogen atmosphere and the manner in which oxygen aids polymer break-down has been discussed[52, 54]. Organotin stabilisers are more effective in a nitrogen atmosphere[48].

An indication of the types of compounds which have been mentioned in recent patent specifications as stabilisers for PVC and related compounds is given in Table 1. It is seen that most of these compounds are organotin esters or mercaptides or derivatives of mercapto acids such as thioglycollic acid in which the two functions are combined.

TABLE I

Organotin stabilisers for PVC and other chlorinated polymers

Stabiliser[a]	Reference
$R_2Sn(OCOR')_2$ ($R'COOH = 2,5\text{-}(C_8H_{17}OCO)_2C_6H_3COOH$, $2,4\text{-}(C_8H_{17}OCO)_2C_6H_3COOH$; $R = alkyl$)	227
$Me_2Sn(OCOR)_2$ ($RCOOH = $ mixed fatty acids from vegetable oils)	32
Dibutyltin esters of aromatic acids	130
Dialkyltin derivatives of polyesters	239
$RCOO(SnBu_2O)_2COR$ ($R = alkyl$)	23
$Bu_2Sn(OCOCH\!\!=\!\!CHCOOR\text{-}cis)_2$ ($R = PhCH_2$)	170
$Bu_2Sn(OCOCH\!\!=\!\!CHCOOR\text{-}cis)_2$ ($R = $ cyclohexyl, 2-methylcyclohexyl)	88
$Bu_2Sn(OCOCH\!\!=\!\!CHCOOCH_2CH_2OR\text{-}cis)_2{}^b$ ($R = alkyl$)	161
$Bu_nSn(OCOCH\!\!=\!\!CHCOOR\text{-}cis)_{4-n}$ ($n = 1,2,3$; $R = $ tetrahydrofurfuryl)	162
Mixture of dibutyltin oxide and maleic anhydride	81
$Bu_2RSnOCOCH\!\!=\!\!CHCOOSnRBu_2\text{-}cis$ ($R = OCOCH\!\!=\!\!CHCOOC_8H_{17}\text{-}cis$, $SCH_2COOC_8H_{17}$)	101
$HOSnBu_2OSnBu_2OCOCH\!\!=\!\!CHCOOR\text{-}cis$ ($R = $ Me, Bu)	171, 177
$Bu_2Sn{\overset{\displaystyle OCOCH_2S}{\underset{\displaystyle OCOCH_2S}{\big\langle}}}CHR$ ($R = CH_2\!\!=\!\!CH$, Ph, $PhCH\!\!=\!\!CH$)	147
$Bu_2Sn{\overset{\displaystyle OCOCH_2CH_2S}{\underset{\displaystyle OCOCH_2CH_2S}{\big\langle}}}SnBu_2$	91
$Bu_2ClSnSCH_2CH_2COO(Bu_2SnSCH_2CH_2COO)_nSnClBu_2$	122
$Bu_2Sn{\overset{\displaystyle OCOCH_2CH_2S}{\underset{\displaystyle OCOCH_2CH_2S}{\big\langle}}}CMe_2$	86, 222
$Bu_2SnSCH_2CH_2COOSnBu_2OCOCH\!\!=\!\!CHCOOC_8H_{17}\text{-}cis$ $\quad\mid$ $OCOCH\!\!=\!\!CHCOOC_8H_{17}\text{-}cis$	100
$Bu_2Sn[OCO(CH_2)_{10}Me]S(CH_2)_{11}Me$	143
$(C_{11}H_{23}COOSnBu_2)_2S$	200
$[CH_2CH_2Sn(OCOR)_2]_n$	99
$[RCH(SCH_2CH_2COO)_2]_3(R'Sn)_2$ (R and $R' = alkyl$)	118, 216
$(PhCH_2)_2Sn(SCOR)_2$ ($R = $ alkyl or aryl)	207
$Bu_2Sn(SC_{12}H_{25})SCOPh$	93
$Bu_2Sn(SCH_2COOC_8H_{17})_2$	144
$(PhCH_2)_2Sn(SCH_2COOC_8H_{17}\text{-}i)_2$	208
$Bu_nSn(SCH_2COOR)_{4-n}$ ($n = 1,2,3$; $R = $ tetrahydrofurfuryl)	163
$Bu_2Sn[SCH(COOBu)CH_2COOBu]_2$	138
$(NCCH_2CH_2)_2Sn(SCH_2COOC_8H_{17}\text{-}i)_2$	205
$Bu_2Sn(SCH_2CH_2OCOC_8H_{17})_2$	139

TABLE I—*cont.*

Stabiliser[a]	Reference
Bu_2Sn $\begin{cases} SCH_2CH_2OCOCH_2 \\ SCH_2CH_2OCOCH_2 \end{cases}$ $\begin{matrix} CH_2 \\ | \\ CH_2 \end{matrix}$	137
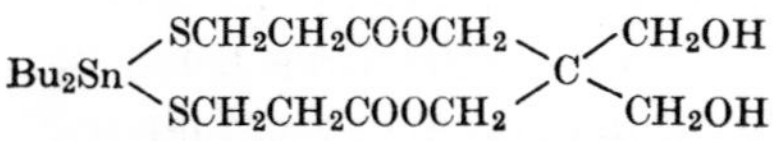 Bu_2Sn $\begin{cases} SCH_2CH_2COOCH_2 \\ SCH_2CH_2COOCH_2 \end{cases}$ C $\begin{matrix} CH_2OH \\ CH_2OH \end{matrix}$	95
$B(OSnBu_2SCH_2COOCHMeCH_2OCOCH_2SH)_3$	172, 190
$Bu_2Sn(SC_{12}H_{25})_2$ and triphenyltin hypophosphite	165
Product obtained by heating $Bu_2SnF_2 + Bu_2SnS$ at 100–220°	204
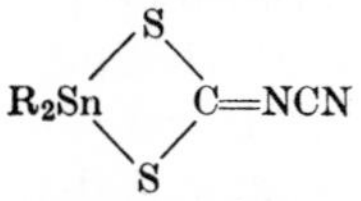 $R_2Sn \begin{matrix} S \\ \diagup \diagdown \\ \diagdown \diagup \\ S \end{matrix} C{=}NCN$ $\quad$ (R = Bu, C_8H_{17})	242
$[SCH_2C_6H_4CH_2SSnBu_2]_n$	218
$[CH_2CH_2Sn(SR)_2]_n$	99
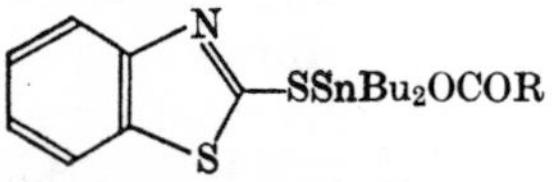 benzothiazole—$SSnBu_2OCOR$ $\qquad$ (R = alkyl or aryl)	213
$BuSn(Cl)(SC_8H_{17})_2$	191
$R_2Sn_2S_3$	136
Product from reaction between $RSnCl_3$, Na_2S and R'SNa	201

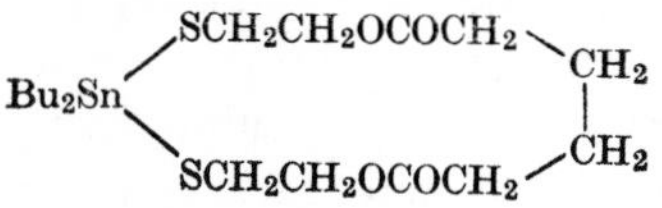

[a] An indication, only, is given of the type of compound, usually by a single example. In many cases the evidence that the stabilisers have the structures assigned to them is incomplete.

[b] This stabiliser also has plasticiser properties.

The choice of stabiliser will be influenced by a number of factors particularly by the use to which the polymer will be put and the relative importance of resistance to pyrolytic and photolytic degradation. Differential thermal analysis has been used to evaluate compounds which stabilise against thermal decomposition[43] and the protection afforded by stabilisers to pyrolytic degradation of PVC in solution has been studied[37]. Certain stabilisers, particularly dioctyltin maleate derivatives, disproportionate at 180° by the reaction below leading to loss of stabiliser activity[18]. Organotin

$$2R_2SnX_2 \longrightarrow R_3SnX + RSnX_3$$

maleates are superior to mercaptides for protection against colour development in clear rigid PVC exposed to sunlight[5, 6]. The relative

20

efficiency of various stabilisers added to rigid PVC does however depend upon climatic conditions[63].

An increasing number of patents are concerned with synergistic effects which are observed when an organotin stabiliser and a tin-free compound are used. The latter are usually antioxidants such as highly hindered phenols[74, 75, 76, 108, 187, 206] but organic sulphides [97], alkyl hydrogen maleates[105], epoxy resins[22, 24] and various other compounds[35, 102, 199] have been advocated. Addition to PVC of epoxidised soya-bean oil, as a lubricant, at concentrations below 3%, had no undesirable effects on the migration of organotin stabilisers[65].

Among stabilisers which have been proposed for use with specific types of poly(vinyl chloride) are organotin sulphur compounds for impact resisting PVC (which contains 10–20% of a synthetic rubber)[44], dibutyltin *bis*(nonyl thioglycollate) for PVC coating compositions[78], dibutyltin distearate for PVC foams[79] and dibutyltin dilaurate for dyed polymer pigment particles[89]. Transparent, nonflammable, thermoplastic material with good weather resistance is made from poly(methyl methacrylate) and PVC, the latter being stabilised with dibutyltin maleate[72]. Dibutyltin oxide is added to the 14% or so of PVC which is incorporated into acrylonitrile film- and fibre-forming polymers [69]. The presence of organotin compounds lowers the extent of radiation-induced grafting of acrylonitrile and styrene on PVC[28] and stabilises heat-curable adhesive compositions containing PVC[87]. Polypropylene is rendered flame-proof by the addition of chlorinated polymers and organotin esters are included in these compositions to prevent discoloration[195]. Chemical, chromatographic and polarographic methods have been used to assay PVC products for stabiliser residues[27, 33] (see also p. 273).

In an entirely different approach to the problem some of the chlorine atoms in PVC have been replaced by R_3Sn groups ($R = Bu$ or Ph) by treatment with R_3SnLi to give a polymer with increased thermal stability[36, 49].

Another group of polymers for which organotin additives may be used for the prevention of degradation are the polyolefins. These polymers, on exposure to heat or light, suffer oxidative degradation which causes discoloration and eventual loss of mechanical properties. Phosphorus-containing stabilisers have been proposed of which the following examples are typical, tributyltin phosphate[154], triphenyltin phosphate[155], diphenylpropyltin phosphate[156] and thiophosphinates such as $Bu_2Sn[S(S)PPh_2]_2$[235, 245]. Several other types of stabiliser for polyolefins have been described including organo-tin-esters[15, 92, 125, 129, 153, 164, 166, 167, 169, 179] and -phenoxides[219], organo-distannanes[159], sulphur-containing compounds such as $PhCH(SCH_2CH_2COOSnBu_3)_2$[116], $Bu_2Sn(SCH_2COOC_8H_{17})_2$[66] and the cyclic compounds *I* and *II*[220]. The manner in which dibutyltin maleate stabilises polyethylene against radiation-induced degradation has been investigated by Mössbauer spectroscopy[1].

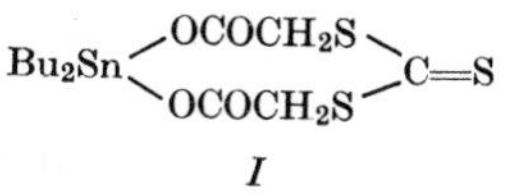

$$Bu_2Sn \diagdown_{OCOCH_2S}^{OCOCH_2S} \diagup C{=}S \qquad Bu_2Sn \diagdown_{OCOCH_2CH_2}^{OCOCH_2CH_2} \diagup S$$

$$I \qquad\qquad II$$

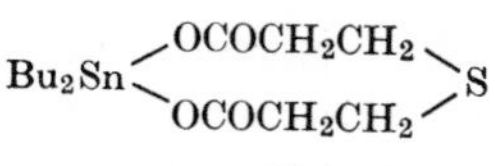

Patents have been taken out covering the use of organotin compounds for the stabilisation of other polymeric materials including phenol-formaldehyde resins[248], polycarbonates[117], polyethers[228], polyamides[176] and a polysulphide sealant compound[236]. Dibutyltin esters were effective in stabilising fluorinated polystyrenes against thermal decomposition[20]. The resistance of polyimide insulating materials to high temperature corona discharge is improved by organotin additives[246].

Other suggested applications in polymer technology include the addition of organotin phenoxides to synthetic resins for the absorption of ultraviolet light[131] and of high molecular weight organotin esters to a number of plastics as anti-static agents[173, 174, 178].

15.2. ORGANOTIN COMPOUNDS AS CATALYSTS

15.2.1. *The formation of urethans*

Organotin compounds catalyse the formation of urethans from isocyanates and alcohols.

$$RNCO + R'OH \xrightarrow[\text{compound}]{\text{organotin}} RNHCOOR'$$

There have been a number of recent kinetic studies of this reaction[16, 17, 29, 41, 42, 50] and the manner in which the catalyst functions has already been discussed (p. 77). This reaction is the basis for the preparation of polyurethans from appropriate difunctional compounds.

$$nOCNXNCO + nHOYOH \longrightarrow$$
$$-CONHXNHCO[OYOCONHXNHCO]_{n-1}OYO-$$

Polyurethan foams are formed when a controlled amount of water reacts with some of the isocyanate groups to give carbamic acid residues which decompose giving carbon dioxide and it is this gas which causes foam formation. The amine groups which are formed react with the isocyanates to give substituted ureas. The problem in

$$OCNXNCO + 2H_2O \longrightarrow [HOCONHXNHCOOH]$$
$$[HOCONHXNHCOOH] \longrightarrow 2CO_2\uparrow + NH_2XNH_2$$
$$mOCNXNCO + mNH_2XNH_2 \longrightarrow -CONHX[NHCONHX]_{2m-1}NH-$$

foam production is the control of the rates of these reactions. Thus if the isocyanate-alcohol reaction (= reaction 1) is too fast an unfoamed gel results which is then simply split by the carbon dioxide evolved. If the isocyanate-water reaction (= reaction 2) is too fast the gel is not strong enough to retain the gas bubbles. Although both reactions are susceptible to catalysis by organotin compounds[2, 217] it is usual to use an organotin compound for reaction 1 and a tertiary amine for reaction 2[47].

The most generally used organotin catalyst for urethan preparation is dibutyltin dilaurate[61, 71, 84, 90, 232, 233]. Alternative compounds which have been advocated include dialkyltin esters of other fatty acids[26] and of chlorinated acids[140, 182], mixed esters containing both laurate and maleate residues[194], dialkyltin oxides[225], oxides of the type $R_2XSnOSnXR_2$ (X = acetate, halogen, pseudo-halogen, etc.)[103, 160, 184, 185] and sulphonium salts such as $Ph_3S^{\oplus}$ $[Me_2SnCl_2Br]^{\ominus}$[240]. Polyurethan foams with improved physical properties and increased resistance to oxygen and moisture were prepared by using adducts of alkyltin trihalides with dimethylformamide as catalysts[141]. There are advantages in adding a polysiloxane surface active agent with the organotin catalyst to increase the tensile strength of the polyurethan foam[214, 215]. It appears that, under certain circumstances, the presence of organotin residues may promote the degradation of polyurethans[30, 45].

The related reaction between phenyl isocyanate and a series of substituted 1,1-pentamethylene-3-phenyl ureas, in the presence of butyltin trichloride, occurs by a different mechanism. The rates of these reactions are proportional to the concentrations of the $BuSnCl_3$ and this catalyst is 28 times more active than $EtSnCl_3$, compounds with 2, 3 or 4 Sn—C bonds are without catalytic effect[14].

15.2.2. Ziegler catalysts

Ziegler catalysts are widely used for olefin polymerisation since reaction is possible under mild conditions to give stereoregular polymers with high molecular weights. The essential components of a Ziegler catalyst system are a transition metal halide and a main-group organometallic compound. In the initial formulations the latter was always an aluminium derivative but organotin compounds are now being used commercially for polyolefin production[53].

Ziegler catalyst systems which are described in the current patent literature vary considerably in the number of components and in the type of organotin compound specified. There are simple formulations containing titanium(IV) chloride and an organotin hydride[114, 119, 158] although the latter may be generated *in situ* and the presence of an organotin phenoxide[107, 181] or of a phenol[180] may be beneficial. The reactions between organotin compounds and titanium(IV) chloride have been studied, the relative stabilities of the organotitanium compounds formed have been examined[46, 57] and there is spectroscopic evidence for products containing Ti—H bonds[56]. The polymerisation of ethylene in aqueous suspension using a mixture of titanium(III) chloride or oxide and tetrabutyltin has been described[211].

A number of ternary catalyst systems have been patented in which a transition metal compound is used in conjunction with an aluminium halide (or alkyl) and a tetraalkyl(aryl)tin compound[7, 8, 110, 120, 123, 157, 223]. In ternary systems containing Ph_4Sn, $AlBr_3$ and VCl_4 the function of the tin compound is solely as a phenylating agent for the aluminium bromide and the system is equivalent in catalytic activity to one containing VCl_4 and $PhAlBr_2$[12]. Other catalysts include a combination of a transition metal halide with the product from the reaction of diethyltin chlorohydride with alumina[124]. The optimum conditions for the polymerisation of styrene using a vanadium oxychloride-tetrapropyltin catalyst have been reported[3]. The reaction between a number of tetraalkyltin compounds and titanium(IV) chloride has been investigated using gas chromatography[58].

15.2.3. *Other catalytic uses*

Organotin compounds have been used alone, or with other catalyst components, to promote the polymerisation of vinyl chloride[51, 149, 189] and a stannic chloride-tetrabutyltin catalyst for the polymerisation of acrylonitrile has been described[221].

Tributyltin lithium is an effective catalyst for the solution polymerisation of conjugated dienes giving particularly transparent polymers[188]. When tetraethyltin together with aluminium bromide and a cobalt carbonyl compound are used to polymerise butadiene the product has $\geqslant 70\%$ cis-1,4-configuration[106]. Improved yields in the stereospecific polymerisation of butadiene catalysed by organolithium compounds are obtained in the presence of organotin compounds such as $Ph_2Sn(CH_2CH=CH_2)_2$[77]. However, the more usual function of organotin additives in such polymerisations is to improve the processing characteristics of the polymer[70, 197, 234]. Organotin derivatives of mercapto esters, of the type $R_2Sn(SCH_2COOR')_2$, are used to accelerate the peroxide-catalysed polymerisation of olefins[80, 133]. A catalyst prepared from dibutyltin dichloride and diethylaluminium chloride promotes the dimerisation of olefins[212].

Catalyst systems containing (−)-menthoxytriethyltin and aluminium trichloride promote the polymerisation of benzofuran giving an optically active product; the effective catalyst is the active menthoxyaluminium dichloride dimer[25, 59]. Binary mixtures of tetraalkyltin compounds and aluminium trichloride are used to give crystalline polymers of vinyl ethers.

Compounds such as dibutyltin-oxide or -diacetate are excellent esterification catalysts and many patents cover the use of these and related compounds for the preparation of terylene and related polyesters[67, 85, 112, 192, 210, 229, 237]. Dibutyltin esters are also used to catalyse the esterification of epoxy resins with fatty acids[202].

The formation of polyamides from diamines and dicarboxylic acids or from caprolactam is catalysed by alkyltin compounds[128, 148].

15.3. MISCELLANEOUS APPLICATIONS

Organopolysiloxanes having terminal hydroxyl groups can be 'cold cured' by the addition of dibutyltin dilaurate or a similar organotin ester. Usually other additives such as tetraethyl silicate are specified; the process involves increasing the molecular weight of the polymers and, in many cases, the introduction of cross-links[68, 96, 135, 145, 183]. These catalysts are used in the production of organopolysiloxane elastomers[94, 98, 127, 132, 203], surface coatings[82, 126, 146] and for a co-polymer based on ethyl acrylate and vinyltrimethoxysilane which is partially hydrolysed before vulcanisation[196]. A kinetic study of the catalysed, room-temperature, vulcanisation of silicones has been reported[196].

Catalysts containing organotin components have also been proposed for the hardening of epoxy resins[113, 115].

The addition of various types of organotin compound to hydrocarbon-based lubricants improves resistance to oxidation at high temperatures [13, 224, 243] and also permits a higher fuel to oil ratio in the mixtures used in two-stroke engines[230]. The antiwear properties of poly(phenyl ether) lubricants and ability to withstand high pressures is increased by organotin additives[73, 226] as is the resistance to oxidation of perfluoroalkoxytriazine- and phosphonitrile-based lubricating fluids[241].

New scintillation detectors for γ- and X-rays have been described which are based on polystyrene containing a tetraaryltin compound[9, 62] or on a copolymer of vinyltoluene and p-triphenylstannyl-styrene[11, 60].

Among patents covering miscellaneous applications of organotin compounds may be mentioned their use as films to prevent corrosion of metals[231, 238], as water-repellants for fibres and other materials[34, 104, 198], as additives for improving the properties of paints and varnishes[134, 193] and as corrosion-inhibitors for chloro- and bromo-hydrocarbons stored in fire extinguishers[83]. If a solution of dibutyltin oxide is sprayed on to glass heated at 600° a layer of stannic oxide results, it is then possible to coat the glass with a firmly-adhering, transparent, light-reflecting layer of cobalt oxide[186]. Tetraalkyl(aryl)tin compounds are useful as ballistic additives for solid rocket engine fuels[111] and hexalkyldistannanes improve the combustion characteristics of gasoline[247]. Organotin compounds are more effective in flame-proofing polyesters than are equimolar quantities of phosphorus compounds[142]. The addition of 2% dibutyltin oxide improves the adhesion of polychloroprene bonding composi-tions[109]. Polymers resulting from the homopolymerisation of

$$Sn(CH_2CH_2COOCH_2\overline{CHCHO})_4$$ or from the copolymerisation of this compound with other epoxides are useful in adhesives, laminates and other applications[244]. Organotin maleates are effective in controlling particle size in the emulsion polymerisation of vinyl compounds[152].

References

1. Aleksandrov, A. Yu., Berlyant, S. M., Karpov, V. L., Leshchenko, S. S., Okhlobyatin, O. Yu., Finkel, E. E. and Shpinel, V. S., *Vysokomol. Soedin.*, 1964, **6**, 2105 (*Chem. Abs.*, 1965, **62**, 41616).
2. Aleksandrova, Yu. V. and Lakosina, T. A., *Plasticheskie Massy*, 1965, 15 (*Chem. Abs.*, 1965, **63**, 11305h).
3. Anand, L. C., Deshpande, A. B. and Kapur, S. L., *Indian J. Chem.*, 1967, **5**, 186.

4. Ayrey, G., Poller, R. C. and Siddiqui, I. H., *Abstracts, 4th International Conference on Organometallic Chemistry*, Bristol, 1969.
4a. Bamford, C. H. and Fenton, D. F., *Polymer*, 1969, **10**, 63.
5. Baum, G. A., *Appl. Polym. Symp. No. 4*, 1967, 189 (*Chem. Abs.*, 1968, **68**, 30476).
6. Baum, G. A., *Mod. Plastics*, 1967, **44**, 148.
7. Biritz, L. F., *Am. Chem. Soc. Div. Petrol Chem. Preprints*, 1963, **8**, B57 (*Chem. Abs.*, 1965, **62**, 644b).
8. Carrick, W. L., *Macromol. Syn.*, 1966, **2**, 33 (*Chem. Abs.*, 1966, **65**, 18692h).
9. Chernobai, A. V. and Kolesnikov, L. N., *Pribory i Tekhn. Eksperim.*, 1964, **9**, 120 (*Chem. Abs.*, 1966, **65**, 11711f).
10. Chevassus, F. and de Broutelles, R. *The Stabilization of Polyvinyl Chloride*, Edward Arnold, London, 1963.
11. Dannin, J., Sandler, S. R. and Baum, B., *Intern. J. Appl. Radiation Isotopes*, 1965, **16**, 589.
12. de Liefde Meijer, H. J., van den Hurk, J. W. G. and van der Kerk, G. J. M. *Rec. Trav. Chim.*, 1966, **85**, 1018; 1025.
13. Dolle, R. E., *Ind. Eng. Chem., Prod. Res. Develop.*, 1967, **6**, 177 (*Chem. Abs.*, 1967, **67**, 92530g).
14. Dyer, E. and Pinkerton, R. B., *J. App. Polymer Sci.*, 1965, **9**, 1713.
15. Egorova, Z. S., Slovokhotova, N. A., Leshchenko, S. S., Karpov, V. L., Finkel, E. E. and Mitrofanova, L. V., *Tr. Komis. po Spektroskopii, Akad. Nauk S.S.S.R.*, 1964, 503 (*Chem. Abs.*, 1965, **63**, 10119b).
16. Entelis, S. G. and Nesterov, O. V., *Kinet. Katal.*, 1966, **7**, 469 (*Chem. Abs.*, 1966, **65**, 10462e).
17. Entelis, S. G., Nesterov, O. V. and Tiger, R. P., *J. Cell. Plast.*, 1967, **3**, 360 (*Chem. Abs.*, 1967, **67**, 83441q).
18. Franzen, V., *Ernaehrungsforschung*, 1966, **11**, 368 (*Chem. Abs.*, 1967, **66**, 95774e).
19. Frye, A. H., Horst, R. W. and Paliobagis, M. A., *J. Polymer Sci. A*, 1964, **2**, 1765; 1785; 1801.
20. Gashtol'd, N. S., Dokukina, A. F., Kocheshkov, K. A. and Talalaeva, T. V., *Vysokomol. Soedin. Ser. A*, 1967, **9**, 1489 (*Chem. Abs.*, 1968, **68**, 30345).
21. Geddes, W. C., *Rubber and Plastics Research Association* (*Shawbury, Shrewsbury, Shropshire*), *Technical Review*, No. 31, 1966.
22. Gel'fman, Ya. A., Lauris, I. V. and Kuskova, V. P., *Sb. Tr. Vses. Nauch.-Issled. Inst. Novykh Stroit Mater.*, 1966, 54 (*Chem. Abs.*, 1968, **68**, 79010).
23. Gel'fman, Ya. A., Zemlyanskii, N. N., Lauris, I. V., Syutkina, O. P., Kuskova, V. P. and Panov, E. M., *Plasticheskie Massy*, 1966, **9**, 10 (*Chem. Abs.*, 1967, **66**, 11378t).
24. Gel'fman, Ya. A., Zemlyanskii, N. N., Lauris, I. V., Syutkina, O. P., Kuskova, V. P. and Panov, E. M., *Sb. Tr. Vses. Nauch.-Issled. Inst. Novykh Stroit. Mater*, 1966, 58 (*Chem. Abs.*, 1968, **68**, 79011).
25. Hayakawa, Y., Fueno, T. and Furukawa, J., *J. Polymer Sci. Pt. A-1*, 1967, **5**, 2099.
26. Kafengauz, A. P., Kafengauz, I. M. and Murashova, V. I., *Plasticheskie Massy*, 1965, 13 (*Chem. Abs.*, 1965, **63**, 16547b).

27. Korn, O. and Woggon, H., *Ernaehrungsforschung*, 1965, **10**, 57 (*Chem. Abs.*, 1965, **63**, 4462c).
28. Lagner, H., *Plaste Kautshuk*, 1966, **13**, 76 (*Chem. Abs.*, 1966, **64**, 14352a).
29. Lipatova, T. E., Bakalo, L. A., Sirotinskaya, A. L. and Syutkina, O. P., *Vysokomol. Soedin. Ser. A*, 1968, **10**, 859 (*Chem. Abs.*, 1968, **69**, 19559).
30. Mack, G. P., *Mod. Plastics*, 1964, **42**, 148 (*Chem. Abs.*, 1965, **62**, 6629e).
31. Marks, G. C., Benton, J. L. and Thomas, C. M., *Soc. Chem. Ind.* (*London*) *Monogr. No.* 25, 1967, 204.
32. Matsuda, S., Matsuda, H. and Yamane, Y., *Kogyo Kagaku Zasshi*, 1964, **67**, 467 (*Chem. Abs.*, 1965, **62**, 16288b).
33. Mazur, H., *Roczniki Panstwowego Zakladu Hig.*, 1965, **16**, 275 (*Chem. Abs.*, 1965, **63**, 11779b).
34. Migdal, S., Gertner, D. and Zilka, A., *Israel J. Chem.*, 1967, **5**, 163 (*Chem. Abs.*, 1967, **67**, 100624h).
35. Minsker, K. S., Fedoseeva, G. T., Zavarova, T. B. and Malysheva, I. P., *Vysokomol. Soedin.*, Ser. B, 1968, **10**, 454 (*Chem. Abs.*, 1968, **69**, 59786).
36. Minsker, K. S., Purinson, Yu. A., Zavarova, T. B., Plate, N. A., Fedoseeva, G. T. and Kargin, V. A., *Vysokomol. Soedin. Ser. A*, 1968, **10**, 1336 (*Chem. Abs.*, 1968, **69**, 36562).
37. Morikawa, T., *Kagaku To Kogyo* (*Osaka*), 1964, **38**, 672 (*Chem. Abs.*, 1965, **62**, 14892b).
38. Morikawa, T. and Yoshida, K., *Kagaku To Kogyo* (*Osaka*), 1964, **38**, 667 (*Chem. Abs.*, 1965, **62**, 14891h).
39. Mufti, A. S. and Poller, R. C., *J. Chem. Soc.* (*C*), 1967, 1767.
40. Nagy, J. and Borbely-Kuszmann, A., *Intern. Symp. Organosilicon Chem.*, *Sci. Commun.*, Prague, 1965, 201 (*Chem. Abs.*, 1966, **65**, 7422h).
41. Nesterov, O. V., Chirkov, Yu. N. and Entelis, S. G., *Kinet. Katal.*, 1967, **8**, 1371 (*Chem. Abs.*, 1968, **68**, 86516).
42. Nesterov, O. V., Zabrodin, V. B., Chirkov, Yu. N. and Entelis, S. G., *Kinet. Katal.*, 1966, **7**, 805 (*Chem. Abs.*, 1967, **66**, 37000p).
43. O'Leary, R. K., Foy, J., Guess, W. L. and Autian, J., *J. Pharm. Sci.*, 1967, **56**, 494 (*Chem. Abs.*, 1967, **66**, 116215r).
44. Ondracek, J. and Manas, J., *Plaste und Kautschuk*, 1966, **13**, 587.
45. Orlov, V. A. and Tarakanov, O. G., *Vysokomol. Soedin.*, 1966, **8**, 1139 (*Chem. Abs.*, 1966, **65**, 12352e).
46. Osipov, O. A. and Kashireninov, O. E., *Zh. Obshch. Khim.*, 1962, **32**, 1717 (*Chem. Abs.*, 1963, **58**, 4590f).
47. Overmars, H. G. J. and van der Kerk, G. J. M., *Chimia*, 1965, **19**, 126 (*Chem. Abs.*, 1965, **62**, 13326d).
48. Popova, Z. V., Tikhova, N. V. and Vyazankin, N. S., *Vysokomol. Soedin.*, *Khim. Svoistva i Modifikatsiya Polimerov, Sb. Statei*, 1964, 175 (*Chem. Abs.*, 1965, **62**, 14892c).
49. Purinson, Yu. A., Plate, N. A., Davydova, S. L., Nurkeeva, Z. S. and Kargin, V. A., *Vysokomol. Soedin.*, Ser. B, 1968, **10**, 257 (*Chem. Abs.*, 1968, **69**, 11020).
50. Rand, L., Thir, B., Reegen, S. L. and Frisch, K. C., *J. Appl. Polmer Sci.*, 1965, **9**, 1787.

304   THE CHEMISTRY OF ORGANOTIN COMPOUNDS

51. Razuvaev, G. A., Minsker, K. S. and Sangalov, Yu. A., *Dokl. Akad. Nauk S.S.S.R.*, 1964, **159**, 158 (*Chem. Abs.*, 1965, **62**, 5337g).
52. Rieche, A., *Kunstoffe*, 1964, **54**, 428 (*Chem. Abs.*, 1965, **62**, 2844a).
53. Ross, A., *Ann. N.Y. Acad. Sci.*, 1965, **125**, 107.
54. Scott, G., *Atmospheric Oxidation and Antioxidants*, Elsevier, London and New York, 1965, 296.
55. Smith, H. V., *Organotin Stabilizers*, The Tin Research Institute, Greenford, 1959.
56. Sorokin, G. V., Pozdnyakova, M. V., Ter-Asaturova, N. I., Perchenko, V. N. and Nametkin, N. S., *Dokl. Akad. Nauk S.S.S.R.*, 1967, **174**, 376.
57. Takami, Y., *Kogyo Kagaku Zasshi*, 1962, **65**, 234 (*Chem. Abs.*, 1963, **58**, 1482g).
58. Takami, Y., *Tokyo Kogyo Shikensho Hokoku*, 1962, **57**, 234 (*Chem. Abs.*, 1965, **62**, 2826b).
59. Takeda, Y., Hayakawa, Y., Feuno, T. and Furukawa, J., *Makromol. Chem.*, 1965, **83**, 234.
60. Tsou, K. C., *Trans. Nucl. Sci.*, 1965, **12**, 28 (*Chem. Abs.*, 1965, **63**, 2597b).
61. Verdol, J. A., Ryan, P. W., Carrow, D. J. and Kuncl, K. L., *Rubber Age*, 1966, **98**, 57 (*Chem. Abs.*, 1966, **65**, 10772f).
62. Vershinina, S. P., Zaplesnichenko, G. P., Kolesnikov, L. N., Skuratovskaya, Zh. V., Chernobai, A. V. and Tsirlin, Yu. A., *Med. Radiol.*, 1965, **10**, 73 (*Chem. Abs.*, 1965, **63**, 9398d).
63. Weisfeld, L. B., Thacker, G. A. and Nass, L. I., *Soc. Plastics Engrs. J.*, 1965, **21**, 649 (*Chem. Abs.*, 1965, **63**, 13491d).
64. Winkler, D. E., *J. Polymer Sci.*, 1959, **35**, 3.
65. Woggon, H., Koehler, U. and Uhde, W. J., *Nahrung*, 1967, **11**, 809 (*Chem. Abs.*, 1968, **68**, 69831).
66. Yoshida, Z. and Miyoshi, H., *Kogyo Kagaku Zasshi*, 1965, **68**, 580 (*Chem. Abs.*, 1966, **64**, 9890c).
67. Zavaglia, E. A., Mosher, W. A. and Billmeyer, F. W., *Offic. Dig. J. Paint Technol. Eng.*, 1965, **37**, 229 (*Chem. Abs.*, 1965, **63**, 3054f).
68. Zherdev, Yu. V., Korolev, A. Ya. and Leznov, N. S., *Plasticheskie Massy*, 1964, 16 (*Chem. Abs.*, 1965, **62**, 2875b).
69. Belgium Patent 638,363, April 8, 1964 (*Chem. Abs.*, 1965, **62**, 7944g).
70. Belgium Patent 644,681, Sept. 4, 1964 (*Chem. Abs.*, 1965, **63**, 13542d).
71. Belgium Patent, 645,019, July 1, 1964 (*Chem. Abs.*, 1965, **63**, 10142c).
72. Belgium Patent 645,883, July 16, 1964 (*Chem. Abs.*, 1965, **63**, 11809g).
73. Belgium Patent 653, 242, Jan. 18, 1965 (*Chem. Abs.*, 1966, **64**, 9495d).
74. Belgium Patent 658,003, July 7, 1965 (*Chem. Abs.*, 1966, **64**, 3802b).
75. Belgium Patent 661,478, July 16, 1965 (*Chem. Abs.*, 1966, **64**, 19904c).
76. Belgium Patent 661,480, July 16, 1965 (*Chem. Abs.*, 1966, **64**, 19904d).
77. Belgium Patent 663,321, Nov. 3, 1965 (*Chem. Abs.*, 1966, **65**, 4079f).
78. Belgium Patent 665,078, Oct. 1, 1965 (*Chem. Abs.*, 1966, **64**, 20000f).
79. Belgium Patent 671,001, Jan. 31, 1966 (*Chem. Abs.*, 1966, **65**, 20315g).
80. British Patent 979,995, Jan. 6, 1965 (*Chem. Abs.*, 1966, **64**, 9901h).
81. British Patent 985,721, March 10, 1965 (*Chem. Abs.*, 1965, **62**, 13334e).
82. British Patent 992,025, May 12, 1965 (*Chem. Abs.*, 1965, **63**, 5855f).
83. British Patent 994,167, June 2, 1965 (*Chem. Abs.*, 1965, **63**, 4090h).

84. British Patent 994,348, June 2, 1965 (*Chem. Abs.*, 1965, **63**, 5904g).
85. British Patent 999,805, July 28,1965 (*Chem. Abs.*, 1965, **63**, 16522h).
86. British Patent 1,004,663, Sept. 15, 1965 (*Chem. Abs.*, 1965, **63**, 16555c).
87. British Patent 1,005,248, Sept. 22, 1965 (*Chem. Abs.*, 1965, **63**, 15065h).
88. British Patent 1,009,368, Nov. 10, 1965 (*Chem. Abs.*, 1966, **64**, 8240f).
89. British Patent 1,016,534, Jan. 12, 1966 (*Chem. Abs.*, 1966, **64**, 8489h).
90. British Patent 1,018,035, Jan. 26, 1966 (*Chem. Abs.*, 1966, **64**, 12933c).
91. British Patent 1,018,111, Jan. 26, 1966 (*Chem. Abs.*, 1966, **64**, 12906h).
92. British Patent 1,019,916, Feb. 9, 1966 (*Chem. Abs.*, 1966, **64**, 11413d).
93. British Patent 1,020,291, Feb. 16, 1966 (*Chem. Abs.*, 1966, **64**, 15926c).
94. British Patent 1,020,501, Feb. 16, 1966 (*Chem. Abs.*, 1966, **64**, 19949a).
95. British Patent 1,027,781, April 27, 1966 (*Chem. Abs.*, 1966, **64**, 19910d).
96. British Patent 1,033, 903, June 22, 1966 (*Chem. Abs.*, 1966, **65**, 10757a).
97. British Patent 1,038,723, Aug. 10, 1966 (*Chem. Abs.*, 1966, **65**, 20304e).
98. British Patent 1,041,851, Sept. 7, 1966 (*Chem. Abs.*, 1966, **65**, 20337c).
99. British Patent 1,047,949, Nov. 9, 1966 (*Chem. Abs.*, 1967, **66**, 19249x).
100. British Patent 1,061,747, March 15, 1967 (*Chem. Abs.*, 1967, **67**, 12144w).
101. British Patent 1,069,165, May 17, 1967 (*Chem. Abs.*, 1967, **67**, 32778j).
102. British Patent 1,071,397, June 7, 1967 (*Chem. Abs.*, 1967, **67**, 44424d).
103. British Patent 1,078,897, Aug. 9, 1967 (*Chem. Abs.*, 1967, **67**, 100705k).
> 104. British Patent 1,081,504, Aug. 31, 1967 (*Chem. Abs.*, 1967, **67**, 101119j).
105. British Patent 1,084,344, Sept. 20, 1967 (*Chem. Abs.*, 1968, **68**, 3565).
106. British Patent 1,100,933, Jan. 24, 1968 (*Chem. Abs.*, 1968, **68**, 60365).
107. British Patent 1,107,188, March 20, 1968 (*Chem. Abs.*, 1968, **68**, 105644).
108. British Patent 1,110,709, April 24, 1968 (*Chem. Abs.*, 1968, **69**, 11139).
109. British Patent 1,112,408, May 8, 1968 (*Chem. Abs.*, 1968, **69**, 11089).
110. British Patent 1,122, 157, July 31, 1968 (*Chem. Abs.*, 1968, **69**, 67906).
111. Czech. Patent 125,310, Dec. 15, 1967 (*Chem. Abs.*, 1968, **69**, 68696).
112. French Patent 83,866, Oct. 30, 1964 (*Chem. Abs.*, 1965, **62**, 10380b).
113. French Patent 1,357,974, April 10, 1964 (*Chem. Abs.*, 1965, **62**, 7943f).
114. French Patent 1,361,845, May 22, 1964 (*Chem. Abs.*, 1965, **62**, 6588b).
115. French Patent 1,364,456, June 19, 1964 (*Chem. Abs.*, 1965, **62**, 6632e).
116. French Patent 1,368,785, Aug. 7, 1964 (*Chem. Abs.*, 1965, **62**, 16460e).
117. French Patent 1,373,716, Oct. 2, 1964 (*Chem. Abs.*, 1965, **62**, 11973b).
118. French Patent 1,375,723, Oct. 23, 1964 (*Chem. Abs.*, 1965, **62**, 11975d).
119. French Patent 1,375,867, Oct. 23, 1964 (*Chem. Abs.*, 1965, **62**, 10541c).
120. French Patent 1,379,988, Jan. 27, 1964 (*Chem. Abs.*, 1965, **62**, 16407d).
121. French Patent 1,381,947, Dec. 11, 1964 (*Chem. Abs.*, 1965, **62**, 10542c).
122. French Patent 1,386,988, Jan. 22, 1965 (*Chem. Abs.*, 1965, **63**, 10135b).
123. French Patent 1,388,997, Feb. 12, 1965 (*Chem. Abs.*, 1965, **62**, 16404d).
124. French Patent 1,399,045, May 14, 1965 (*Chem. Abs.*, 1966, **64**, 5228a).
125. French Patent 1,406,055, July 15, 1965 (*Chem. Abs.*, 1966, **65**, 5587).
126. French Patent 1,411,819, Sept. 24, 1965 (*Chem. Abs.*, 1966, **64**, 2278g).
127. French Patent 1,424,599, Jan. 14, 1966 (*Chem. Abs.*, 1966, **65**, 13928h).
128. French Patent 1,474,509, March 24, 1967 (*Chem. Abs.*, 1967, **67**, 55093w).
129. French Patent 1,476,628, April 14, 1967 (*Chem. Abs.*, 1968, **68**, 22676).
130. French Patent 1,480,479, May 12, 1967 (*Chem. Abs.*, 1967, **67**, 117764b).
131. French Patent 1,487,752, July 7, 1967 (*Chem. Abs.*, 1968, **68**, 60272).

132. French Patent 1,506,186, Dec. 15, 1967 (*Chem. Abs.*, 1968, **69**, 106878).
133. German Patent 1,150,814, June 27, 1963 (*Chem. Abs.*, 1966, **64**, 8403f).
134. German Patent 1,189,220, March 18, 1965 (*Chem. Abs.*, 1965, **62**, 16512e).
135. German Patent 1,223,548, Aug. 25, 1966 (*Chem. Abs.*, 1967, **66**, 56412r).
136. German Patent 1,227,658, Oct. 27, 1966 (*Chem. Abs.*, 1967, **66**, 46483a).
137. German Patent 1,232,737, Jan, 19, 1967 (*Chem. Abs.*, 1967, **66**, 86273p).
138. German Patent 1,232,739, Jan. 19, 1967 (*Chem. Abs.*, 1967, **66**, 86274q).
139. German Patent 1,232,740, Jan. 19, 1967 (*Chem. Abs.*, 1967, **66**, 86276s).
140. German Patent 1,239,092, April, 20, 1967 (*Chem. Abs.*, 1967, **67**, 33314s).
141. German Patent 1,249,833, Sept. 14, 1967 (*Chem. Abs.*, 1967, **67**, 100703h).
142. German Patent 1,266,497, April 18, 1968 (*Chem. Abs.*, 1968, **69**, 3448).
143. German Patent 1,270,799, June 20, 1968 (*Chem. Abs.*, 1968, **69**, 59838).
144. German Patent 1,271,389, June 27, 1968 (*Chem. Abs.*, 1968, **69**, 59836).
145. German (East) Patent 36,060, May 25, 1965 (*Chem. Abs.*, 1965, **63**, 15066c).
146. German (East) Patent 39,200, May 25, 1965 (*Chem. Abs.*, 1965, **63**, 18480b).
147. German (East) Patent 49,295, July 20, 1966 (*Chem. Abs.*, 1967, **66**, 29616d).
148. Japanese Patent 20,526, Sept, 19, 1964 (*Chem. Abs.*, 1965, **62**, 16411b).
149. Japanese Patent 7,064, April 7, 1965 (*Chem. Abs.*, 1965, **63**, 703c).
150. Japanese Patent 10,588, May 28, 1965 (*Chem. Abs.*, 1965, **63**, 16497f).
151. Japanese Patent 10,589, May 28, 1965 (*Chem. Abs.*, 1965, **63**, 16497g).
152. Japanese Patent 12,181, June 16, 1965 (*Chem. Abs.*, 1965, **63**, 18293a).
153. Japanese Patent 12,375, June 17, 1965 (*Chem. Abs.*, 1965, **63**, 18376d).
154. Japanese Patent 13,259, June 26, 1965 (*Chem. Abs.*, 1966, **65**, 9116h).
155. Japanese Patent 13,260, June 26, 1965 (*Chem. Abs.*, 1966, **65**, 9116f).
156. Japanese Patent 13,261, June 26, 1965 (*Chem. Abs.*, 1966, **65**, 9116g).
157. Japanese Patent 21,435, Sept. 22, 1965 (*Chem. Abs.*, 1966, **64**, 2189c).
158. Japanese Patent 29,025, Dec. 23, 1965 (*Chem. Abs.*, 1966, **64**, 9838h).
159. Japanese Patent 6,755, April 18, 1966 (*Chem. Abs.*, 1966, **65**, 13894f).
160. Japanese Patent 19,414, Nov. 10, 1966 (*Chem. Abs.*, 1967, **66**, 46485c).
161. Japanese Patent 21,625, Dec. 17, 1966 (*Chem. Abs.*, 1967, **67**, 44433f).
162. Japanese Patent 864, Jan. 18, 1967 (*Chem. Abs.*, 1967, **67**, 33316u).
163. Japanese Patent 865, Jan. 18, 1967 (*Chem. Abs.*, 1967, **67**, 33317).
164. Japanese Patent 4,275, Feb. 22, 1967 (*Chem. Abs.*, 1967, **67**, 44419f).
165. Japanese Patent 5,345, March 4, 1967 (*Chem. Abs.*, 1968, **68**, 40551).
166. Japanese Patent 5,554, March 7, 1967 (*Chem. Abs.*, 1968, **68**, 60275).
167. Japanese Patent 7,687, March 29, 1967 (*Chem. Abs.*, 1968, **68**, 14003).
168. Japanese Patent 12,529, July 17, 1967 (*Chem. Abs.*, 1967, **67**, 109337z).
169. Japanese Patent 13,302, July 28, 1967 (*Chem. Abs.*, 1968, **68**, 49767).
170. Japanese Patent 18,816, Sept. 26, 1967 (*Chem. Abs.*, 1968, **68**, 60282).
171. Japanese Patent 18,818, Sept. 26, 1967 (*Chem. Abs.*, 1968, **69**, 11133).
172. Japanese Patent 19,177, Sept. 28, 1967 (*Chem. Abs.*, 1968, **68**, 79165).
173. Japanese Patent 24,044, Nov. 20, 1967 (*Chem. Abs.*, 1968, **68**, 87885).
174. Japanese Patent 24,047, Nov. 20, 1967 (*Chem. Abs.*, 1968, **68**, 87887).
175. Japanese Patent 615, Jan. 10, 1968 (*Chem. Abs.*, 1968, **68**, 115182).
176. Japanese Patent 1,102, Jan. 16, 1968 (*Chem. Abs.*, 1968, **69**, 3667).
177. Japanese Patent 2,329, Jan. 27, 1968 (*Chem. Abs.*, 1968, **69**, 27528).
178. Japanese Patent 3,005, Feb. 3, 1968 (*Chem. Abs.*, 1968, **69**, 20022).
179. Japanese Patent 3,012, Feb. 3, 1968 (*Chem. Abs.*, 1968, **69**, 44396).

180. Japanese Patent 12,813, May 30, 1968 (*Chem. Abs.*, 1968, **69**, 87585).
181. Japanese Patent 13,961, June 13, 1968 (*Chem. Abs.*, 1968, **69**, 107267).
182. Netherlands Patent 6,406,534, Dec. 10, 1965 (*Chem. Abs.*, 1966, **64**, 16092e).
183. Netherlands Patent 6,407,761, March 31, 1965 (*Chem. Abs.*, 1965, **63**, 10132h).
184. Netherlands Patent 6,411,204, March 29, 1965 (*Chem. Abs.*, 1966, **65**, 4083f).
185. Netherlands Patent 6,412,532, May 3, 1965 (*Chem. Abs.*, 1966, **64**, 5267h).
186. Netherlands Patent 6,413,375, June 24, 1965 (*Chem. Abs.*, 1966, **64**, 420e).
187. Netherlands Patent 6,413,688, May 26, 1966 (*Chem. Abs.*, 1966, **65**, 20303h).
188. Netherlands Patent 6,500,606, July 23, 1965 (*Chem. Abs.*, 1966, **64**, 5276a).
189. Netherlands Patent 6,503,797, Oct. 4, 1965 (*Chem. Abs.*, 1966, **64**, 8340c).
190. Netherlands Patent 6,504,150, Oct. 4, 1965 (*Chem. Abs.*, 1966, **64**, 8406h).
191. Netherlands Patent 6,512,796, April 4, 1966 (*Chem. Abs.*, 1966, **65**, 7393c).
192. Netherlands Patent 6,513,108, April 12, 1966 (*Chem. Abs.*, 1966, **65**, 12359g).
193. Netherlands Patent 6,513,952, May 2, 1966 (*Chem. Abs.*, 1966, **65**, 13955c).
194. Netherlands Patent 6,514,097, May 2, 1966 (*Chem. Abs.*, 1966, **65**, 12371g).
195. Netherlands Patent 6,515,354, May 26, 1966 (*Chem. Abs.*, 1966, **65**, 17154c).
196. Netherlands Patent 6,516,388, June 20, 1966 (*Chem. Abs.*, 1966, **65**, 15613a).
197. Netherlands Patent 6,602,265, Aug. 24, 1966 (*Chem. Abs.*, 1967, **66**, 3523k).
198. Netherlands Patent 6,603,742, Sept. 23, 1966 (*Chem. Abs.*, 1967, **66**, 86538d).
199. Netherlands Patent 6,604,743, Oct. 17, 1966 (*Chem. Abs.*, 1967, **66**, 66219x).
200. Netherlands Patent 6,604,827, Oct. 13, 1967 (*Chem. Abs.*, 1968, **68**, 22547).
201. Netherlands Patent 6,606,681, Nov. 22, 1966 (*Chem. Abs.*, 1967, **66**, 86266p).
202. Netherlands Patent 6,609,502, Jan. 8, 1968 (*Chem. Abs.*, 1968, **68**, 88279).
203. Netherlands Patent 6,612,421, March 3, 1967 (*Chem. Abs.*, 1967, **67**, 44618v).
204. Netherlands Patent 6,615,781, May 10, 1968 (*Chem. Abs.*, 1968, **69**, 44398).
205. Netherlands Patent 6,700,013, July 4, 1967 (*Chem. Abs.*, 1967, **67**, 108761w).
206. Netherlands Patent 6,700,014, July 4, 1967 (*Chem. Abs.*, 1967, **67**, 109344z).
207. Polish Patent 48,178, May 18, 1964 (*Chem. Abs.*, 1965, **62**, 1688f).
208. Polish Patent 49,815, June 28, 1965 (*Chem. Abs.*, 1966, **64**, 16082c).
209. U.S. Patent 2,219,463, Oct. 29, 1940 (*Chem. Abs.*, 1941, **35**, 1145).
210. U.S. Patent 3,160,609, Dec. 8, 1964 (*Chem. Abs.*, 1965, **62**, 5402h).
211. U.S. Patent 3,166,547, Jan. 19, 1965 (*Chem. Abs.*, 1965, **62**, 9259b).
212. U.S. Patent 3,168,590, Feb. 2, 1965 (*Chem. Abs.*, 1965, **62**, 9006b).
213. U.S. Patent 3,183,238, May 11, 1965 (*Chem. Abs.*, 1965, **63**, 1816b).
214. U.S. Patent 3,194,770, July 13, 1965 (*Chem. Abs.*, 1965, **63**, 8578c).
215. U.S. Patent 3,194,773, July 13, 1965 (*Chem. Abs.*, 1965, **63**, 8577g).
216. U.S. Patent 3,196,129, July 20, 1965 (*Chem. Abs.*, 1966, **64**, 12906f).
217. U.S. Patent 3,198,757, Aug. 3, 1965 (*Chem. Abs.*, 1965, **63**, 11821b).
218. U.S. Patent 3,208,969, Sept. 28, 1965 (*Chem. Abs.*, 1965, **63**, 18376g).
219. U.S. Patent 3,208,970, Sept. 28, 1965 (*Chem. Abs.*, 1965, **63**, 16554g).
220. U.S. Patent 3,209,017, Sept. 28, 1965 (*Chem. Abs.*, 1965, **63**, 16382e).
221. U.S. Patent 3,210,329, Oct. 5, 1965 (*Chem. Abs.*, 1966, **64**, 2187e).
222. U.S. Patent 3,217,004, Nov. 9, 1965 (*Chem. Abs.*, 1967, **67**, 90939y).
223. U.S. Patent 3,225,022, Dec. 21, 1965 (*Chem. Abs.*, 1966, **64**, 8340g).
224. U.S. Patent 3,236,772, Feb. 22, 1966 (*Chem. Abs.*, 1966, **64**, 15658g).
225. U.S. Patent 3,240,730, March 15, 1966 (*Chem. Abs.*, 1966, **64**, 16087f).

226. U.S. Patent 3,244,629, April 5, 1966 (*Chem. Abs.*, 1966, **64**, 19296f).
227. U.S. Patent 3,244,737, April 5, 1966 (*Chem. Abs.*, 1966, **65**, 4054e).
228. U.S. Patent 3,245,907, April 12, 1966 (*Chem. Abs.*, 1966, **65**, 5402h).
229. U.S. Patent 3,245,959, April 12, 1966 (*Chem. Abs.*, 1966, **64**, 19830f).
230. U.S. Patent 3,251,661, May 17, 1966 (*Chem. Abs.*, 1966, **65**, 3649h).
231. U.S. Patent 3,252,215, May 24, 1966 (*Chem. Abs.*, 1966, **65**, 3506d).
232. U.S. Patent 3,261,813, July 19, 1966 (*Chem. Abs.*, 1966, **65**, 18849b).
233. U.S. Patent 3,274,134, Sept. 20, 1966 (*Chem. Abs.*, 1967, **66**, 3916c).
234. U.S. Patent 3,278,508, Oct. 11, 1966 (*Chem. Abs.*, 1966, **65**, 20334g).
235. U.S. Patent 3,296,193, Jan. 3, 1967 (*Chem. Abs.*, 1967, **66**, 55981p).
236. U.S. Patent 3,297,630, Jan. 10, 1967 (*Chem. Abs.*, 1967, **66**, 56671z).
237. U.S. Patent 3,297,651, Jan. 10, 1967 (*Chem. Abs.*, 1967, **66**, 47232e).
238. U.S. Patent 3,305,388, Feb. 21, 1967 (*Chem. Abs.*, 1967, **66**, 97827y).
239. U.S. Patent 3,306,920, Feb. 28, 1967 (*Chem. Abs.*, 1967, **66**, 95815u).
240. U.S. Patent 3,311,648, March 28, 1967 (*Chem. Abs.*, 1967, **67**, 11590b).
241. U.S. Patent 3,313,731, April 11, 1967 (*Chem. Abs.*, 1967, **67**, 3103c).
242. U.S. Patent 3,316,284, April 25, 1967 (*Chem. Abs.*, 1967, **67**, 64539y).
243. U.S. Patent 3,322,671, May 30, 1967 (*Chem. Abs.*, 1967, **67**, 55952a).
244. U.S. Patent 3,347,833, Oct. 17, 1967 (*Chem. Abs.*, 1967, **67**, 117696f).
245. U.S. Patent 3,358,006, Dec. 12, 1967 (*Chem. Abs.*, 1968, **68**, 30705).
246. U.S. Patent 3,389,111, June 18, 1968 (*Chem. Abs.*, 1968, **69**, 36706).
247. U.S. Patent 3,397,969, Aug. 20, 1968 (*Chem. Abs.*, 1968, **69**, 79047).
248. U.S.S.R. Patent 174,354, Aug. 28, 1965 (*Chem. Abs.*, 1966, **64**, 879b).

INDEX

Compounds are listed either as 'organotin' or 'stannyl' derivatives, therefore both these headings should be examined when looking for a particular type of compound.